Shaping Space

Exploring Polyhedra in Nature, Art, and the Geometrical Imagination

Marjorie Senechal
Editor

Shaping Space

Exploring Polyhedra in Nature, Art, and the Geometrical Imagination

with George Fleck and Stan Sherer

 Springer

Editor
Marjorie Senechal
Department of Mathematics and Statistics
Smith College
Northampton, MA, USA

ISBN 978-1-4939-3948-0   ISBN 978-0-387-92714-5 (eBook)
DOI 10.1007/978-0-387-92714-5
Springer New York Heidelberg Dordrecht London

Mathematics Subject Classification: 51-01, 51-02, 51A25, 51M20, 00A06, 00A69, 01-01

Printed on acid-free paper

Springer is part of Springer Science+Business Media (www.springer.com)

Contents

Part III   Polyhedra in the Geometrical Imagination

Preface

Molecules, galaxies, art galleries, sculpture, viruses, crystals, architecture, and more: *Shaping Space: exploring polyhedra in nature, art, and the geometrical imagination* is an exuberant survey of polyhedra in nature and art. It is at the same time hands-on, mind-turned-on introduction to one of the oldest and most fascinating branches of mathematics. In these pages you will meet some of the world's leading geometers, and learn what they do and why they do it. In short, *Shaping Space* is as many-faceted as polyhedra themselves.

Shaping Space is a treasury of ideas, history, and culture. For students and teachers, from elementary school to graduate school, it is a text with context. For the multitude of polyhedra hobbyists, an indispensable handbook. *Shaping Space* is a resource for professionals—architects and designers, painters and sculptors, biologists and chemists, crystallographers and physicists and earth scientists, engineers and model builders, mathematicians and computer scientists. If you are intrigued by the exquisite shapes of crystals and want to know how nature builds them, if you marvel at domes and wonder why most stay up but some fall down, if you wonder why Plato thought earth, air, fire and water were made of polyhedral particles, if you wonder what geometry is and are willing to try it yourself, this book is for you. In *Shaping Space* you will see that polyhedra are as new as they are old, and that they continue to shape our spaces in new and exciting ways, from computer games to medical imaging.

The computer revolution has catalyzed new research on polyhedra. A quarter century ago, discrete and computational geometry (the branch of mathematics to which polyhedra belong) was less a field in its own right than—in the eyes of many people, even many mathematicians—a grab-box of mathematical games. Today an international journal, *Discrete and Computational Geometry*, publishes six issues a year with the latest research on configurations and arrangements, spatial subdivision, packing, covering, and tiling, geometric complexity, polytopes, point location, geometric probability, geometric range searching, combinatorial and computational topology, probabilistic techniques in computational geometry, geometric graphs, geometry of numbers, and motion planning, and papers with a distinct geometric flavor in such areas as graph theory, mathematical programming, real algebraic geometry, matroids, solid modeling, computer graphics, combinatorial optimization, image processing, pattern recognition, crystallography, VLSI design, and robotics.

Figure 1. An icosahedron built and decorated by elementary school children. Photograph by Stan Sherer.

Figure 2. Sculptures by Morton Bradley. Photograph by Stan Sherer.

Yet, it is also true, as the saying goes, *plus ça change, plus c'est la même chose*. The more things change, the more they stay the same, especially in school mathematics curricula. Despite its central importance in the sciences, the arts, in mathematics and in engineering, solid geometry has all but vanished from the schools, plane geometry is being squeezed to a minimum, and model-building is relegated to kindergarten. Reasons for this unfortunate (and, unfortunately, long-term) trend include a lack of teacher training and pressures to teach testable skills. But educators will realize, sooner rather than later, that "technology in the classroom" is more than clicking the latest gadgets, it means understanding our technological world. Geometry will reappear as a blend of model-building, engineering and fundamental math and science.

Meanwhile, the internet is helping to bring geometry back to life and with it a community of geometers. You can explore polyhedra in nature, art, and the geometrical imagination on the world wide web by yourself, with *Shaping Space* as your guide, and share your findings and frustrations with the like-minded through chat groups. Keep pencils, paper, a ruler, scissors, and tape handy: Confucius got it right 2500 years ago:

> I hear and I forget,
> I read and I remember,
> I do and I understand.

Shaping Space will evolve as the subject grows. The notes and references at the end of this book are also on my website, http://www.marjoriesenechal.com. Authors will post updates there; you will also find links to instructional and recreational materials, and to websites of polyhedra-minded scientists, artists and hobbyists. Visit often!

Indeed, *Shaping Space* has grown already. Its ancestor, *Shaping Space: a polyhedral approach* was inspired by a three-day festival of workshops, exhibitions and lectures on polyhedra held at Smith College in 1984. *Shaping Space: Exploring Polyhedra in Nature, Art, and the Geometrical Imagination* includes the best of the past and new chapters by Robert Connelly, Erik Demaine (with Martin Demaine and Vi Hart), George Hart, Joseph O'Rourke, Ileana Streinu, and Günter Ziegler (with Moritz Schmidt).

Figure 3. H. S. M. Coxeter (1907–2003). Photograph by Stan Sherer.

Figure 4. Arthur L. Loeb (1923–2002). Photograph by Stan Sherer.

Shaping Space: exploring polyhedra in nature, art, and the geometrical imagination is dedicated to the memory of two friends and colleagues, the legendary geometer H. S. M. Coxeter and the many-faceted design scientist Arthur L. Loeb. Without their enthusiasm, encouragement, support and participation, the Shaping Space Conference could not have been held and the first edition of this book might never have appeared. They continue to inspire us.

Northampton, MA, USA   Marjorie Senechal

1

Introduction to the Polyhedron Kingdom

Marjorie Senechal

What is a polyhedron? The question is short, the answer is long. Although you may never have heard of the Polyhedron Kingdom before, it is nearly as vast and as varied as the animal, mineral, and vegetable kingdoms (and it overlaps all three of them). There are aristocrats and workers, families and individuals, old polyhedra with long and interesting histories and young polyhedra born yesterday or the day before. In this kingdom you can take a walking tour of polyhedral architecture, visit a nature preserve and an art gallery and an artisans' polyhedra fair. As you stroll along you may even glimpse polyhedral ghosts from four-dimensional space.

The boundaries of the Polyhedron Kingdom are in dispute (as are those of most kingdoms) but it is safe to visit the border areas. You need not worry about the nature of the disputes until the last part of this book.

The language of the Polyhedron Kingdom is mathematics but, for this brief first visit, you can get by if you learn three important words: face, edge, and vertex. The word *polyhedron* comes from the Greek word for "many" and an Indo-European word for "seat." To geometers, it means an object with many faces. In Figures 1.1 and 1.2 we see polyhedra with faces. But this is not what we mean when we speak of the faces

Figure 1.1. Cube with face, by a fifth-grade student at the Smith College Campus School.

of a polyhedron. For our purposes the *faces* of a polyhedron are the polygons from which its surface is constructed. The *edges* of a polyhedron are the lines bounding its faces; its *vertices* are the corners where three or more faces (and thus three or more edges) meet (Figure 1.3). You will see as we go along that these terms can have more general meanings, but these definitions will do for the moment. As you tour the Polyhedron Kingdom you will become more comfortable with an increasing vocabulary and a wider range of common usages.

We begin our tour, of course, with a visit to the rulers of the Kingdom.

M. Senechal
Department of Mathematics and Statistics,
Smith College,
Northampton, MA 01063, USA
e-mail: senechal@smith.edu; http://math.smith.edu/~
senechal; http://www.marjoriesenechal.com

M. Senechal (ed.), *Shaping Space*, DOI 10.1007/978-0-387-92714-5_1,
© Marjorie Senechal 2013

Figure 1.2. A polyhedral monster, also by Campus School student.

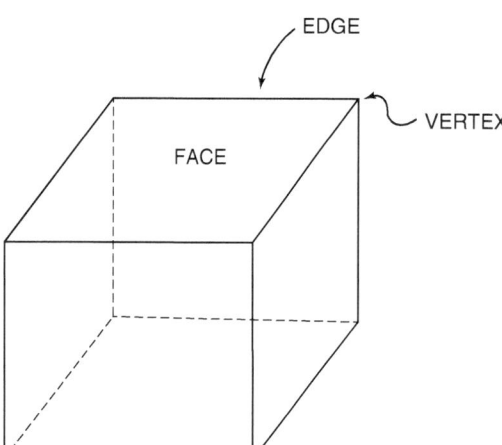

Figure 1.3. The cube has six faces, twelve edges, and eight vertices.

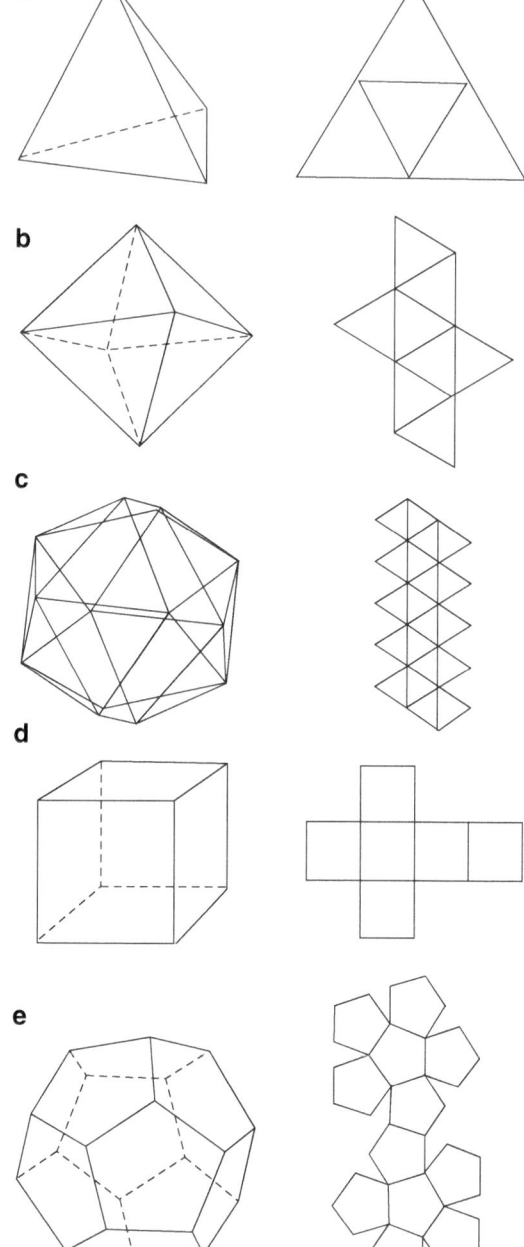

Figure 1.4. *Left*: the five regular polyhedra. *Right*: the same, "unfolded" into planar nets (For more about nets for polyhedra and some unsolved problems concerning them, see Chapters 6 and 22).

The Regular "Solids"

At the gates of the Kingdom live its rulers, the famous and venerable regular "solids" pictured in Figure 1.4. Each of these polyhedra is called *regular* because of certain very special properties: its faces are identical regular polygons, and the same number of polygons meet at each vertex.

(Remember that regular polygons are polygons whose edges have equal lengths and whose angles have equal measure: a regular polygon of three edges is an equilateral triangle, of four edges a square, and so on.) So the faces of each regular polyhedron are all alike and their vertices (or, more precisely, the arrangements of polygons at their vertices) are all alike too.

If we try to build polyhedra with the regularity property just described, we will quickly find that there are only five possibilities. We start by constructing polyhedra whose faces are equilateral triangles. First, we can put three triangles together to form one vertex of a polyhedron. If we continue this pattern at all the other corners, we obtain a pyramid that has four triangular faces, four vertices, and six edges; this is the regular *tetrahedron* (Figure 1.4a). If we put four triangles at each vertex, we can build an *octahedron* (Figure 1.4b); if we put five together then we get the *icosahedron* (Figure 1.4c). Six equilaterial triangles fit together around a point to form a flat surface, so that arrangement is out. And if we try to fit seven or more together—well, try it and see what happens! So these three polyhedra are the only regular ones that can be built out of equilateral triangles.

Now let us try to build a regular polyhedron out of squares. We see that there is just one possibility, the *cube* (Figure 1.4d), in which three faces meet at each vertex, because four squares in a plane lie flat around a point. (What happens if we try to fit five?) If we use regular pentagons, we can again build just one solid, the *dodecahedron* (Figure 1.4e). We cannot continue this procedure with regular polygons with a greater number of sides because three regular hexagons lie flat, three or more heptagons or octagons buckle, and so forth. We conclude that there are no other regular polyhedra.

The regular polyhedra are also known as the "Platonic solids" because the Greek philosopher Plato (427–347 B.C.E.) immortalized them in his dialogue *Timaeus*. In this dialogue Plato discussed his ideas about the "elements" of which he believed the universe to be composed: earth, air, fire, and water. Today when we think of "element," we usually think of the chemical elements

in the Periodic Table. (We recognize the solid, gas, plasma, and liquid *states* of matter.) But notice that we still speak of needing protection from the "elements," and when we say this we mean snow, wind, lightning, and rain. In *Timaeus*, Plato argued that the geometric forms of the smallest particles of these elements are the cube, the octahedron, the tetrahedron, and the icosahedron, respectively. (The fifth regular solid, the dodecahedon, was assigned to the Great All, the cosmos.) This association of the regular solids with the elements captured the imagination of many people from Plato's time to our own. The twentieth century artist M.C. Escher presented them in various ways; Figure 1.5 might be subtitled "Platonic Puzzle," because all of the five Platonic solids appear in it in one form or another! Figure 1.6 shows an icosahedral candy box decorated by Escher.

Plato aside, do the regular polyhedra have any special significance outside the Polyhedra Kingdom? Maybe not. The astronomer Johannes Kepler (1571–1630) believed that he had at last discovered their true meaning: the spheres in which they can be inscribed, nested one inside another, are the divine model for the orbits of the six planets! This explained why there could be only six! (Kepler's ideas are discussed in detail by H.S.M. Coxeter in Chapter 3.) The beauty of the regular polyhedra has led scientists astray in our own time as well. In 1936 Dorothy Wrinch

Figure 1.5. *Reptiles.* Woodcut by M.C. Escher.

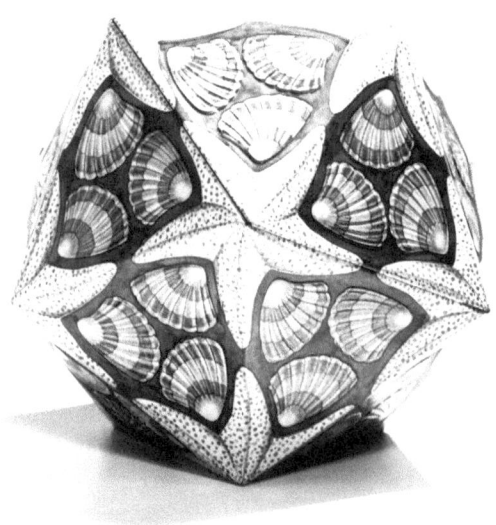

Figure 1.6. *Icosahedron with Starfish and Shells*, a candy box by M.C. Escher.

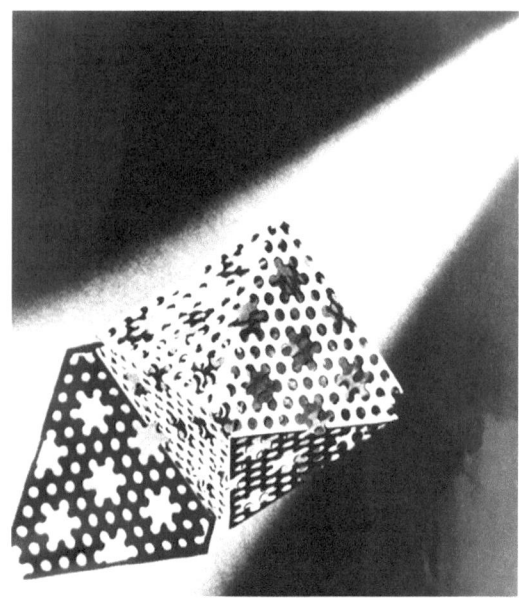

Figure 1.8. The model for protein structure proposed by Dorothy Wrinch in 1936.

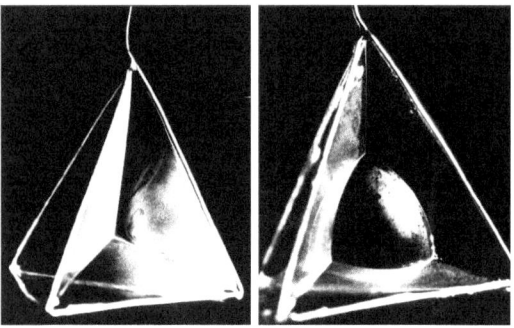

Figure 1.7. Soap films, made by dipping a tetrahedral wire frame into a soapy solution. Notice that the tetrahedral bubble has curved faces.

Figure 1.9. The icosahedron and other polyhedra often appear as decorative elements in Baroque architecture; here, the church of Santissimi Apostoli by Borromini.

proposed a patterned octahedron as the first model for the molecular structure of proteins (Figure 1.8); unfortunately the structures of proteins have turned out to be much less elegant.

The regular polyhedra may not solve the riddle of the universe or reveal the secret of life, but they do crop up in the most unexpected places: for example, in the soap films shown in Figure 1.7 (if we agree that a polyhedron can have curved faces and edges), in decorative ornament (Figure 1.9), and as the shapes of many viruses.

The shapes of many molecules are thought to be closely related to the regular polyhedra (Figure 1.10). Many crystals have cubic, octahedral, or dodecahedral forms; others are tetrahedral or icosahedral. But most dodecahedral (and icosahedral) crystals, like the pyrite crystals in Figure 1.11, are not regular. (Indeed, until November 1984, it was believed that regular dodecahedral and icosahedral crystals could not exist, because their symmetry is theoretically impossible for a crystal. Then some crystals with this symmetry were discovered, posing some challenging problems for symmetry theory!) Perhaps to make up for its limited role in the mineral kingdom, the regular dodecahedron with its twelve faces has been used by people in imaginative ways, such as street corner recycling bins in France (Figure 1.12).

Today we believe that it is not the classical form of the regular polyhedra that is significant: instead it is the high degree of order which they represent. Indeed, as Figure 1.4 suggests, the regular "solids" are not always found in solid form. In some contexts, they have hollow interiors; in others, they have perforated surfaces; in yet others they have no faces, but appear as skeletons made of edges and vertices. Still, they are usually recognizable because of their high degree of *symmetry*. For example, all of the regular polyhedra have *mirror symmetry:* they can be divided into mirror-image halves in many different ways. They also have *rotational symmetry*: there are many ways in which they can be rotated without changing their apparent position. Both mirror symmetry and rotational symmetry are due to the fact that, for each of these polyhedra, every face, every vertex, and every edge is like every other. In other words, they are repetitively organized; this is one of the reasons that they are found so often in nature. This organization is also

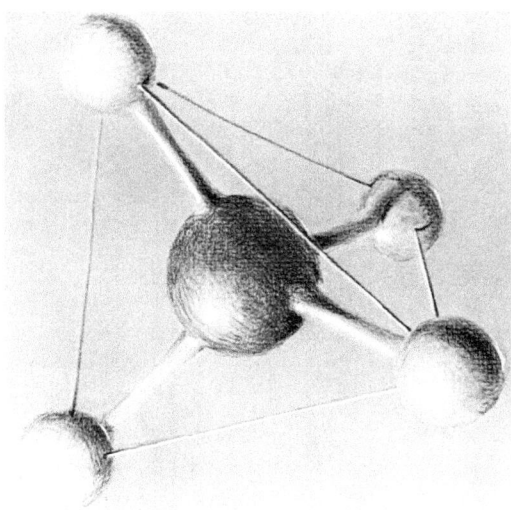

Figure 1.10. An artist's conception of a methane molecule.

Figure 1.11. Pyrite crystals.

Figure 1.12. Dodecahedral recycling bin for glass, on a street corner in Paris, France. Photograph by Marjorie Senechal.

aesthetically pleasing, and it is largely because of their symmetry that they are considered to be beautiful. The regular solids have the highest possible symmetry among polyhedra that are finite in extent. This is one reason why we can justly say that the regular solids are the rulers of the Polyhedron Kingdom. As you read through this book you will learn a great deal about symmetry.

Direct Descendants

There are many variations on the theme of the regular polyhedra. First let us meet the eleven (in Figure 1.13) which can be made by cutting off (*truncating*) the corners, and in some cases the edges, of the regular polyhedra so that all the faces of the faceted polyhedra obtained in this way are regular polygons. These polyhedra were first discovered by Archimedes (287–212 B.C.E.) and so they are often called Archimedean solids. Notice that vertices of the Archimedean polyhedra are all alike, but their faces, which are regular polygons, are of two or more differ-

ent kinds. For this reason they are often called *semiregular*. (Archimedes also showed that in addition to the eleven obtained by truncation, there are two more semiregular polyhedra: the snub cube and the snub dodecahedron. (Also shown in Figure 1.13.)

By this definition, *prisms* (see Figure 1.14) with regular polygonal bases and square sides are semiregular solids too. Prisms are quite common in nature and in architecture, as we will see later (Chapter 7). *Antiprisms* also have two identical polygonal faces, but the "top" face is rotated relative to the "bottom" one, so that the two polygons are joined by triangles (see Figure 1.15); when its faces are regular polygons, an antiprism is a semiregular polyhedron.

Perhaps the most elaborate variations on the theme of the regular polyhedra are those of the sixteenth-century Nuremberg goldsmith Wenzel Jamnitzer, who engraved a fascinating and extensive series of polyhedra in honor of Plato's theory of matter. In his book *Perspectiva Corporum Regularium*, published in 1568, each of the five regular solids is presented in exquisite variation. Can you tell which solid is being varied in Figure 1.16? Jamnitzer's figures show us that polyhedra need not be *convex*; that is, they can have indentations. Regular *polygons* that are not convex, such as the famous pentagram (Figure 1.17), are familiar to most of us. Such "star polygons"

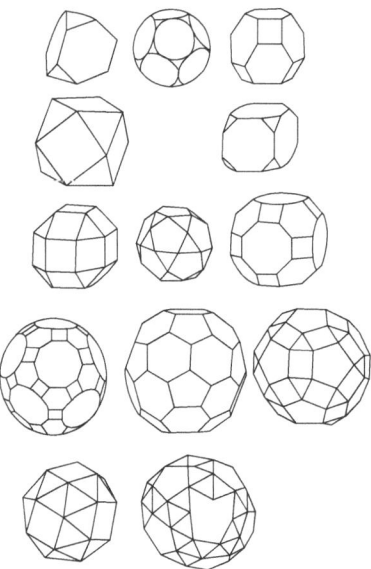

Figure 1.13. The Archimedean or semiregular polyhedra; The first eleven can be obtained from the regular polyhedra by truncation.

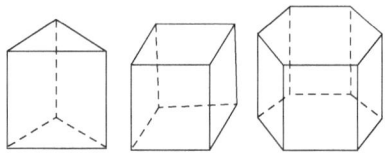

Figure 1.14. The three semiregular prisms.

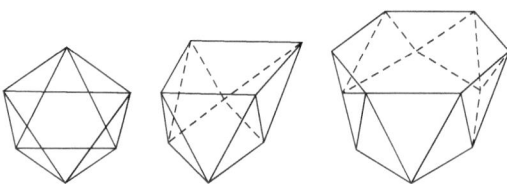

Figure 1.15. Three semiregular antiprisms.

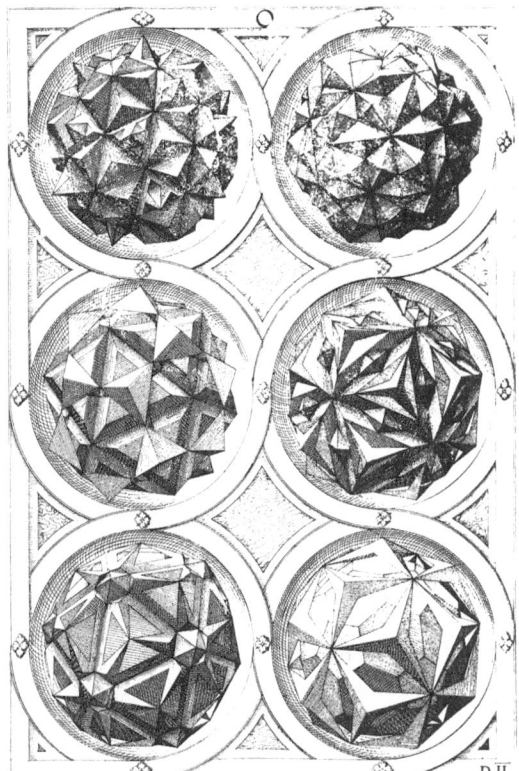

Figure 1.16. Plate D.II. from Wenzel Jamnitzer, *Perspectiva Corporum Regularium*, 1568.

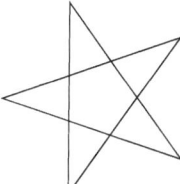

Figure 1.17. The pentagram has equal sides and equal angles.

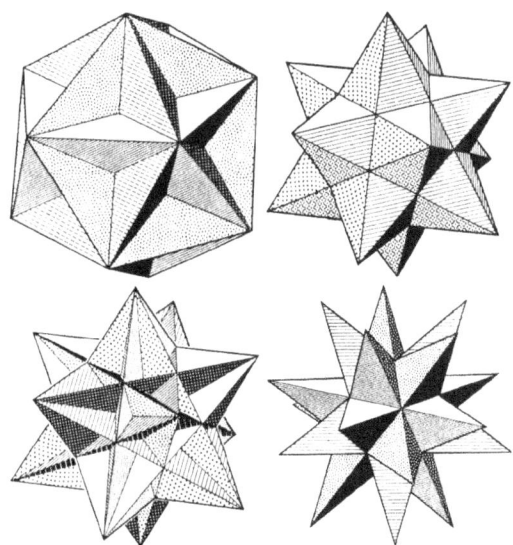

Figure 1.18. The four regular star polyhedra.

Figure 1.19. Marble tarsia (1425–1427) in the Basilica of San Marco, Venice, attributed to Paolo Uccello.

can be used to build regular "star polyhedra." There are exactly four regular star polyhedra (see Figure 1.18). Notice that all their faces are regular polygons and the same number of faces meet at each vertex. In this case, however, either the faces or the vertex arrangements are pentagrams. The lineage of these polyhedra can be traced to fifteenth-century Venice (see Figure 1.19), but no general theory seems to have been developed at that time. Later Kepler investigated regular star polyhedra and found two of them; after that

star-shaped polyhedra (not necessarily regular) became ubiquitous (see for instance Figure 1.20). But it was not until the early nineteenth century that two more regular star polyhedra were found and the French mathematician Augustin-Louis Cauchy (1789–1857) showed that there are no others (see Chapter 4).

The *uniform* polyhedra are polyhedra, star or otherwise, whose vertices are all symmetrically

Figure 1.20. Courtyard of Borromini church.

equivalent. (They are generalizations of the Archimedean polyhedra.) Perhaps the most spectacular uniform polyhedron is the Yog-Sothoth, shown in Figure 1.21. Although its existence had been predicted (on theoretical grounds) for many years, no one had ever seen it before Bruce Chilton's was presented to society for the first time at the Shaping Space Conference. The debut was a spectacular success. The Yog-Sothoth has 112 faces: 12 are pentagrams, 40 are triangles of one type, and 60 are triangles of another. Yet despite its complexity its symmetry is that of the icosahedron and dodecahedron, no more no less!

There are many other interesting lines of descent from the regular solids. For example, there are polyhedra whose faces are all alike but whose vertices are not. Closely related to the semiregular solids, these polyhedra are especially important in the study of crystal forms.

Impossible Polyhedra

Despite its diversity, the Polyhedra Kingdom is exclusive. You will not find polyhedra with any number of faces, edges, and vertices you might think up; only certain combinations are permitted. In the eighteenth century a Swiss Mathematician named Leonhard Euler discovered why. He found a curious relation among the numbers of faces, edges, and vertices of any convex polyhedron. (*Convex* means that the surface has no bumps or dents.) For example, a cube has six faces, twelve edges, and eight vertices; a tetrahedron has four faces, six edges, and four vertices. In both cases, the sum of the numbers of faces and vertices is two more than the number of edges. If we write F for the number of faces of a given polyhedron, V for the number of its vertices, and E for the number of its edges, we have a simple formula: $E = F + V + 2$.

This means, for example, that there is no polyhedron with four faces, six vertices, and nine edges. Nor—though this is harder to prove—can you build a "soccer ball" out of hexagons.

Next Steps

But before losing yourself in contemplation of the impossible, you should build some *possible* polyhedra with your own hands.

Figure 1.21. Three plan views of the Yog-Sothoth, along five-, three-, and twofold axes, drawn by Bruce L. Chilton.

2

Six Recipes for Making Polyhedra

Marion Walter, Jean Pedersen, Magnus Wenninger, Doris Schattschneider, Arthur L. Loeb, Erik Demaine, Martin Demaine, and Vi Hart

This chapter includes six "recipes" for making polyhedra, devised by famous polyhedrachefs. Some recipes are for beginners, others are intermediate or advanced. You can use these recipes, or devise your own. Building models is fun, and will give you a deeper understanding of the chapters that follow.

M. Walter (✉)
Department of Mathematics, University of Oregon, Eugene, OR 97401, USA
e-mail: mwalter@uoregon.edu

J. Pedersen
Department of Mathematics and Computer Science, Santa Clara University, 500 El Camino Real, Santa Clara, CA 95053, USA
e-mail: jpedersen@scu.edu

M. Wenninger
Saint John's Abbey, 31802 County Road 159, Collegeville, MN 56321-2015, USA
e-mail: mwenninger@csbsju.edu

D. Schattschneider • A.L. Loeb (1923–2002)
2038 Sycamore St., Bethlehem, PA 18017, USA
e-mail: schattdo@moravian.edu

E. Demaine • M. Demaine
MIT CSAIL, 32 Vassar St, Cambridge, MA 02139, USA
e-mail: edemaine@mit.edu, mdemaine@mit.edu

V. Hart
e-mail: vi@vihart.com

Constructing Polyhedra Without Being Told How To!

Marion Walter

Getting Started: How to Attach Polygons

Put some cut-out regular polygons on a table. Put a little glue on a flat tile, a plastic lid, or a piece of plastic, and spread out the glue a little so that you can dip a whole edge of a polygon into the glue.

Choose two polygons that you want to glue together along an edge, and dip one of these edges in the glue. Dip lightly; if polygons don't stick well it is usually because there is too much glue (Figure 2.1).

Hold the two edges together firmly. The joint will remain flexible but the polygons will stick together (Figure 2.2).

If you find later that you need extra glue on an edge of a polygon that you have already attached, you can (lightly) dip a toothpick or applicator stick in the glue to smear some along an edge.

What Shape Are You Going to Make?

It is most fun and most rewarding to make a shape you yourself create rather than following someone else's plans. How can you do this?

Figure 2.1.

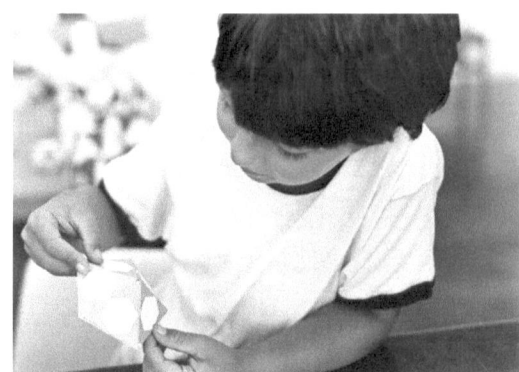

Figure 2.4.

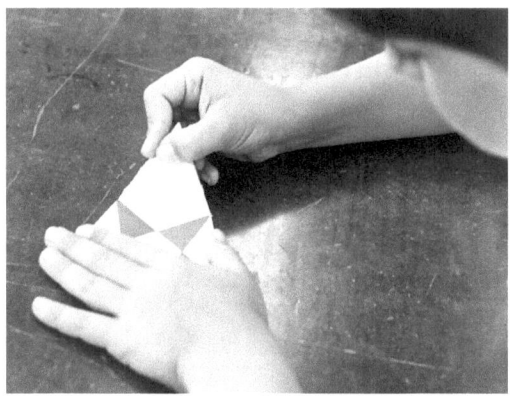

Figure 2.2.

Figure 2.5.

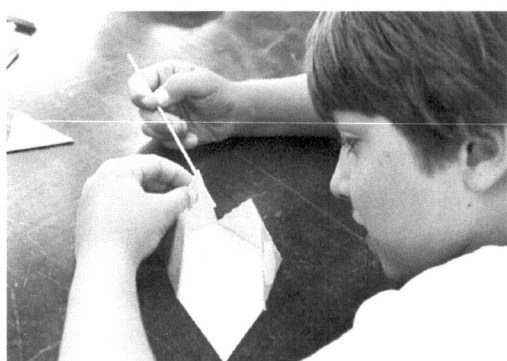

Figure 2.3.

tagons, or triangles and squares. What shapes can you make using triangles and only one pentagon? (See Figure 2.3).

The first shape the boy shown in Figure 2.4 made has a pentagon for its base and triangles for sides. It is called a pentagonal pyramid. Now make up another question of your own. What will your first shape look like? When you experiment freely, you may get a few surprises and you will learn a lot. For example, six triangles lie flat.

What a surprise: the shape in Figure 2.5 lies flat too! Notice that the twelve triangles that surround the hexagon help to make a bigger hexagon. The student shown in the photograph

There are many ways to start. One way is to limit yourself to using only one or two different shapes — say triangles, or triangles and pen-

Figure 2.6.

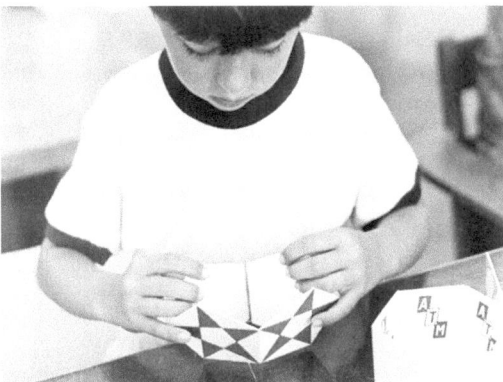

Figure 2.8.

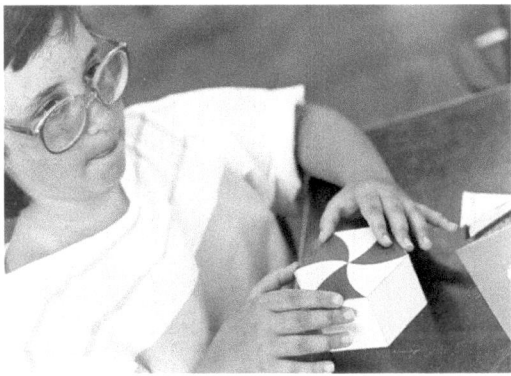

Figure 2.7.

Figure 2.9.

also had a surprise after she attached only six triangles to the hexagon. Do you think it will make a pyramid with a hexagonal base?

What shapes can you make with hexagons and squares? (See Figures 2.6 and 2.7).

Making shapes requires thinking ahead. Try to make a shape using only pentagons. What a relief: the two edges in Figure 2.8 really do seem to meet! How will the boy shown go on? Do the girls in Figures 2.9 and 2.10 seem to be making the same shape?

The shape in Figure 2.11 is made entirely of pentagons: how many of them were used? Turn it around and look at it. How many edges does it have? How many corners? How many edges meet at one corner? How many faces meet at a corner? This shape is a *dodecahedron*.

When you are experimenting, don't expect that your shape will always close! (Figure 2.12). Some shapes may have holes that you cannot fill with the shapes that we have; remember that we are using only regular polygons.

Shapes You Can Make with Triangles

The shape in Figure 2.13 is only one of the many you can make using just triangles. It is an *icosahedron*. Look at it from many sides. How many faces, edges, and corners does it have? Compare these numbers to the corresponding numbers you found for the dodecahedron.

In Figure 2.14 the girl is placing one five-sided pyramid over the base of another one. How many

Figure 2.11.

Figure 2.10.

faces will this polyhedron have? What other
shapes can you make with triangles?

A Note to the Teacher

Every problem leads to new observations and
questions. For example, even the simple problem
"Make all possible convex shapes using only
equilateral triangles" is very rich in possibilities.
These shapes are called *deltahedra*, after the
triangular Greek letter Δ. Usually after some
experimentation, students will discover the tetrahedron, the octahedron, the triangular and pentagonal bipyramids, and the icosahedron. Later
the search also yields the 12-, 14-, and 16-sided
deltahedra. Figure 2.15 shows a 14-sided deltahedron made of applicator sticks. Use sticks all of
the same length. Some drugstores sell applicator
sticks which are ideal; be sure to get the kind
without cotton at each end. Hobby and craft

Figure 2.12.

stores often sell small-diameter wooden dowel
rods which work well. Put a small amount of
contact glue on the ends of the sticks and let it
dry for about 15 minutes, until the glue is tacky.
Then the sticks will join well and yet stay flexible.
Don't be surprised if a cube or dodecahedron
made of applicator sticks won't stand up, however. Unlike structures built entirely of triangles,
these structures are nonrigid.

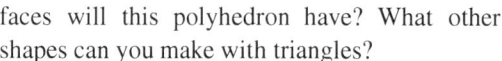

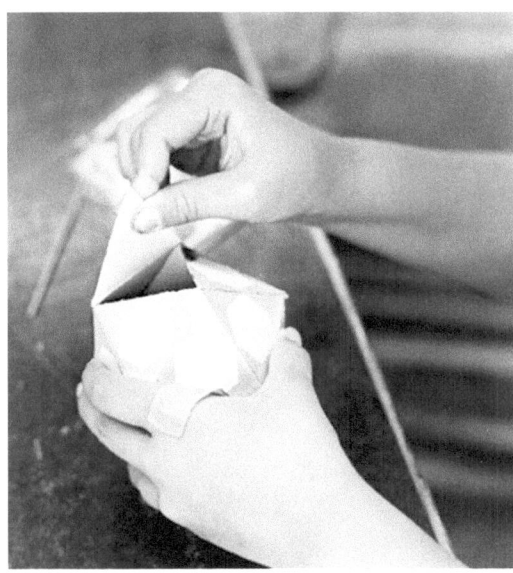

Figure 2.13.

Figure 2.15.

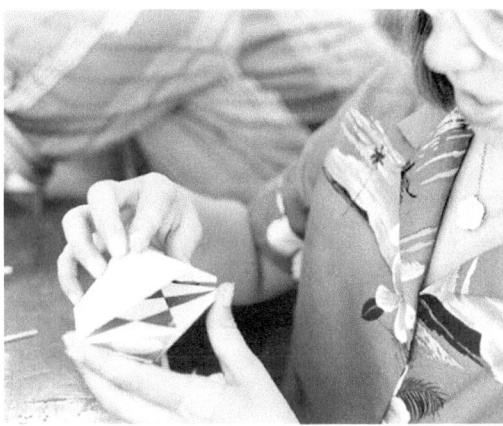

Figure 2.14.

The observation that each deltahedron has an even number of faces leads to the question of why this should be so. The reason is straightforward once one sees it! Each triangle has three edges. If the shape has F faces, then there are $3F$ edges altogether. These $3F$ edges are glued in pairs, so there must be an even number of edges. Hence $3F$ and therefore F must be be even. Noticing that there exist 4-, 6-, 8-, 10-, 12-, 14-, 16-, and 20-sided deltahedra immediately sets off a search for an 18-sided one. Can an 18-sided deltahedron

be made? It was not until 1947 that the answer was proved to be *no*.

Looking at deltahedra is one thing; visualizing them without models is quite another. I found it difficult to close my eyes and visualize the 12-, 14-, 16-sided deltahedra. One day while I was looking at a cube made from applicator sticks and glue, I decided to pose problems by using the What-If-Not Strategy. The idea is that one starts with a situation, a theorem, a diagram, or in our case an object, lists as many of its attributes as one can, and then asks, "What if not?" For example, among the many attributes (not necessarily independent) of a cube that I had listed were the following:

1. All edges are equal.
2. All faces are squares.
3. The object is not rigid.
4. The top vertices are directly above the bottom ones.
5. Opposite faces are parallel.

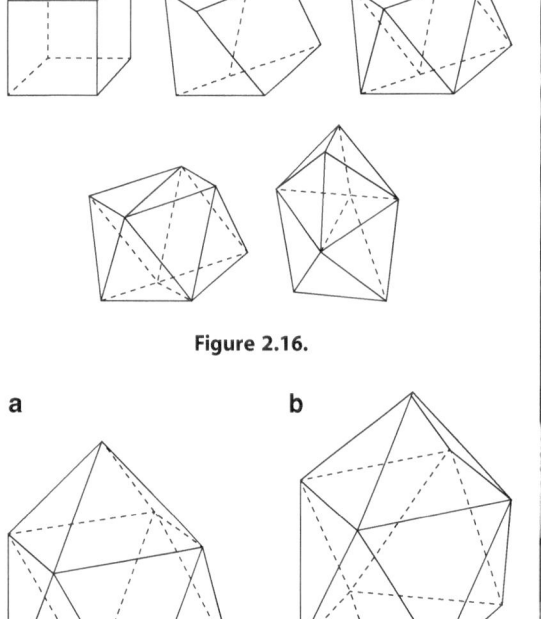

Figure 2.16.

a   b

Figure 2.17.

Figure 2.18. Alice Shearer beginning construction of a model.

While working on attribute 4, I asked myself: "What if the top vertices were *not* directly above the bottom ones?" And because the contact glue gives movable joints, it was easy to give the top square a twist. As my twist approached 45°, I began to see an antiprism emerge. I attached sticks to complete the antiprism, but the shape wasn't rigid. The obvious thing to do to make it rigid was to add diagonals to the top and bottom squares. Since all the applicator sticks are of the same length, I had to squeeze the squares into "diamonds." The resulting shape was rigid- and was built of 12 equilateral triangles! (See Figure 2.16).

How else could I have made the antiprism rigid? I hastily removed the top diagonal, and added four sticks that meet above the square to form a square pyramid (Figure 2.17a.) Lo and behold, I had made a 14-sided deltahedron! From there it was a quick step to remove the bottom diagonal also, build another four-sided pyramid, and thus obtain the 16-sided deltahedron (Figure 2.17b).

Not only have these deltahedral "villains" now become friends, I see now that they are closely related to one another. One can also place the icosahedron in this family, since it is a pentagonal antiprism capped with two pentagonal pyramids. (Indeed the octahedron itself is an antiprism, and the tetrahedron can be viewed as an antiprism in which the two bases have shrunk to an edge. Two opposite edges may be considered degenerate polygons, which are here in antiprism orientation.) That leaves us only with the 6- and 10-sided deltahedra as "odd ones out," but they are both bipyramids and are easy to visualize.

Figure 2.19. Jane B. Phipps contemplating a polyhedron constructed from MATs.

A Word About Materials

Cardboard always works well; you should experiment with different weights. I prefer MATs, described in the next paragraph. All the polygons shown in these photographs are MATs. A glue used for carpets, such as Flexible Mold Compound – Mold It® is excellent, as is the English Copydex.

Adrien Pinel found that hexagonal cardboard beer mats (used in English pubs) were excellent for making polyhedra with holes and, when augmented by triangles and squares cut from the hexagons, became even more useful. It was not long before the Association of Teachers of Mathematics of Great Britain had regular polygons of three, four, five, six and eight sides produced from the same easy-to-glue material as the beer mats. They call them Mathematics Activity Tiles (MATs for short). They also produce rectangles and isosceles triangles. The polygons may be ordered separately or in two different kits: Kit A has 100 each of equilateral triangles, squares, pentagons, and hexagons, and Kit B has 200 each of triangles and squares and 50 each of pentagons, hexagons and octagons.

Constructing Pop-Up Polyhedra

Jean Pedersen

Required Materials

- One 22 × 28 inch piece of brightly colored heavyweight posterboard
- Six rubber bands
- One yard stick or meter stick
- One ballpoint pen
- One pair of scissors

General Instructions for Preparing the Pattern Pieces

Begin by drawing the pattern pieces on the posterboard as shown in Figure 2.20. Press hard with the ballpoint pen so that the posterboard will fold easily and accurately in the final assembly. Label the points indicated. Be certain to put the labels on what will become the cube (or octahedron) when the model is finished — not on the paper that surrounds it. Cut out the pattern

pieces and snip the notches at A and B (but not the notches at C and D).

Constructing the Cube

1. Crease the pattern piece with square faces on all of the indicated fold lines, remembering that the unmarked side of the paper should be on the outside of the finished cube. Thus each individual fold along a marked line should hide that marked line from view.
2. Position the pattern piece so that it forms a cube with flaps opening from the top and the bottom, as shown in Figure 2.21.
3. Temporarily attach the two rectangles together inside the cube with paper clips. Then, with the cube still in its "up" position, cut through both thicknesses of paper at once to produce the notches at the positions which you already labeled C and D.
4. Connect three rubber bands together, as shown in Figure 2.22.
5. Slide one end-loop of this chain of rubber bands through the slot which you labeled A,

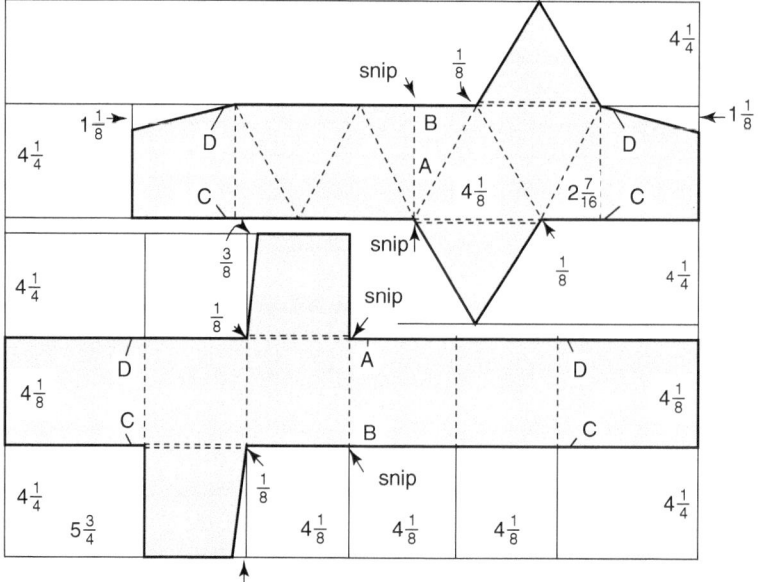

Figure 2.20.

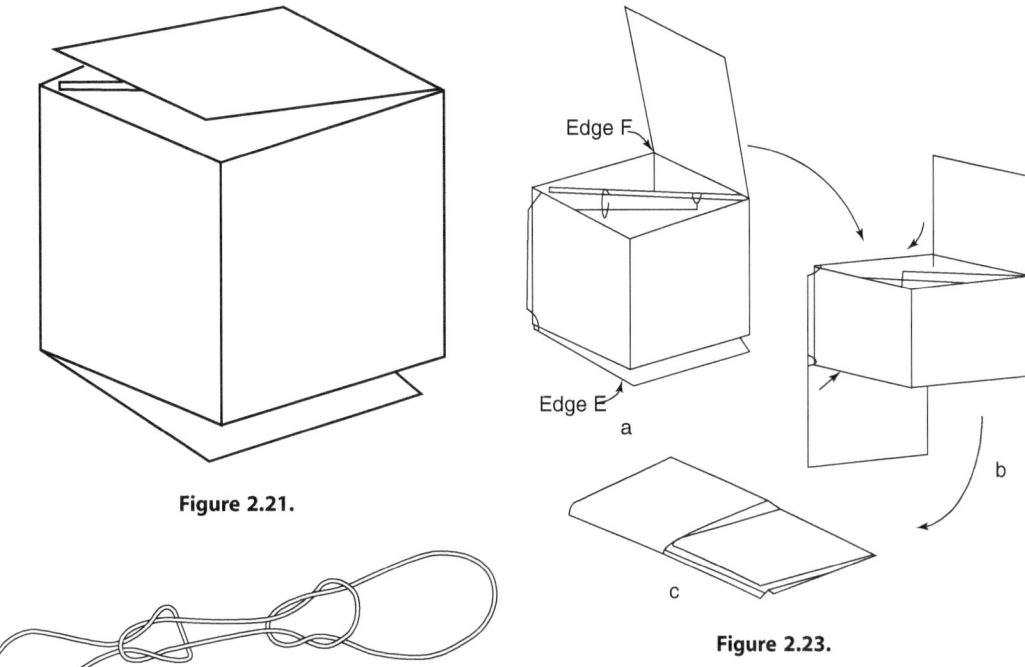

Figure 2.21.

Figure 2.22.

Edge F

Edge E

a

b

c

Figure 2.23.

and the other end-loop through the slot labeled B, leaving the knots on the outside of the cube.

6. Stretch the end loops of the rubber bands so that they hook into slots C and D, as shown in Figure 2.23. The bands must produce the right amount of tension for the model to work. If they are too tight the model will not go flat and if they are too loose the model won't pop up. You may need to do some experimenting to obtain the best arrangement.

7. Remove the paper clips when you are satisfied that the rubber bands are performing their function.

8. To flatten the model push the edges labeled E and F toward each other as shown in Figure 2.23b and wrap the flaps over the flattened portion as in Figure 2.23c.

9. Holding the flaps flat, toss the model into the air and watch it *pop up*. If you want it to make a louder noise when it snaps into position, glue an additional square onto each visible face of the cube in its "up" position. This also allows you to make the finished model very colorful.

Constructing the Octahedron

1. Crease on all the indicated fold lines so that the marked lines will be on the inside of the finished model.

2. Position the pattern piece so that it forms an octahedron with triangular flaps opening on the top and bottom, as shown in Figure 2.24a. Don't be discouraged by the complicated look of the illustration; the construction is so similar to the cube that once you have the pattern piece in hand, it becomes clear how to proceed.

3. Secure the quadrilaterals inside the octahedron with paper clips and cut through both thicknesses of paper to make the notches at C and D. Angle these cuts toward the center of the octahedron (so that the rubber bands will hook more securely). Gluing the quadrilaterals inside the model to each other in their proper position produces a sturdier model.

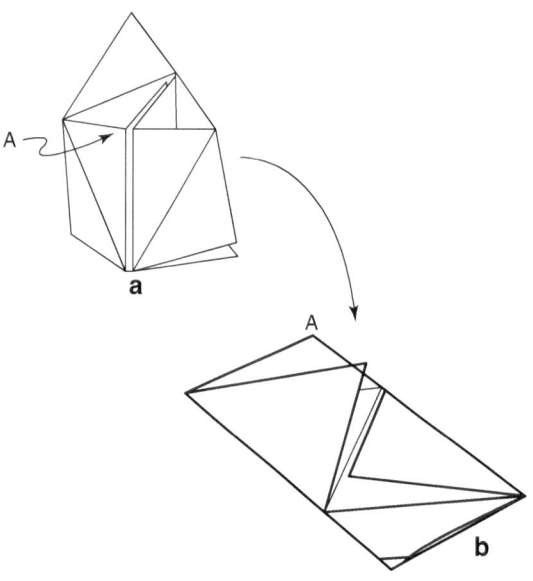

Figure 2.24.

be necessary, so experiment to find the best arrangement.

7. Remove the paper clips when you have a satisfactory arrangement of rubber bands.

8. To flatten the model put your fingers inside and pull at the vertices nearest A and D so that you are pulling those opposite faces away from each other until each is folded along an altitude of that triangular face. Then wrap the triangular flaps over the flattened portion so that it looks like Figure 2.24b.

9. Holding the triangular flaps flat, toss the model and watch it *pop up*. Just as with the cube, this model will make more noise if you glue an extra triangle on the exposed faces. Of course, if you use colored pieces the resulting model is more interesting.

A helpful hint:

4. Connect three rubber bands together, as shown in Figure 2.22.

5. Slide one loop-end of the rubber band arrangement through the slot A and the other loop-end through the slot B, leaving both knots on the outside of the octahedron.

6. Stretch the end loops of the rubber bands so that they hook into the slots at C and D. Some adjustment in the size of the rubber bands may

If you store either the cube or the octahedron in its flattened position for several hours, or days, it may fail to pop up when tossed in the air. This is because the rubber bands lose their elasticity when stretched continuously for long periods of time. If the rubber bands have not begun to deteriorate, the model will behave normally as soon as you let the rubber bands contract for a short while.

The Great Stellated Dodecahedron

Magnus Wenninger

The great stellated dodecahedron (Figure 2.25) makes a lovely decoration or an interesting ornament for any time or place. It is very attractive, and when made as suggested here it is also very sturdy and rigid even though it is entirely hollow inside. The pattern to use for making this model is simply an isosceles triangle with base angles of 72° and a vertex angle of 36°. The length of the base should be between 1 and 2 inches (between 2.5 and 5.0 cm). The angular measures just given will automatically make the equal sides of the triangle τ times longer than the base, where τ is 1.618034 (the golden section number). You will need 60 such triangles to complete one model. Very attractive results can be obtained by using different colors of index card, namely ten triangles of each of six different colors. Astonishingly beautiful results can be obtained by using glitter film with pressure sensitive adhesive backing to cover the index card.

Figure 2.25. Katherine Kirkpatrick studying models made in Magnus Wenninger's workshop.

Getting Started

Begin the work by first cutting all the glitter film triangles to exactly the same size. You can lay out a tessellated network of such triangles by marking the back or waxy side of a sheet of glitter film with a scoring instrument and then cutting out the triangles with scissors. Next peel off one corner of the waxy backing from the film and attach this to a piece of index card. Finally, remove the entire backing while you smooth out the film on the card. Now trim the card with scissors, leaving a border of card all around the film. A quarter inch or so is suitable (about 7 or 8 mm). Next trim the vertices of the triangle as suggested in Figure 2.26. You will now find it easy to bend or fold the card down along the edges of the film even without scoring the card. This edging of card serves as a tab for joining the triangles together. Use ordinary white paper glue (such as Elmer's Glue-All® for this purpose.

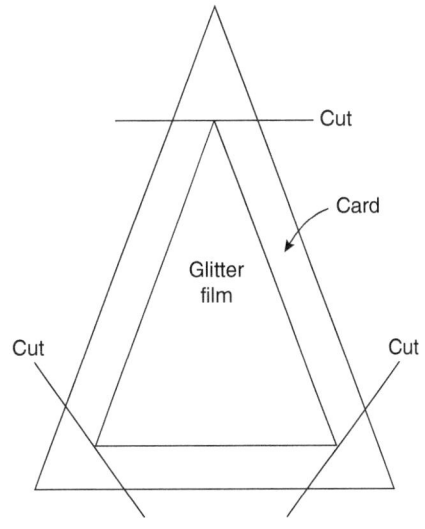

Figure 2.26.

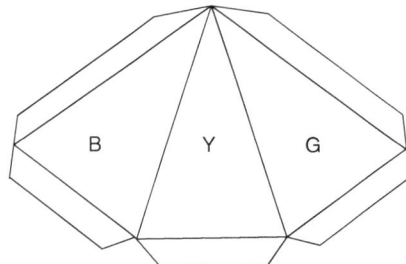

Figure 2.27.

Figure 2.28.

Assembling the Model

Glue three triangles together as shown in Figure 2.27. Shape this part into a triangular pyramid without a base. This will then form one trihedral vertex of the great stellated dodecahedron. The color arrangement for ten vertices is as follows:

(1) B Y G	(6) B W G	Y = yellow or gold
(2) O B Y	(7) O W Y	B = blue
(3) R O B	(8) R W B	O = orange
(4) G R O	(9) G W O	R = red
(5) Y G R	(10) Y W R	G = green
		W = white or silver

The first five vertices or triangular pyramids are joined in a ring with the bottom edges of the middle Y, B, O, R, G of (1), (2), (3), (4), (5), forming an open pentagon. Then the next five parts are added to each edge of this pentagon, so that the W of (6) is glued to the Y of (1) and so on around. This completes half the model. You may find it a bit tricky to get the colors right at first, but the arrangement suggested here makes each star plane the same color. The triangles are star arms, so once you get started right it is not hard to continue.

The remaining ten vertices or parts have their colors in reverse order. They are the mirror image arrangement of the first ten. To make them, just read the color table in reverse order and from right to left. For example, vertex (11) will be R W Y, the reverse of (10) which is Y W R. And this is glued in place diametrically opposite to its counterpart on the model. Watching the colors of the star arms will help you get all the remaining parts in their proper places. As the model closes up it is helpful to use tweezers to get the tabs to adhere. The secret is to do only one pair of tabs at a time. On the last part glue one pair of tabs first. Then, when this has set firmly, put glue on the remaining two sets of tabs and close the triangular opening. The model now has sufficient rigidity so that the tabs will adhere by applying gentle pressure from the outside with your hands. An extra drop of glue at the base of each pyramid corner will provide extra strength where you may perceive a small opening remaining.

You should now see, if you have not already noticed this, that parallel star planes are the same color. Hence twelve star planes complete this model, two of each of the six colors. The twelve stars give this model its name: stellated dodecahedron (Figure 2.28). It is called "great" because it is the final stellation of the dodecahedron, truly a beautiful thing to behold!

"A thing of beauty is a joy forever."

Figure 2.29. Magnus Wenninger leading a workshop.

Creating Kaleidocycles and More

Doris Schattschneider

The transition from a flat pattern to a three-dimensional form can be fascinating to explore. Even the youngest child can shape a simple basket by cutting squares from the corners of a rectangular piece of construction paper and folding it up. But except for this well-known pattern learned as a preschool exercise, the two-dimensional pattern of an unfamiliar three-dimensional object often seems to yield little information about the object. Perhaps part of the reason for this is that we are rarely asked to imagine what shape will result from folding up a flat pattern. The following exercises provide hands-on exploration of some of the relations between flat nets and three-dimensional forms and provide an extra surprise in the creation of kinetic forms.

Folding Strips of Triangles

Begin by constructing strips of four connected congruent triangles like the one sketched in Figure 2.30. You should construct several different kinds of strips — those whose triangles are (1) equilateral triangles, (2) isosceles acute triangles, (3) isosceles right triangles, (4) isosceles obtuse triangles, (5) scalene triangles (a strip of acute triangles, or of right, or obtuse). *Note:* If you are in a hurry, use graph paper for rapid layout of the strips of congruent triangles. If more time is available, carry out rule-and-compass construction of the strips (brush up on the congruence theorems!).

Question: For each of the constructed nets, what three-dimensional shape will be formed

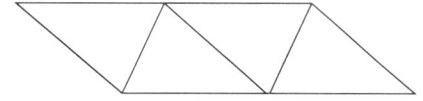

Figure 2.30.

when the net is folded along the common edges of its triangles?

First, guess answers to this question. Then score the connecting edges of the triangles (use a medium ballpoint pen held against a straight edge), cut out the strips of triangles, and fold each of them to see what happens. (All folds should be of the same type, folding the pattern back-to-back.)

After this, other strips of four triangles can be explored: for instance, a strip of four triangles, all acute, but not all congruent; a strip of triangles with some triangles right, others acute, and so forth. Exploring what happens when these nets are folded up leads to some natural questions:

1. When will four congruent triangles form a tetrahedron?
2. What must be true of four triangles if they are to form a tetrahedron?
3. Are there different flat nets (other than the strips of four triangles) that will fold up to make the same shapes as those formed by the strips of four triangles?

Kaleidocycles

Next, we will create and explore nets of connected strips of triangles. For ease and accuracy of construction, large paper and long (18 inch) rulers should be used. Graph paper can be purchased in size 17×22 inch, just right for two constructions. Drawing paper can easily be purchased in large sizes. Lay out each of the two grids shown in Figures 2.31a and 2.32. The grid in Figure 2.31a is made up of six connected vertical strips of congruent isosceles triangles that are characterized by the property that base equals altitude. The grid is easily laid out using graph paper; it is also easily constructed with ruler and compass because of the simple defining property of the triangles. (There is a grid of squares which underlies the triangular grid; this is shown in Figure 2.31b).

The grid in Figure 2.32 is made up of twelve connected vertical strips of isosceles right trian-

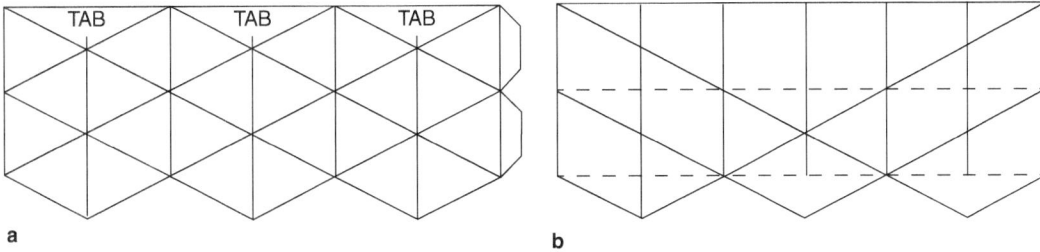

Figure 2.31.

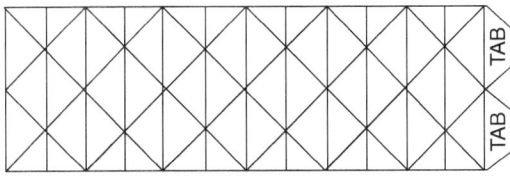

Figure 2.32.

gles, where the top and bottom triangles have been cut in half. This grid is obviously based on a grid of squares, and is easily laid out on graph paper or constructed with ruler and compass.

Before the grids are turned into three-dimensional objects, ask yourself the question that was asked earlier for the single strips. *Question:* From each of the constructed nets (as in Figures 2.31a and 2.32), what three-dimensional shape will be formed when they are folded along the common edges of the triangles?

Of course, you will need to use the earlier answers to the question in attempting to answer the question for the more complex nets. An auxiliary question that is worth asking is: Will all of the lines in the grid (which are common edges of pairs and triangles) play the same role in the three-dimensional form?

Now score all the lines in each grid (use a medium ballpoint pen), and cut out the nets around the outline (be sure to cut around the tabs). Fold the nets as follows:

1. Fold the net face-to-face (valley fold) on all vertical lines, including those to which the tabs are attached.
2. And fold the net back-to-back (mountain fold) on all diagonal lines.

Then cup the folded net in both hands, and gently squeeze it to encourage the top and the bottom to come together.

The net in Figure 2.31a should come together easily, with the half-triangles labeled as tabs completely covered. A chain of linked tetrahedra is formed. (Glue or tape the edges of the tetrahedra fitted over the tabs.) Holding the ends of the chain, bring the ends of the chain together, fitting the tabs at one end of the chain into the open edge at the other end of the chain. (If the chain does not come together easily, turn it until it does.) Glue or tape these last two edges to the tabs, completing the model.

The ring of six linked tetrahedra is a (carefully) crinkled torus (doughnut), and has the property that it can be *endlessly turned through its center hole.* Simply grasp the model in both hands and turn the tetrahedra inward, pushing the points through the center hole!

The Isoaxis

The net in Figure 2.32 will not come together to form a closed three-dimensional form, but rather it will form a (carefully) crinkled cylinder that will also turn through its center hold, changing its shape and appearing to "bloom" as it is turned. This form was discovered by graphic designer Wallace Walker, and is called Isoaxis®. Assemble Isoaxis as follows. Gently squeeze the scored and folded net so that it begins to curl and collapse along the fold lines. When fully collapsed, it will look like an accordion-folded paper with

Figure 2.33. Corraine Alves and Diana Weimer making kaleidocycles.

square cross section. (One method of achieving this state of the model is to begin at one end of the net, collapsing the net along the folds to form a square cross section, and holding the collapsed part between thumbs and forefingers, "gathering" the rest of the net into the collapsed state with the middle fingers.) The accordion-folded net should be pressed firmly; it is best if it can be pressed under a heavy object for 12 hours or more to set the folds fully. The two ends of the folded net are then joined (use tape or glue), matching tabs to the inside of opposite triangles. Join one tab at a time; the model will be tight, and so turn it through its center to join the second tab. To rotate this model, hold it in both hands and bring points to the center; push on the points. The crinkled cylinder will turn continuously through its center hole!

Further Exploration

There are many avenues for follow-up. A few are suggested below.

1. Explore the symmetry properties of the three-dimensional models. This can be enhanced by decorating the faces of the models to display various symmetries. One question that will

need to be answered is: What faces in the flat net are adjacent (or become adjacent during rotation) in the three-dimensional forms?

2. Create other similar nets, varying the kinds of triangles chosen, and the number of triangles in the net. Fold in the same manner to see what three-dimensional forms result. A good challenge that can be met using only a knowledge of elementary geometry is: create other rings of tetrahedra having more tetrahedra, but such that the center hole in the ring is (in theory) a point, as is the case for the model in Figure 2.31. The model in Figure 2.31 has been called a "hexagonal kaleidocycle" by Walker and Schattschneider, because the center cross section of the assembled form is a regular hexagon.

Information on Construction Materials

The basic necessities for the above constructions are

- Paper
- 18-inch ruler
- Medium ballpoint pen
- Scissors
- White glue or tape

If the models are to be decorated, then coloring materials that will not weaken or warp the paper should be used. Since the models rotate, the paper chosen must not easily tear or break when bent repeatedly. Ordinary construction paper is not suitable. In addition, the paper should be heavy enough so that the three-dimensional models have suitable firmness. Medium-weight drawing paper, 100% rag, is excellent, and takes decoration well. Ordinary graph paper is too thin, but there are excellent heavier drafting and design papers that come in large sizes. The nets should not be made too small, or they become very difficult to put together and manipulate. A good size for the nets is 2.5 to 3 inches width for each "panel" of linked triangles for Figure 2.31, and 1.25 to 1.75 inches for each "panel" of linked triangles for Figure 2.32. The overall width of the nets should be in the range of 15 to 22 inches.

White glue seems to be best for assembling the models; in any case, the glue chosen should not warp the paper, nor should it be the "instant hold" variety since tabs need to be manipulated into place before the glue sets. If tape is used, then it must be the type used for hinging; ordinary clear plastic tape will break after a few turns of the model.

Giftwrap paper which has a pattern based on a square grid can be laminated to drawing paper to create a nicely decorated Isoaxis with an all-over pattern. Use a spray glue that will not be brittle when dry to affix the gift paper. The square grid of the pattern must be carefully followed for the lines of the net of Isoaxis.

The Rhombic Dodecahedron

Arthur L. Loeb

This is a recipe for constructing modules that generate the rhombic dodecahedron in two fundamentally different ways. The first construction stellates a cube with six square pyramids; the second stellates a regular octahedron with eight triangular pyramids.

The Pyramids

The first step is the construction of the sides of the pyramids. The square pyramids have an apex angle whose cosine equals $1/3$, while the triangular pyramids have an apex angle whose cosine equals $-1/3$. In order to produce mutually congruent dodecahedra by both methods, we construct the template shown in Figure 2.34 by the following steps:

1. Draw two mutually perpendicular lines. Call their intersection O.
2. Choose a point A, different from O, on one of the mutually perpendicular lines.
3. Draw a circle having radius equal to three times the distance OA, whose center is located on A.
4. Call the intersections of this circle with the extension of the line OA C and B, as shown. Call an intersection of the circle with the line perpendicular to OA D, as shown.
5. Connect C and D, as well as B and D.

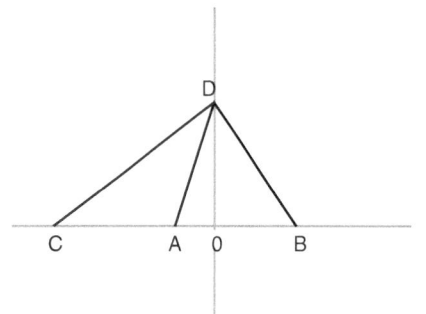

Figure 2.34.

The resulting template furnishes the following linear and angular dimensions:

- The length of the line segment CD is the edge length of the octahedron to be built.
- The length of the line segment BD is the edge length of the cube to be built.
- The triangle CAD is the shape of the sides of the triangular pyramids to be built.
- The triangle BAD is the shape of the sides of the square pyramids to be built.

Construction of the Polyhedra

1. Construct a regular octahedron whose edge length equals the length of line segment CD.
2. Construct a cube whose edge length equals the length of line segment BD.
3. Construct eight triangular pyramids whose bases are equilateral triangles having edge length equal to the length of line segment CD, and whose sides have the shape of triangle CAD.
4. Construct six square pyramids whose bases are square having edge length equal to the length of line segment BD, and whose sides have the shape of triangle BAD.

Juxtaposition of Polyhedra

Arrange the six square pyramids so that their square bases are in the configuration shown in Figure 2.35. Hinge them together so that they can rotate with respect to each other around their shared edges. When the pyramids are folded inward until their six apexes touch, they will form a cube congruent with the cube also constructed.

Arrange four of the triangular pyramids with their triangular bases in the configuration shown in Figure 2.36, and hinge them together as above. When folded in until their apices touch, they will form a regular tetrahedron. Repeat for the remaining four tetrahedra.

Place the six square pyramids around the cube, square faces joined to square faces. Place the

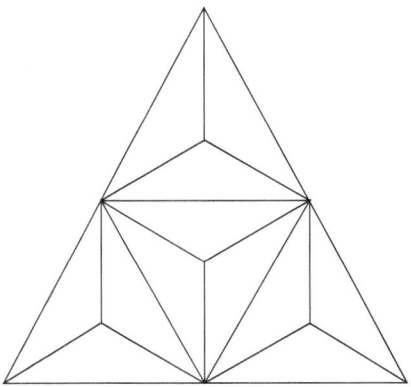

Figure 2.35.

Figure 2.36.

- Cube
- Rhombic dodecahedron
- Square pyramid
- Irregular octahedron
- Regular octahedron combined with eight triangular pyramids

Combinations of these (say cube in combination with square pyramids) are, of course, also possible.

A Note on Materials

Contributed by Jack Gray

Any useful polyhedral model is formed on the spectrum between "a rough sketch" and "a long-lasting work of art." The position on the spectrum is determined by the choices of materials, tools, and techniques as well as by the time and care used in the construction process. A rough sketch is always a valid precursor to a work of art. Expect to make a few mistakes on the sketch, and then try to conquer those in a second model.

Transparent tape is a good hinging material, while paper tape is thicker and more cumbersome. Use permanent tape, taking care to position it as follows. Place a strip of tape, sticky side up, on a flat surface. The strip should be longer by a good amount than the edge to be hinged. Weight down the ends of the tape so that it cannot move while you are connecting the polyedron to it.

Carefully lower the edge of the first polyhedron to the tape. Before letting it make contact, make sure that it is in the center of the width and the length of the tape. Make contact along the whole length of the edge.

Orient the second polyhedron to the first. Slide the second polyhedron down the face of the first until its edge touches the tape; then rotate it about that edge, so that contact with the tape is made along the entire edge. Trim off the excess tape with an X-Acto® knife or a single-edge razor blade.

Flip over the pair of joined polyhedra and inspect the tape hinge. Burnish it with your finger to complete contact along the full surface of the tape.

eight triangular pyramids around the octahedron, with the equilateral triangles joined. The result should be two mutually congruent rhombic dodecahedra. Note that the cube edges constitute the shorter, the octahedron edges the longer, diagonals of the rhombic faces.

Place two square pyramids with their square bases joined. The result is an octahedron that is not regular, because its faces are not equilateral. Six of these irregular octahedra can be put together to form a rhombic dodecahedron. (*Note:* this would require *twelve* square pyramids rather than the six already constructed).

Space-Fillers

Of the polyhedra constructed, the following will fill space without interstitial spaces.

Figure 2.37. Arthur L. Loeb demonstrating his models.

Place another piece of tape of the same length on the flat surface. This will be used to tape the other side of the joined edges, creating a hinge that is equally strong on both sides. The tape should be weighted down as before, and care should be taken in centering the already-joined edges on the strip of tape before making contact. Let the joined faces lie flat against the tape to make contact along the full width of tape surface. Remove the excess tape and burnish as before.

If your sketch looks like a work of art, plan a finished model. Visit a local art store or hobby shop to examine the sheet materials that are available.

Various colored art papers, colored and transparent acetate, mirrored Mylar®, oak tag, and construction paper can be found. Fine rice papers are good for finishes, though they are too flexible for the body of such models. (In adding a surface finish of thin sheet material to your model, cut out each polygonal face so that it will not go across the hinge. Otherwise, the finish material will buckle when flexed.) Your experience making a sketch model will prepare you to pick materials "by feel."

Thicker sheet material like mat board and Plexiglas® need to have their edges mitered to half the dihedral angle between faces to prevent the thickness of the material from creating inaccuracies. Great care should be used in gluing such joints, so that glue does not spill onto the surfaces.

On a finished model, the hinging should be done with transparent polyester hinging tape. Cloth tape can be used on larger models.

Balloon Polyhedra

Erik Demaine, Martin Demaine, and Vi Hart

You have probably seen a long balloon twisted into a dog as in Figure 2.38. But did you know that the same balloon can be twisted into a regular octahedron like Figure 2.39? Here are two one-balloon constructions and their associated networks of edges and vertices (the technical term for such a network is a *graph*).

Balloon twisting offers a great platform for making, exploring, and learning about both polyhedra and graphs. Even the classic balloon dog can be thought of as a graph, with edges corresponding to balloon segments between the twists. Children of all ages can thus enjoy the physicality of the balloon medium while learning about mathematics.

Here we give a practical guide to making polyhedra and related geometric constructions with balloons, while briefly describing the mathematics and computer science related to balloon construction.

Figure 2.38. Classic dog (one balloon).

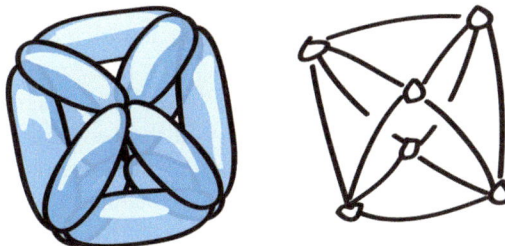

Figure 2.39. Octahedron (one balloon).

If Euler Were a Clown

What polyhedra can be made by twisting a single balloon, like the octahedron in Figure 2.39? The balloon must traverse every edge of the polyhedron in sequence, and can traverse each edge only once, though it can (and will) visit each vertex multiple times. Mathematicians will recognize this structure as an *Eulerian path*. We review the classic mathematics of this structure in the context of balloons by wondering: what is the simplest polyhedron that can be made from a single balloon?

The simplest (nonflat) polyhedron is a tetrahedron, but a tetrahedron cannot be made from a single balloon. Such a balloon would start at some vertex and end at some other vertex, but for every remaining vertex, each time the balloon enters the vertex it also exits the vertex. Thus, for a balloon twisting of a graph to possibly exist, every vertex except possibly two (the starting and ending vertices) must have an even number of edges incident to it, or even *degree*. But in the tetrahedron, all four vertices have odd degree. Thus the tetrahedron cannot be made from one balloon.

The four-dimensional analog of a tetrahedron (called the 4-*simplex*) consists of four tetrahedra glued together face-to-face. One 3D projection of a 4-simplex is a regular tetrahedron with all four vertices joined to a fifth vertex added in the center. Every vertex is thus joined to all four other vertices, giving it even degree. So the 4-simplex does not have the "too many odd-degree vertices" obstruction that the tetrahedron had. Indeed, Figure 2.40 shows a 4-simplex made from one balloon. Building one is somewhat difficult for practical reasons; we suggest you try it.

In fact, a single balloon can be twisted into any connected graph in which all vertices have even degree. The starting point is to traverse the graph naïvely: start the balloon at any vertex, route it along any incident edge, and keep going, at each step following any edge not already visited. Because the vertex degrees are all even, whenever the balloon enters a vertex, there is

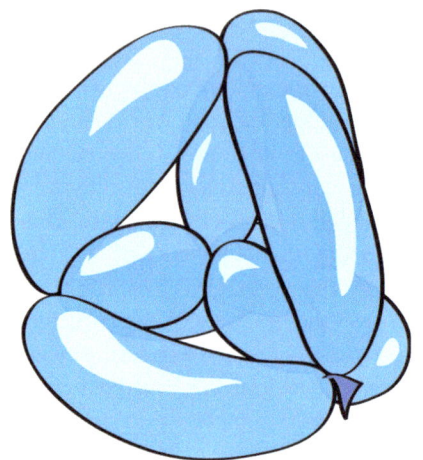

Figure 2.40. The 4-simplex, made from one balloon, is a puzzle to twist.

always another unvisited edge along which it can exit. The only exception is the starting vertex, where the balloon's initial exit left an odd number of unvisited edges. This vertex is the only place where the balloon might have to stop, and eventually this must happen because the balloon will run out of edges to visit. When this happens, the balloon forms a loop that visits some edges once, but possibly does not visit some edges at all.

Now consider just the graph of unvisited edges. It too has even degree at every vertex, because the balloon exited every vertex it entered, including the start vertex. Hence we can follow the algorithm again with a second balloon, and a third balloon, and so on, until the balloons cover all the polyhedron's edges (and no two balloons cover the same edge). Now we take any two balloons that visit the same vertex and merge them into one balloon by the simple switch shown in Figure 2.41. This also forms a loop, and visits all the same edges. Repeating this process, we end up with one balloon forming a loop that visits every edge exactly once.

Now we have a construction for twisting many graphs from a single balloon. We have seen one 4D polyhedron to which this construction applies, but what about 3D polyhedra? One example is the octahedron from Figure 2.39. The octahedron has

six vertices, each of even degree 4. Therefore the general construction applies; Figure 2.42 shows a practical construction. The octahedron is actually the only Platonic solid twistable from a single balloon: for all the others, every vertex has odd degree (3 or 5).

Are there simpler 3D polyhedra than the octahedron that are twistable from one balloon? If we glue two tetrahedra together, we get a *triangular dipyramid*. The two apexes have odd degree 3, but the remaining vertices all have even degree 4. We know that a single balloon cannot tolerate more than two odd-degree vertices, but this polyhedron has just two, putting it right at the borderline of feasibility. Figure 2.43 shows that it indeed can be made from one balloon.

More generally, a single balloon can be twisted into any connected graph with exactly two vertices of odd degree. Just imagine adding an extra edge to the graph, connecting the two odd-degree vertices. This addition changes the degrees of the two odd-degree vertices by 1, making them even, and does not change the degree of any other vertices. Hence this modified graph has all vertices of even degree, so it can be twisted from a single balloon forming a loop. We can shift the loop of the balloon so that it starts and ends at one of the odd-degree vertices. Then we remove the added edge from both the graph and the balloon. We are left with a single balloon visiting every edge of the original graph exactly once. Naturally, the balloon starts at one odd-degree vertex and ends at the other.

Summarizing what we know, a one-balloon graph has at most two odd-degree vertices, and every connected graph with zero or two odd-degree vertices can indeed be twisted from one balloon. What about graphs with one odd-degree vertex? Don't worry about them; they don't exist. Euler showed that every graph has an even number of odd-degree vertices. To see why, imagine summing up the degrees of all the vertices. We can think of this sum as counting the edges of the graph, except that each edge gets counted twice, once from the vertex on either end. Therefore the sum is exactly twice the number of edges, which is an even number. The number of odd terms in

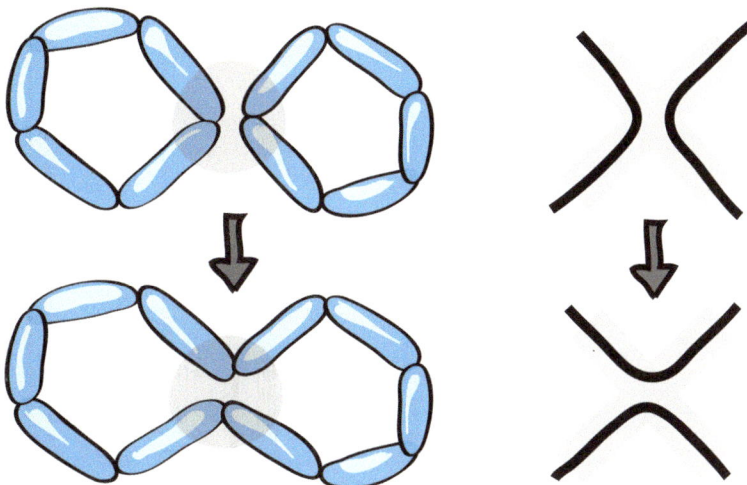

Figure 2.41. Joining two balloon loops (*top left*) into one balloon loop (*bottom right*) when the two loops visit a common vertex (*shaded*, and abstracted on the *right*).

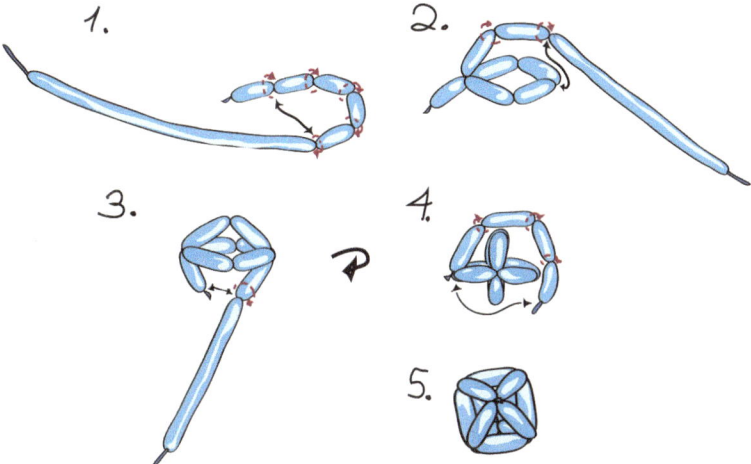

Figure 2.42. The octahedron is like the balloon dog of balloon polyhedra. With practice, it can be twisted quickly from one balloon.

the sum (the number of odd-degree vertices) must thus be even.

We conclude that we know all one-balloon graphs. In fact, we have just rediscovered the classic characterization of Eulerian paths, common in graph-theory textbooks, but in the context of balloons. If you are looking for a good challenge for making a polyhedron from one balloon, we recommend the cuboctahedron. Every vertex has degree 4, but there are twenty-four edges and sharper dihedral angles, making

it difficult to twist except from especially narrow balloons.

Cheating with One Balloon

One trick for transforming graphs into one-balloon graphs is to double every edge: whenever two vertices are connected by an edge, add a second edge alongside it. This change doubles the degree of every vertex, so all resulting vertices

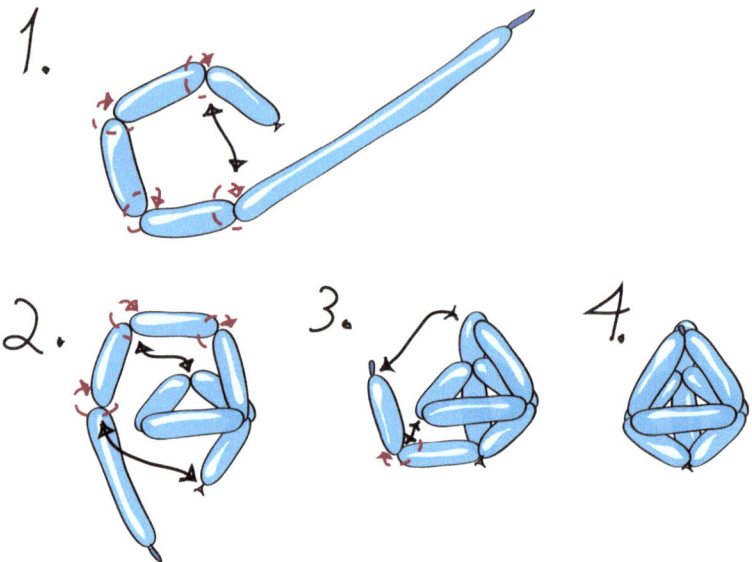

Figure 2.43. The triangular dipyramid is perhaps the simplest polyhedron twistable from one balloon, and it is easier than the octahedron.

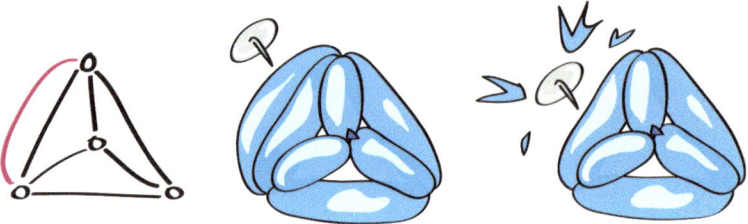

Figure 2.44. Pop-twisting a tetrahedron from one balloon.

have even degree. Therefore we can make the doubled graph from one balloon. In other words, one balloon visits every edge in the original graph exactly twice instead of once.

In fact, we do not need to double every edge: we just need to double edges along paths connecting the odd-degree vertices in pairs, except for one pair of odd-degree vertices that we can leave alone. Minimizing the number of edges we double is the *Chinese postman problem*, a well-studied problem in computer science. Balloons offer a fun context for studying efficient algorithms for this problem.

After we have made a balloon with some of the edges doubled, we could *pop* the extra edges using a sharp object. The result is a more uniform

aesthetic. For example, Figure 2.44 shows how a tetrahedron, which would normally require two balloons, can be made from one balloon with one popped segment. (We leave the construction of the tetrahedron with a doubled segment as an exercise for you. Remember to start and end at the two odd-degree vertices.) In practice, be careful to twist the ends of a segment extensively before popping it to prevent affecting the incident segments.

On the topic of balloon popping, a challenging type of puzzle is this: given an already-twisted balloon polyhedron, can you pop some of the balloon segments to make another desired graph?

This problem is known as *subgraph isomorphism*, and is among the family of computa-

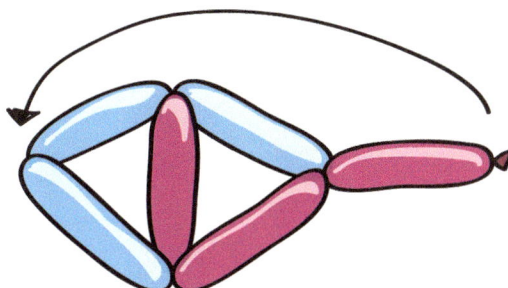

Figure 2.45. The tetrahedron is easy to twist from two balloons.

tionally intractable "NP-complete" problems, so there is likely no good algorithm to solve it.

Polyballoon Constructions

If we cannot make a graph with just one balloon, how many do we need? The minimum number of balloons that can make a particular graph, with each edge covered by exactly one segment, is the graph's *bloon number*.

There turns out to be a very simple formula for the bloon number: it is half the number of odd vertices (unless, of course, the graph has no odd vertices; then the bloon number is 1, not 0).

On the one hand, we cannot hope for fewer balloons: each odd-degree vertex must be the start or end of some balloon, so each balloon can "satisfy" only two odd-degree vertices. On the other hand, there is a construction with just this many balloons, using a simple construction similar to the arguments above. First, we add edges to the graph, connecting odd-degree vertices in disjoint pairs. The number of added edges is half the number of odd-degree vertices. The resulting graph has all vertices of even degree, because we added one edge incident to each vertex formerly of odd degree. Therefore it can be made from one balloon forming a loop. Now we remove all the edges we added. The number of removed edges, and hence the number of resulting balloons, is half the number of odd-degree vertices.

It now becomes an easy exercise to figure out how many balloons we need for our favorite polyhedron. For example, a tetrahedron requires only two balloons, as shown in Figure 2.45. An icosahedron requires six balloons, which can be made into identical two-triangle units as shown in Figure 2.46. The same unit can make a snub cube from twelve balloons, as shown in Figure 2.47, as well as a snub dodecahedron from thirty balloons (another exercise).

Note that, in both cases, there are two fundamentally different ways as to how the balloons can be assembled, right-handed and left-handed. This is a nice lesson on chirality (handedness).

For all Platonic and Archimedean solids, there is a construction out of balloons that

1. Uses the fewest possible balloons,
2. Uses balloons all of the same length, and
3. Preserves all or most of the symmetry of the polyhedron.

Achieving all of these properties together can be a fun puzzle. In fact, achieving just the first two properties is a computationally difficult problem: decomposing a graph into a desired number of equal-length balloons is a special case of "Holyer's problem" in graph theory, and it turns out to be among the family of difficult "NP-complete" problems, even for making polyhedra.

Tangles

Balloons can make many more graphs than just polyhedra. A simple extension is to look at disconnected graphs, made from multiple shapes. In *Orderly Tangles*, Alan Holden introduced *regular polylinks*, symmetric arrangements of identical regular polygons, which make a good subject for balloon twisting. Tangles have recently been explored with the aid of (freely available) computer software to find the right thicknesses of the pieces, which may be especially useful for determining the best-size balloons for twisting.

Perhaps the simplest example of a tangle is a model of the Borromean rings assembled from three rectangles, each lying in a coordinate plane. This model is fairly easy to construct from three balloons, as shown in Figure 2.48. Figure 2.49 shows a more challenging construction made

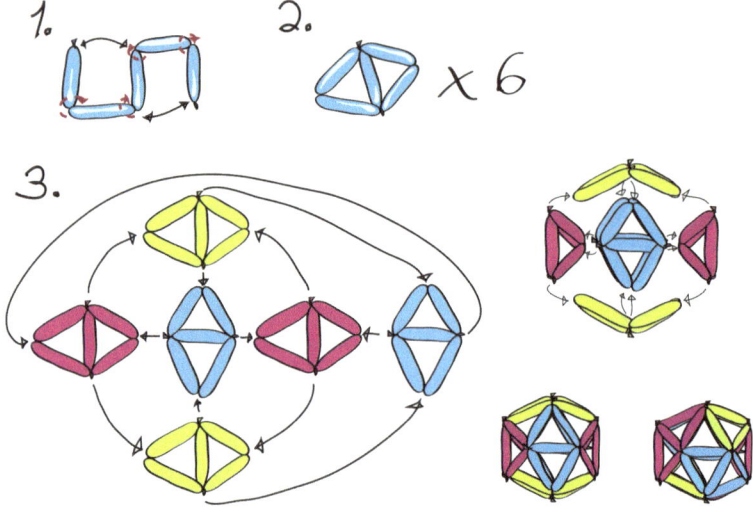

Figure 2.46. The icosahedron is a good example of joining several (six) identical balloon units. The tied balloon ends make it easy to attach vertices together.

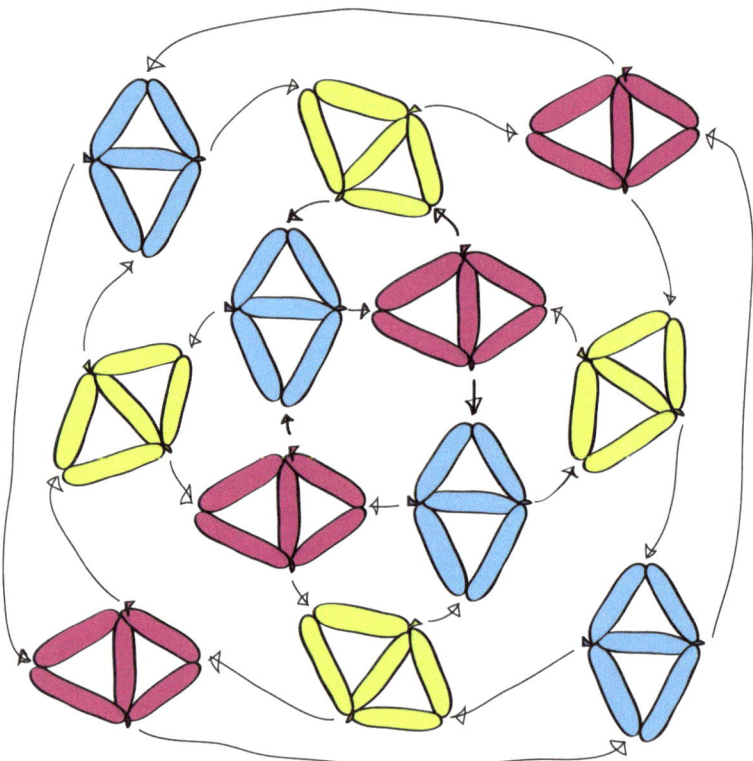

Figure 2.47. The snub cube is a bigger example of joining several (twelve) identical balloon units, recommended for groups of polyhedral balloon twisters. We leave the final form as a surprise.

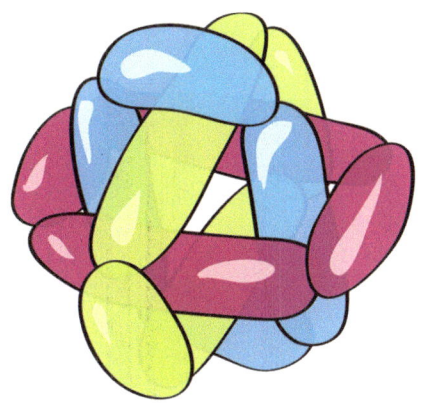

Figure 2.48. The Borromean rings are easy to make from three balloons.

from six squares; the difficulty is not twisting the squares, of course, but interlinking them in the correct over-under pattern. Harder still are the six-pentagon tangle and the four-triangle tangle.

A further generalization of regular polylinks are *polypolyhedra*, which allow the symmetrically arranged shapes to be Platonic solids in addition to regular polygons. Some examples such as the famous "five intersecting tetrahedra" have been around for many years, and recently even the subject of balloon twisting. But a thorough enumeration of all polypolyhedra is relatively recent. For a longer project, ideally with a group of mathematical balloon twisters, we recommend looking through this catalog of polypolyhedra.

Practical Guide for Twisting Balloon Polyhedra

Twisting balloons into polyhedra and related structures has been explored by several others. Several websites listed in the Notes for this chapter include video instructions. In the rest of this section, we give some practical tips for twisting your own balloons into polyhedra.

Long skinny balloons come in two main sizes: 160s (1-inch diameter, 60-inch length) and 260s (2 inches by 60 inches). For making one-balloon polyhedra, or complicated tangles, 160s are better. For most multiple-balloon polyhedra, the extra thickness of the 260s is better for stability. Of course, by not inflating all the way, or by attaching multiple balloons together, the balloon can reach any width-to-length ratio you like.

One of the biggest challenges is to get the balloon segments to be of equal length. Twisting a single balloon into the correct number of equal-length segments is something that just takes practice. If your lengths are off, you can always untwist and start again. When working with multiple balloons, it can be helpful to inflate them all to the same length before twisting them.

You may want to shorten the lengths of all the edges on a finished balloon structure, for example to get a tangle to fit together snugly. One way to do this is to grab all the edges at a vertex, and twist them all together. Performing such a twist at every vertex results in an aesthetically pleasing effect that accents the vertices.

Another problem you may run into is wanting to pass the end of a balloon through a hole that is narrower than the inflated balloon. If you can fit the deflated end of the balloon through, you can then squeeze the air through the deflated portion and inflate the other side. It is better, though, to figure out a twisting order that avoids this, and this is a puzzle in itself.

Enjoy your mathematical twists!

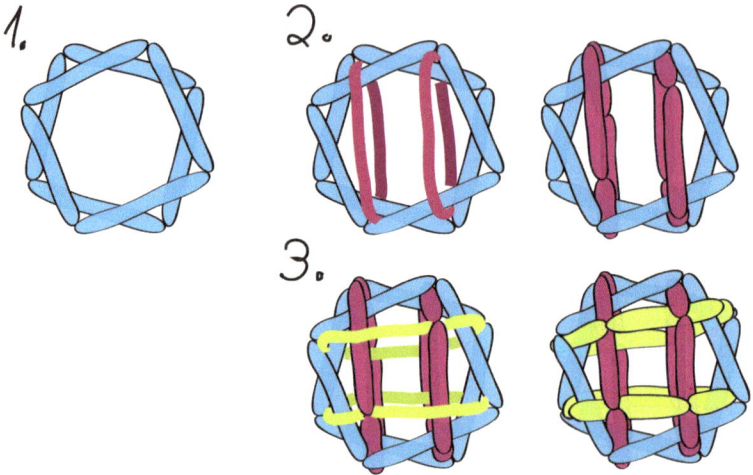

Figure 2.49. The six-square tangle is a good puzzle in assembling balloon tangles.

3

Regular and Semiregular Polyhedra

H.S.M. Coxeter

The cube, the octahedron, and the tetrahedron obviously have been admired for thousands of years. It is impossible to say who first described them. Certainly the Pythagoreans knew all about them. I understand that a dodecahedron was found in Italy which was apparently made in 500 B.C. or perhaps even earlier, and that icosahedral dice were used by the ancient Egyptians. They can be seen in the British Museum, although there is some doubt about their exact date. All the five so-called Platonic solids are described in the later books of Euclid. Subsequent writers have made it much easier to see that the number of Platonic solids is just five.

First of all, perhaps one should define what one means by a regular solid. It is rather strange that not many people realize how very simple the definition can be. If one starts in the plane defining a regular polygon, one cay say that a polygon is regular if it has a circumcircle and an incircle which are concentric. All the vertices lie on a circle and all the sides touch a circle and those two circles have the same center. That is a very obvious way of defining a regular polygon. The same thing works in the analogous situation for a polyhedron in three dimensions. A polyhedron is regular if it has three spheres, all with the same center: one through all the vertices, one touching all the edges and one touching all the faces. And that is all one needs. It is very easy to show from this that the faces are regular polygons and that they are all alike.

Of course, if you are dealing with a honeycomb (that is, a tessellation of the plane with regular polygons), you have to make the definition a little bit different and say that the polygons are regular and all alike. And then you know that the ways of filling the plane with regular polygons are just three: triangles, six at a vertex, which I call $\{3, 6\}$; squares, four at vertex, which I call $\{4, 4\}$; and hexagons, three at a vertex, which I call $\{6, 3\}$. Those are what we call Schläfli symbols. The first entry is the number of sides of a face and the second is the number of faces at a corner. So a cube, for instance is called $\{4, 3\}$. Figure 3.1 gives such symbols for all five Platonic solids.

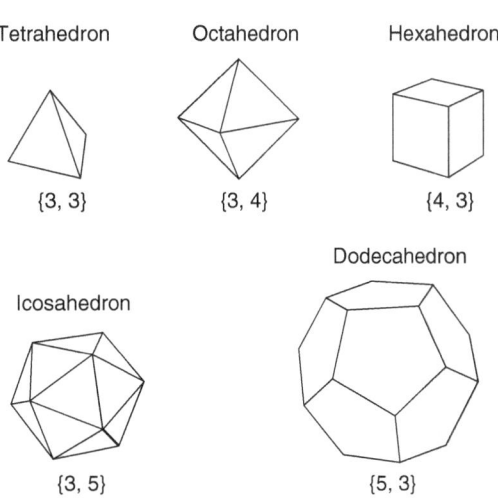

Figure 3.1. Schläfli symbols for the Platonic solids.

H.S.M. Coxeter (1907–2003)

M. Senechal (ed.), *Shaping Space*, DOI 10.1007/978-0-387-92714-5_3,
© Marjorie Senechal 2013

There is a very nice book on the history of these things by van der Waerden called *Scientific Awakening*. He has a fine chapter about Pythagoras, and he says that an Etruscan dodecahedron made of soapstone was found near Padua, dated from before 500 B.C. The faces of a dodecahedron are pentagons. You know that if you draw the diagonals of a pentagon you get a star pentagon inside. The star pentagon is the ancient symbol of the Pythagoreans. The story is told in van der Waerden's book that one of the Pythagoreans was lying on his deathbed in a foreign country, unable to pay the man who had taken care of him. And he advised this man to paint a star pentagon on the door of the house so that any Pythagorean who might enter would make inquiries. And many years later, a Pythagorean did come, and the man was richly rewarded. A rather nice little story.

Coming to much more recent times, René Descartes (1596–1650) wrote a book called *De Solidorum Elementis*. Although the manuscript was soaked for three days after a shipwreck on the river Seine, it was copied in 1676 by Leibniz before it was lost forever. His copy was lost too, but that loss was only temporary; two hundred years later Leibniz's copy was found in Hanover, Germany. Federico gives the details of this story.

Shortly before Descartes came Leonardo da Vinci and Luca Pacioli. Pacioli wrote a book called *Divina Proportione* in which he had pictures of regular and Archimedean solids based on models made by da Vinci.

Johannes Kepler was very interested in these things, and his *Harmonices Mundi* of 1619 contains one of his famous illustrations. In Figure 3.2 you see at the top (Mm) his attempt to fill the plane with polygons of various kinds such as a dodecagon with hexagons and squares around it. Next you see a tetrahedron, and then two halves of an octahedron (Oo) showing that the octahedron is just two square pyramids put base to base. Rr is the dodecahedron and here he has divided it into two parts by cutting along a set of ten edges which form a Petrie polygon. See how they fit together. And in Qq the cube is divided by a skew hexagon, which is *its* Petrie polygon. Those two halves fit together to make

the cube. Pp is the icosahedron with little caps taken off the top and bottom, leaving a pentagonal antiprism between the two pentagonal pyramids. Alternatively, just take the antiprism and stick the two pyramids on top and bottom and there is the icosahedron.

One of the ideas that Kepler got from the ancient Greeks was making four of the five Platonic solids correspond to the four elements: earth, air, fire, and water. In Figure 3.2 you see the tetrahedron with a bonfire drawn on it because the tetrahedron represents *fire*; the octahedron, representing *air*, has birds; the icosahedron has a lobster and fish because the icosahedron represents *water*; and you see a hoe and spade on one face of the cube, a carrot on another, and a tree on a third because the cube represents *earth*. Well of course there was a fifth solid and no fifth element, so the ancients just said that the dodecahedron should correspond to the whole universe. That was curiously echoed by the Japanese; I have a model in which if you look closely, you find that on the twelve faces are drawn the twelve Japanese signs of the Zodiac. It's a little bit different from the Greek Zodiac: one sign is a dragon, one is a doe, one is a dog, one is a chick, and so on.

In Figure 3.2 you see also two *stellated* dodecahedra, each derived from the convex dodecahedron by extending the planes of the faces. Kepler's drawings Tt of the *great* stellated dodecahedron are a little bit inaccurate, but they give us the right idea. In Ss we see two views of the simpler *small* stellated dodecahedron. Somewhere in Italy there is an elaborate floor on which a picture of this polyhedron appears, nicely drawn in 1420 as a mosaic. So the small stellated dodecahedron may have been discovered by Paolo Uccello, two centuries before Kepler. Figure 3.3 shows a different view of the same polyhedron.

Kepler seems to have understood the nature of *reciprocation:* the idea that the cube and the octahedron are reciprocal, and the dodecahedron and the icosahedron are reciprocal. Figure 3.4 shows a shaded cube and white octahedron with corresponding edges crossing each other at right angles. You see that each corner of the cube emerges through a face of the octahedron and vice versa. Curiously enough, although Kepler

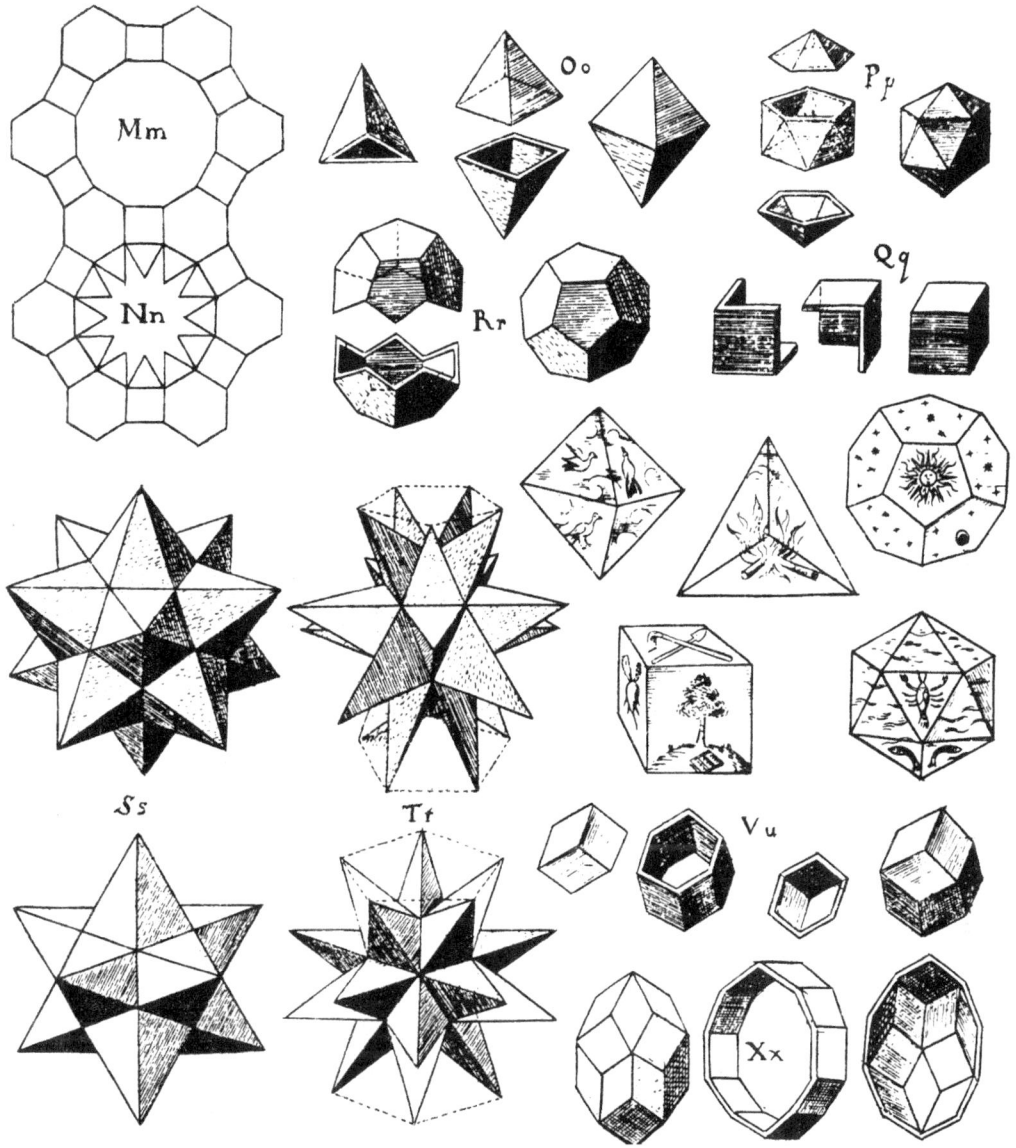

Figure 3.2. Kepler's drawing of the regular solids (and others).

had that idea, he did not pursue it in the matter of the star polyhedra. If you think of a Schläfli symbol, whereby the cube is called {4, 3} and the reciprocal octahedron {3, 4}, then it would be natural to call the small stellated dodecahedron {5/2, 5} because the faces are star pentagons corresponding to the fraction 5/2, and there are five at each corner. Similarly, the great stellated dodecahedron is {5/2, 3}. The faces are pentagrams again and there are three of these at every corner.

Turn these symbols around and you have {5, 5/2} and {3, 5/2}. {3, 5/2}, the great iscosahedron, is shown in Figure 3.5. These were not discovered until two hundred years later, by the Frenchman Louis Poinsot, but when they finally were there, it became clear they were reciprocal to Kepler's {5/2, 5} and {5/2, 3}.

I have a little more to say about Kepler because of his interest in mystical connections between the Platonic solids and astronomy. Of course,

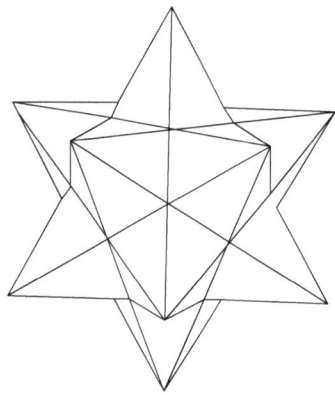

Figure 3.3. The small stellated dodecahedron.

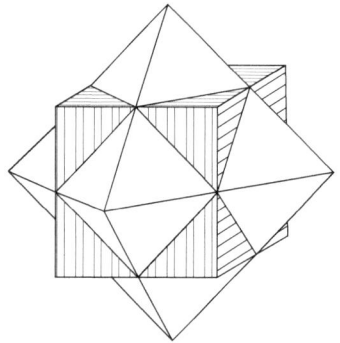

Figure 3.4. The cube and the octahedron.

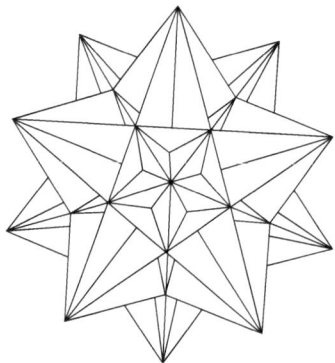

Figure 3.5. The great icosahedron.

he was very interested in astronomy. He had a curious idea about the orbits of the planets which is very nicely described in *The Sleepwalkers* by Arthur Koestler, the man who also wrote *Darkness at Noon*. Koestler said: "Into the orbit or sphere of Saturn he inscribed a cube."

Now let's say exactly what that means. You think of the orbit of Saturn as the equator of a big sphere and similarly the orbit of Jupiter as the equator of a smaller sphere inside. Kepler was a sufficiently good astronomer to know that the orbits were not really circles, but more like ellipses, so that there is a minimum distance and a maximum distance from the sun in each case. What he did was to imagine a sphere in space that was made as a shell, not a mathematical sphere but a solid shell with an outer radius corresponding to the maximum distance of the planet and an inner radius to the minimum distance. So it is a hollow sphere. And he would take the minimum distance of Saturn and divide it by the maximum distance of Jupiter to get the ratio of the circumradius and in radius of a cube. For the six planets known to Kepler, Table 3.1 gives the distances from the sun in millions of miles for comparison with the circumradius $_0R$ and inradius $_2R$ of each Platonic solid. For instance, in the case of Saturn and Jupiter the astronomical result is 1.69 as an approximation to the square root of 3. So let me go on quoting from Koestler:

> Into the orbit or sphere of Saturn he inscribed a cube, and into the cube another sphere which was that of Jupiter. Inscribed in that was the tetrahedron, and inscribed in that, the sphere of Mars. Between the spheres of Mars and Earth came the dodecahedron, between the Earth and Venus, the icosahedron, between Venus and Mercury, the octahedron. Eureka!... This is the ultimate fascination of Kepler, both as an individual and as a case history. For Kepler's misguided belief in the five perfect bodies was not a passing fancy, but remained with him in a modified version to the end of his life, showing all the symptoms of a paranoid delusion; and yet it functioned as the *vigor matrix*, the spur, of his immortal achievements.

Kepler's system is shown in Figure 3.6. I think it is rather nice to see how Koestler acknowledged that, although all this is nonsense, if Kepler hadn't had these curious fantasies he might never have gone on to do all the great things that he did.

Symmetry

One of the most remarkable things about the regular and Archimedean solids is their symmetrical

Table 3.1. Maximum and minimum distances of the planets from the sun in millions of miles, together with ratios needed for examining Kepler's theory of the solar system

Planet	Max. Dist.	Min. Dist.	Ratio	Polyhedron	$_0R/_2R$
Saturn	935	837	$\frac{837}{507} = 1.69$	Cube	$\sqrt{3} = 1.73$
Jupiter	507	459	$\frac{459}{155} = 2.98$	Tetrahedron	3
Mars	155	128	$\frac{128}{94\frac{1}{2}} = 1.35$	Dodecahedron	$\sqrt{(15 - 6\sqrt{5})} = 1.26$
Earth	$94\frac{1}{2}$	$91\frac{1}{2}$	$\frac{91\frac{1}{2}}{68} = 1.35$	Icosahedron	$\sqrt{(15 - 6\sqrt{5})} = 1.26$
Venus	68	67	$\frac{67}{43} = 1.54$	Octahedron	$\sqrt{3} = 1.73$
Mercury	43	28			

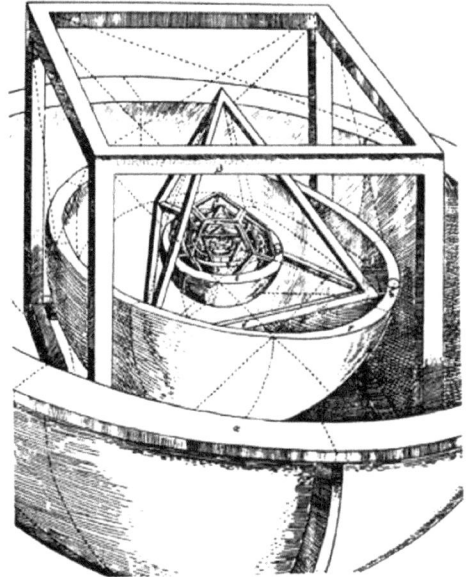

Figure 3.6. The planetary system of Johannes Kepler, detail. From Kepler's *Mysterium Cosmographicum* (1595).

nature. In his little book *Symmetry*, Hermann Weyl mentions that the old problem of enumerating the five kinds of Platonic solids was superseded in the 1870s by the problem of enumerating the five kinds of rotation groups. I would like just to run through the details because this is a beautiful piece of pure mathematics that Felix Klein did a hundred years ago, in his book *Lectures on the Icosahedron*.

Klein considered rotations of the sphere into itself, the way an eyeball rotates in its socket, but continued through a whole turn. There are rotations of various periods. A half turn is a rotation of period 2; a quarter turn is a rotation of period 4, and so on. Suppose that you have a rotation of period p, greater than or equal to 2. The axis of rotation penetrates the sphere at two opposite *poles* of that same period p. Working on a sphere, you may have poles of various periods, and a given group of rotations will transform various poles into one another; you get a certain class of equivalent poles. If the total order of the group (the total number of rotations altogether, including the identity) is N, then in each class of equivalent poles, if they are p-gonal poles, there will be N/p poles. That is because if you take a point close to a pole and just move it by that rotation, you get a little p-gon around that point. It is rotated by a rotation of period p. And so all the rotations in the group, when applied to this point near one of the poles, will give you a lot of p-gons all over the place. As there are N of these points altogether, there are N/p equivalent poles in each class.

The next thing to observe is (thinking of all the rotations in the group), for each axis of period p there are $p-1$ rotations not including the identity. Turn through one pth of a turn and then two $p$ths, and so on. There are $p - 1$ rotations on each axis. And as there are two poles at opposite ends of every axis, there are $(p-1)/2$ for each pole. Now as there are N/p poles in each class of equivalent poles, the number of rotations for each class is

$$\frac{N}{p}\frac{(p-1)}{2} = \frac{N}{2}\left(1 - \frac{1}{p}\right).$$

Not counting the identity, the whole rotation group consists of $N - 1$ rotations, so you simply put them together. Summing over the various classes of equivalent poles, you have

$$\frac{N}{2} \sum \left(1 - \frac{1}{p}\right) = N - 1,$$

where N is the total number of rotations in the group. Just twist that over by a little more algebra and you get

$$2 - \frac{2}{N} = \sum \left(1 - \frac{1}{p}\right).$$

This is summed over the class of equivalent poles.

How many will there be? Well if $N = 1$ that is a trival case, of course, where there is no pole. So from now on we can suppose the number of rotations in the whole group is greater than or equal to 2. And if that is so, then you can get the inequality:

$$1 \le (2 - 2/N) \le 2.$$

Now we have this sum,

$$\sum (1 - 1/p).$$

Could there be only one term? No, there couldn't be only one, because $(1 - 1/p) < 1$, while the summation has to be greater than or equal to 1, and also less than 2. Could there be as many as four terms in this summation? No, because four terms, each $1/2$ or more, would add up to two or more. So we know that this sum has two or three terms.

Take two terms and you have

$$2 - 2/N = (1 - 1/p) + (1 - 1/q).$$

Cancel the $2 = 1 + 1$, multiply all through by N and you get this curious little equation:

$$N/p + N/q = 2.$$

But as we saw right at the beginning, N/p is the number of poles in a class of equivlent poles. So N/p is a whole number. Similarly N/q is a whole number. And those two whole numbers add up to 2. So there is no conclusion except that they

are both 1: $N = p = q$. And that means that if there are only two terms in this sum, you have a group of order p with only one axis but two poles, one at each end of that axis. That is C_p, a cyclic group, the group that you get by having a p-gonal rotation and all its powers, and that is all.

Now suppose there are three terms in this summation, say

$$(1 - 1/p) + (1 - 1/q) + (1 - 1/r) = 2 - 2/N,$$

so that

$$1/p + 1/q + 1/r = 1 + 2/N.$$

Since $1/3 + 1/3 + 1/3 = 1$, the three periods, p, q, r cannot all be 3 or more. So at least one of them is 2, say $r = 2$, and you have

$$1/p + 1/q = 1/2 + 2/N.$$

Multiplying through by $2pq$, you get

$$2q + 2p = pq + 4pq/N,$$

whence

$$(p - 2)(q - 2) = 4 - 4pq/N.$$

So $(p - 2)$ and $(q - 2)$ are two nonnegative integers whose product, being less than 4, can only be 0 or 1 or 2 or 3. Assuming, for convenience, that $p \ge q$, you may have $(q - 2) = 0$, but otherwise the "two nonnegative integers" can only be 1 and 1, or 2 and 1, or 3 and 1. It follows that, apart from the cyclic group C_p, the only finite rotation groups (p, q, r) are the

- Dihedral group $(p, 2, 2)$ with $N = 2p$
- Tetrahedral group $(3, 3, 2)$ with $N = 12$
- Octahedral group $(4, 3, 2)$ with $N = 24$
- Icosahedral group $(5, 3, 2)$ with $N = 60$

If you think of figures possessing this kind of symmetry, you soon see that each $(p, q, 2)$ is the group of rotatory symmetry operations of the polyhedron $\{p, q\}$ and of its reciprocal $\{q, p\}$. The inequality

$$(p - 2)(q - 2) < 4$$

gives you at the same time a proof that there are only five Platonic solids: $\{3,3\}$, $\{4,3\}$, $\{3,4\}$, $\{5,3\}$, and $\{3,5\}$.

The planes of symmetry of such a solid $\{p,q\}$ decompose its circumsphere into a pattern of $2N$ spherical triangles (with angles $\pi/p, \pi/q, \pi/2$) which may be vividly distinguished by blackening N of them. The rotation group permutes the N triangles of either color. For instance, $(4,3,2)$ yields Figure 3.7, which is really simpler than it looks at first sight. Simply draw a circle to indicate the sphere, put in one elipse, then another; then draw their major axes and two diagonal lines bisecting the angles between them. And now you see that the sphere has been divided into little triangles with angles $\pi/4, \pi/3, \pi/2$: 24 white and 24 black, half of them visible and the rest hidden behind. The angle at P is 45° because it belongs to four triangles of each color. Six such points (including the antipodes of P) are the vertices of the octahedron $\{3,4\}$. The angle at Q, which belongs to three triangles of each color, is 60°. Eight such points are the vertices of the reciprocal cube $\{4,3\}$. The twelve points where right angles occur are the vertices of the *cuboctahedron*, an Archimedean solid whose faces consist of six squares and eight triangles.

The *Archimedean solids* are polyhedra which have regular faces of two or three kinds, while all the vertices are transformed into one another by one of the rotation groups described above. (Prisms and antiprisms satisfy this definition, but are not usually considered Archimedean.) Thus the cycle of faces around a vertex is the same for all the vertices of each solid, and the numbers of sides of the polygons in this cycle provide a concise symbol. For instance, the symbol for the cuboctahedron is $3\cdot4\cdot3\cdot4$ or $(3\cdot4)^2$, because each vertex belongs to two triangles and two squares arranged alternately. The p-gonal prism is $4^2\cdot p$ and the p-gonal antiprism is $3^3\cdot p$. The books of Archimedes on this subject were lost. So it was left to Kepler to give names for them. The name "cuboctahedron" is rather natural: it is a combination of the words cube and octahedron. Similarly, he called the common part of the dodecahedron and the icosahedron an *icosidodecahedron*. It has twenty triangles and twelve pentagons. Four others are show in Figures 3.8–3.11. The whole list of thirteen is as follows:

- The truncated tetrahedron $3\cdot6^2$
- The truncated cube $3\cdot8^2$
- Truncated octahedron $4\cdot6^2$
- The truncated dodecahedron $3\cdot10^2$
- The truncated icosahedron $5\cdot6^2$
- The cuboctahedron $(3\cdot4)^2$
- The icosidodecahedron $(3\cdot5)^2$
- The rhombicuboctahedron $3\cdot4^3$
- The rhombicosidodecahedron $3\cdot4\cdot5\cdot4$
- The truncated cuboctahedron $4\cdot6\cdot8$
- The truncated icosidodecahedron $4\cdot6\cdot10$
- The snub cube *(cubus simus)* $3^4\cdot4$
- The snub dodecahedron *(dodecahedron simum)* $3^4\cdot5$

Before Klein there was another German, A. F. Möbius, who observed that the use of Figure 3.7 (as above) to construct the vertices of the cube, octahedron, and cuboctahedron can be extended to yield all the Platonic and Archimedean solids. One of the black triangles in Figure 3.7 is marked PQR in Figure 3.12. Its angles have been bisected so as to yield points S (equidistant from the great

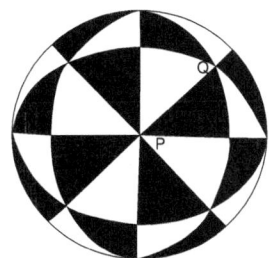

Figure 3.7. The nine circles of symmetry of the cube.

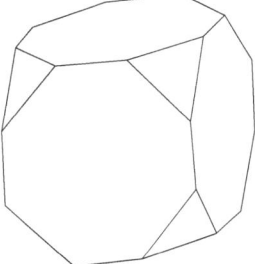

Figure 3.8. The truncated cube $3\cdot8^2$.

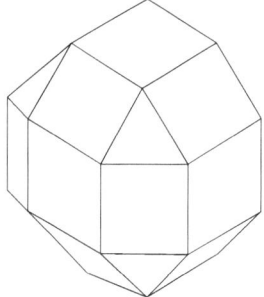

Figure 3.9. The rhombicuboctahedron $3 \cdot 4^3$.

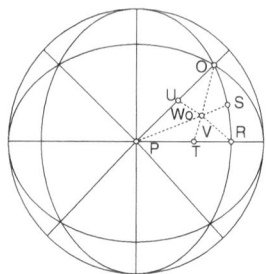

Figure 3.12. Typical vertices of the seven polyhedra.

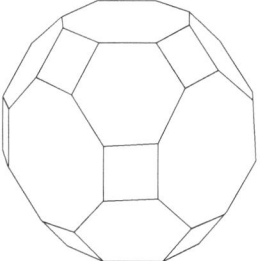

Figure 3.10. The truncated cuboctahedron $4 \cdot 6 \cdot 8$.

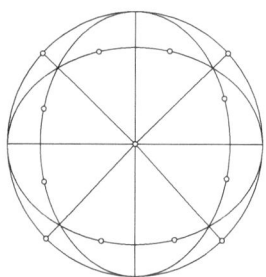

Figure 3.13. Vertices S of the truncated cube.

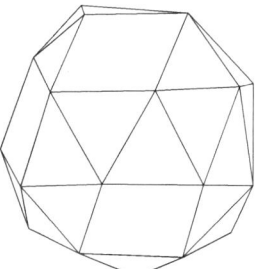

Figure 3.11. The snub cube $3^4 \cdot 4$.

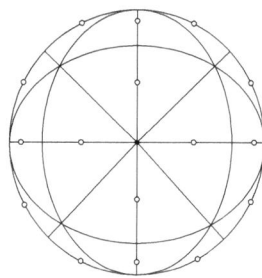

Figure 3.14. Vertices T of the truncated octahedron.

circles RP, PQ), T (equidistant from PQ, QR), U (equidistant from QR, RP), and V (equidistant from all three sides of the triangle PQR, so that V is the center of the inscribed small circle). When the planes of the great circles RP and PQ are taken to be mirrors, the images of S in this two-mirror kaleidoscope are the vertices of a regular octagon. Introducing a third mirror QR, we get Möbius's three-mirror kaleidoscope in which the image of S are the vertices of the truncated cube, as in Figure 3.13. Other Archimedean solids can be derived similarly from the points T, U, V, as in Figures 3.14–3.16. Figure 3.17

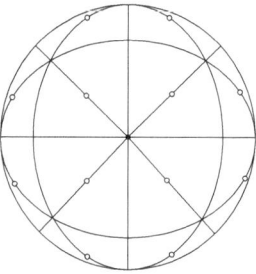

Figure 3.15. Vertices U of the rhombicuboctahedron.

shows points derived from a point W, different from V, by applying the octahedral rotation group $(4, 3, 2)$ of order 24, which is a subgroup of index 2 in the kaleidoscopic group of order 48. In other

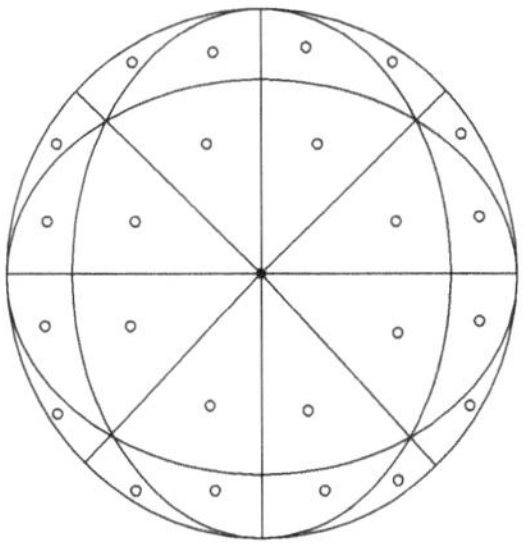

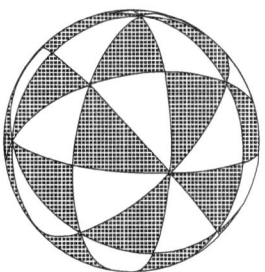

Figure 3.16. Vertices V of the truncated cuboctahedron.

Figure 3.19. Great circles related to the group $(4, 3, 2)$.

Figure 3.17. Vertices W of the snub cube.

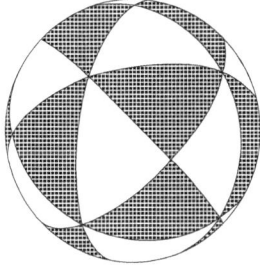

Figure 3.20. Great circles related to the group $(5, 3, 2)$.

Figure 3.18. Six of the great circles of Figure 3.7, related to the group $(3, 3, 2)$.

words, the 24 vertices of the snub cube $3^4 \cdot 4$ are points situated like W in all the *black* triangles.

Figure 3.18 shows six of the nine great circles in Figures 3.7 or 3.19; they yield the tetrahedral rotation group $(3, 3, 2)$ and the truncated tetrahedron. Figure 3.20 shows the analogous set of great circles related to the icosahedral group $(5, 3, 2)$. This "icosahedral kaleidoscope" yields the complicated polyhedra; for instance, points suitably placed in all the black triangles now yield the snub dodecahedron.

It is interesting to see how one could work out coordinates for the vertices of a snub cube. Let us begin with a large cube in its natural position

for Cartesian coordinates, so that the vertices of the cube, are, shall we say, $(1, 1, 1)$, and the same with various changes of sign: one vertex is $(1, 1, 1)$ and another one is $(-1, 1, 1)$ and so on. In a concise notation, the eight vertices are $(\pm 1, \pm 1, \pm 1)$. A smaller square inside the face $(1, \pm 1, \pm 1)$ of the cube may be supposed to have vertices $(1, y, x), (1, x, -y), (1, -y, -x), (1, -x, y)$, where $x > y$, this being in the face where the first coordinate is 1. (In Figure 3.21, $(1, -x, y)$ appears as $1\bar{x}y$.) These four points are the vertices of a square that is inside one square face of the cube, but twisted around through a certain angle. And you ask that these points and others analogously situated in other faces of the cube should be at the same distance from their neighbors, so that the distance from one point to another is the same in those various pairs. One pair gives you $xy + y + x = 1$, whence

$$x = (1 - y)/(1 + y).$$

Another yields

$$x = y(1 + y)/(1 - y).$$

Multiplying them together, you get

$$x^2 = y.$$

Substituting x^2 for y in $xy + y + x = 1$, you are left with the nice cubic equation

$$x^3 + x^2 + x = 1.$$

You can work this out in various ways and find x to be about 0.543689. Then you have to take the point with coordinates $(1, x^2, x)$, apply all the cyclic permutations, put in an even number of minus signs, and put them in a different order with an odd number of minus signs; this gives you all the 24 vertices of the snub cube (Figure 3.21).

The late R. Buckminster Fuller, a great engineer and architect, was very interested in the structures he called *geodesic domes*, which were often modifications of the icosahedron and dodecahedron. In Figure 3.22 you see one of those, in which you simply take a dodecahedron and put a small pentagonal pyramid onto each of its faces, so that you have altogether $5 \times 12 = 60$ triangles. Although they are not equilateral, they are all congruent and so you get an attempt toward

finding a sixth regular solid. Of course it is not regular, because there are five triangles around some points, and six triangles at other points. But at least this is the sphere covered with a large number of triangles that are all nearly equilateral and nearly alike.

Fuller went on doing this in more and more elaborate ways. Figure 3.23 is a slightly different one of the same sort, with nearly equilaterial triangles, nearly the same number around each corner. If you look closely you see that at some corners there are six triangles coming together and at others there are five. And so you can classify these polyhedra by seeing how you can go from one place where there are five triangles to another place where there are also five triangles. To go from one to the other is a sort of modified chess knight's move: two steps this way and then one step that way, and you get another pentagonal point. Everywhere in between there are six triangles at a vertex. It is a rather nice consequence of the theory of polyhedra that just as the icosahedron has twelve vertices, each one of which belongs to five triangles, even after you have put in all these extra triangles it is still true: there are just twelve points on the

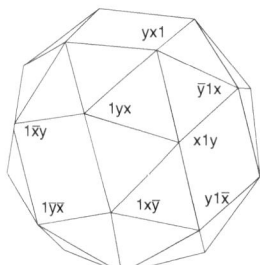

Figure 3.21. The snub cube $3^4 \cdot 4$.

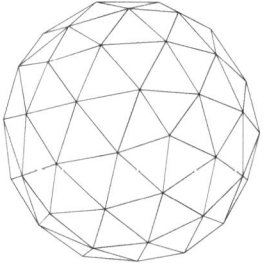

Figure 3.23. $\{3, 5+\}_{2,1}$.

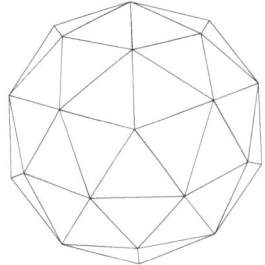

Figure 3.22. Pentakisdodecahedron, $\{3, 5+\}_{1,1}$.

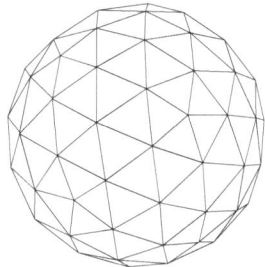

Figure 3.24. $\{3, 5+\}_{3,0}$.

Figure 3.25. H.S.M. Coxeter.

Figure 3.27. Professor Coxeter showing Figure 3.20 during a lecture.

Figure 3.26. A truncated octahedron, by a Campus School student.

sphere where the number of triangles is only five instead of six. Essentially, this is a consequence of Euler's formula. The number of vertices minus the number of edges plus the number of faces of a polyhedron is two. If you fiddle with that formula a little, you can see it always must be true: that if you have a polyhedron whose faces consist entirely of triangles, six coming together at some vertices and five at others, then the number of vertices where there are five triangles coming together is exactly twelve. You might ask the question, how many of the other kind will there be? That was answered by a very able geometer, Branko Grünbaum, who showed that if a convex polyhedron has only triangles for faces, five or six round each vertex, then the number of vertices where you've got six triangles coming together may be any number except one. It can be two or anything greater. Such a polyhedron is a remarkable generalization of the Platonic solids.

Figure 3.28. Professor Coxeter talking with members of the audience after a lecture.

4

Milestones in the History of Polyhedra

Joseph Malkevitch

Considering the fact that polyhedra have been studied for so long, it is rather surprising that there has been no exhaustive study of their history. But we are very lucky that the authors of four modern classics on the theory of polyhedra—Brückner, Coxeter, Fejes-Tóth and Grünbaum—were interested in historical information and provided detailed historical notes in their books. I propose to present an outline of the milestones in the history of the subject, putting together the thread of what happened as the theory developed. I will pay special attention to regularity concepts.

I would like you to imagine what might happen if, a thousand years from now, someone digging at an archaeological site should find a Rubik's cube. What would archaeologists, historians, and mathematicians of that future time deduce about our knowledge of geometry from this one object? We can get some idea from an Ed Fisher cartoon in The New Yorker showing a museum statue labeled "man"; it has three legs, two heads, and misplaced feet and hands. There are two "Martian" creatures looking at it; one is saying: "It certainly is amazing what our scientists can reconstruct from just a few bones and fragments." I think that is the state of our knowledge in studying polyhedra in ancient times.

Let me begin with what perhaps are the most famous of polyhedral objects, the pyramids of Egypt (Chapter 7). They are awesome, and the engineering accomplishment of making them raises the question as to exactly what kinds of information the geometers of Egypt had about polyhedra. Our knowledge of Egyptian geometry comes down to us from two sources: two papyri. One is called the Rhind mathematical papyrus, the other the Moscow mathematical papyrus. There are some problems on the Rhind papyrus showing computations about the relationship between what we would call today the slant height of the pyramid and the height. The Moscow papyrus has a calculation that illustrates the kind of detective work one has to do when one looks at these papyri. The authors did not spell out: "We will now do this calculation." Instead there are some symbols and sometimes an accompanying diagram, and the modern historian has to try to determine what it was that they were trying to do. In this particular instance, it seems that they may have been calculating the volume of the truncated pyramid. This is the *first milestone* in the history of polyhedra.

Milestone 1 ◇ *Volume of Truncated Pyramid*
- *(c.1890 B.C.E.) Problem 14 of the Moscow Papyrus suggests Egyptians may have known how to compute the volume of a truncated square pyramid, using*

$$V = \frac{h}{3}(a^2 + ab + b^2)$$

J. Malkevitch
Professor Emeritus, Department of Mathematics, York College (CUNY), Jamaica, NY 11451, USA
e-mail: jmalkevitch@york.cuny.edu; http://www.york.cuny.edu/~malk

M. Senechal (ed.), *Shaping Space*, DOI 10.1007/978-0-387-92714-5_4,
© Marjorie Senechal 2013

(where a and b are sides of the square bases and h is the height of the pyramid).

The calculation seems to follow the modern formula for the volume of the frustrum of a pyramid. There is quite a bit of scholarly debate as to whether this formula was actually known to the Egyptians, but I will not get involved in that particular thicket. Part of the reason for the controversy is that we have no real basis for trying to determine how the Egyptian geometers might have arrived at that result. Furthermore, the papyri do not contain any problem which calculates the volume of an untruncated pyramid! They did not seem to be in possession of anything like calculus. How else might they have found the formula? We know from modern work that you cannot find the volume of a tetrahedron by the method of cutting it up into a finite number of pieces and reassembling them into something whose volume can be computed easily. (This follows from Max Dehn's solution to Hilbert's problem on equidecomposability, discussed below.) Various proposals have been made for what the Egyptians did do; all are very speculative.

Let me briefly describe what actually seems to be the history of the development of the theory of the volume of a pyramid, because it is certainly not an intuitive result. It was quite an accomplishment for ancient peoples; we can consider it to be the *second milestone*. Archimedes referred to the fact that Democritus, who flourished at the end of the fifth century B.C.E., knew that the volume of the pyramid was one-third the area of the base times the height, and that the proof was devised by Eudoxus. Eudoxus's method is known as the method of "exhaustion" and his approach is the one Euclid followed in the *Elements*.

Milestone 2 ◇ *Volume of a Pyramid*
- *Democritus (fl.end 5th century B.C.E.) discovers that volume of a pyramid is equal to $\frac{1}{3}$(area of base)(height).*
- *Eudoxus (c.409–c.356 B.C.E.) proves the result above using the method of "exhaustion." (Achimedes confirms Eudoxus' role).*

Let us turn now to the origin of regularity concepts. More specifically, we might ask the extent to which ancient peoples had knowledge of regular solids and theories of regular solids. Before attempting to answer this question, let me clarify what I mean by "regular." As the theory of polyhedra has developed, the meaning of the word "regular" has broadened. There are two major uses of the word. One approach to regularity is "local": it requires congruence of faces and/or vertex-figures (that is, the pattern of faces at a vertex) and/or edge-figures. The other approach considers the polyhedron as a whole: a polyhedron is regular if its symmetry operations act transitively on its vertices, edges, or faces. For example, vertex transitivity would mean that any vertex of the polyhedron could be moved to any other vertex by a symmetry operation of the polyhedron. In either approach we are primarily interested in the case where the faces are regular polygons. The second approach is very modern (nineteenth century) and almost certainly was not the approach used by the Egyptians or Greeks. Henceforth, when the word *regular* is used, it will refer to the first approach to regularity. As near as I am able to determine, there was no knowledge of individual regular solids, and no theory of them in ancient Egypt; I will have some more comments about that later.

As you will see in Chapter 10, many very beautiful polyhedral forms occur in nature as crystals. Fluorite crystals often grow as octahedra and there are pyrite (combinatorial) dodecahedra. Examples of such crystals must have been known in ancient times. The oldest man-made dodecahedral object is generally attributed to pre-Pythagorean times. It is a dodecahedral shape with incised markings on it, discovered in an excavation on Mt. Loffa in Italy near Padua. Nobody knows what the symbols actually mean. There is some linguistic evidence that the dodecahedron was not known by the name *dodecahedron* in early days, but instead was referred to as "the sphere of 12 pentagons." That may have had something to do with how the theory actually developed.

There is a tradition, which goes back into Greek history, that assigns knowledge of the five "Platonic solids" to the Pythagoreans. Eudemus of Rhodes referred to Pythagoras himself as having discovered the five regular polyhedra. But modern scholars seem to discount this. In a scholarly thesis in 1917, Eva Sachs gave some cogent arguments suggesting that, in fact, the Pythagoreans did not know all of the five Platonic solids. Many modern scholars argue that the proper history is based on a scholium or comment in an extended edition of Euclid's *Elements* which reads as follows: "In this book, the 13th [the thirteenth book of Euclid], are constructed five figures called Platonic, which do not however belong to Plato. Three of these figures, the cube, pyramid and dodecahedron, belong to the Pythagoreans, while the octahedron and icosahedron belong to Theatetus." But the history is complicated by an understanding of what "belong to" means. The Greeks were very interested, as we know, in ruler-and-compass constructions. The question arises as to whether this quotation means that Theatetus had found ruler and compass constructions for these polyhedra. Perhaps all five solids were known, as objects, earlier.

William Waterhouse gives some very interesting linguistic and other arguments in favor of a later date than the Pythagoreans for the origins of actually thinking of the regular solids as a *family* and singling them out for study. Theatetus (415–369 B.C.E.) seems to have looked at this collection of solids not merely as isolated objects; he considered the question of discussing them as part of a theory (Milestone 3). On the other hand, as we have seen, these polyhedra are often referred to as the Platonic solids. Plato, who was a friend of Theatetus, built them into his cosmology in the dialogue *Timaeus* (Milestone 4). This is important for the history of polyhedra because one of the threads that kept polyhedra alive during the Renaissance was the renewed interest in classical studies. (See, for example, Raphael's painting *The School of Athens,* in which he showed the geometers at work.) The association of Plato with these solids probably helped keep this knowledge alive for a long period of time.

Milestone 3   ◇ *Theatetus (c.415–369 B.C.E.)*
- *Develops a general theory of regular solids, specifically adding the octahedron and icosahedron to solids known earlier.*

Milestone 4   ◇ *Plato (427–347 B.C.E.)*
- *In the dialogue Timaeus Plato incorporates his knowledge of the five regular polyhedra into his philosophical system.*

(His popularity and influence result in their becoming known as "Platonic solids.")

By the time we get to Euclid, we already have a fairly full-blown theory of solid geometry. In Book XI of *Elements*, Euclid gave a full treatment of metric properties of polyhedra, in Book XII he discussed the volume of prisms and pyramids including Eudoxus' proof, and in Book XIII he showed how to construct the regular convex polyhedra and "proved" that there are only five of them (Milestone 5). I use quotation marks because Euclid never told us what a polyhedron was. This raises the question of what people at various times have had in mind when they used the word "polyhedron." It is my contention that throughout the history of the development of regularity concepts, the notions of polygon and polyhedron have diversified. This diversification has been the driving force behind creating a lot of the theory we know today.

Milestone 5   ◇ *Euclid (fl.323–285 B.C.E.)*
- *Book XI of the Elements treats metric properties of polyhedra.*
- *Book XII of the Elements discusses the volume of prisms and pyramids.*
- *Book XIII of the Elements treats the five regular polyhedra, concluding with a "proof" that there are exactly five.*

The sixth milestone is the description by Archimedes, in a manuscript that now is lost, of what we today call the semiregular or Archimedean solids (Milestone 6). By some miracle or another, Pappus (whose works unfortunately don't appear to exist in English) gave an account (Milestone 7) of the lost book of Archimedes which dealt with the semiregular solids. It is significant that he explicitly mentioned that there are thirteen of them, and described them in terms of how many polygons each has with various numbers of sides at a vertex. We will see that this turns out to be rather significant at a later time.

Figure 4.2. Faïence icosahedron with Greek letters incised on the faces.

Figure 4.1. Steatite icosahedron with Greek letters incised on the faces.

Milestone 6 ◇ *Archimedes (c.287–212 B.C.E.)*
- *Describes 13 "semi-regular" solids in a now lost treatise.*

Milestone 7 ◇ *Pappus (fl.4th century)*
- *(320 A.D.) Pappus' Collection (Book V) gives an account of the 13 "semiregular" solids discovered by Archimedes.*

In the centuries between Archimedes' time and the time of Pappus, polyhedral objects of various kinds and types were made. Fortunately some of these objects can still be seen today. I had the big thrill of examining some of them personally at the Metropolitan Museum of Art in New York. There I saw regular icosahedral objects incised with the first 20 letters of the Greek alphabet (Figures 4.1 and 4.2). Four similar icosahedra used to be on display in the Egyptian rooms of the British Museum in London. The exact origins and provenance of these objects is not known. It is, however, fairly certain that they are not indigenous to Egypt, that they are probably not even of Greek origin. They may date from the Roman period. Claims in the literature about Egyptian knowledge of regular solids appear to be false extrapolation from the assumption that these icosahedra were made in Egypt.

There are also other objects, from perhaps a somewhat later period, that have been cited by scholars. Most of these were found in England, France, and Italy and have dodecahedral shapes. Figure 4.3 shows a bronze dodecahedron dug up in 1768 in Carmarthen, typical of about fifty which have been found in the northwestern provinces of the Roman empire. It has been suggested that they were used as surveying instruments. There is considerable controversy about their origins and uses; the best guess is that they were candle holders. I have never actually seen an original of one of these, but from photographs some of them appear to be quite handsome objects. A large collection of such polyhedral objects has been described, including a rhombic triacontahedron, opening to question claims that nobody had seen such a thing until Kepler's time. The best estimate for when these objects were made is about 500 A.D.

There does not seem to have been any systematic account of polyhedra from the time of Pappus until the Renaissance. As we know, a lot of mathematical traditions died during that period. As I mentioned earlier, what ultimately resuscitated these ideas was the renewed interest in Plato. So I will jump to the Renaissance period. I would like to single out the work of Albrecht Dürer

Figure 4.3. Bronze dodecahedron found in Carmarthen 1768, and now in the possession of the Society of Antiquaries of London. Overall height $4\frac{1}{8}$ inches.

for Milestone 8. This is the concept of studying polyhedra by drawing what are today called nets. By folding a planar piece of cardboard along prescribed lines and joining the edges of the figure, the net becomes a polyhedron. Dürer made nets for the dodecahedron and for other regular and semiregular solids (see Chapter 6).

Milestone 8 ◇ *Albrecht Dürer (1471–1528)*
• *Invents the concept of the "net" of a polyhedron.*

The situation during the Renaissance is extremely complicated. A great many scholars, artists, and artisans discovered and rediscovered various classical solids (Milestone 9). Some of them drew what appear to be star-shaped solids; others drew compounds; others drew convex polyhedral solids. In discussing this period it is very difficult to reach any firm conclusions about who discovered what and when. For example there is a picture of a solid in the famous Jamnitzer pictures (published in 1568) that some say was discovered by Kepler about 1619. Professor Coxeter (Chapter 3) alluded to the book of Luca Pacioli's work on the regular and semiregular polyhedra which was illustrated by Leonardo da Vinci. There are

indications that Pacioli actually made polyhedron models of glass; there is a painting which shows Pacioli with a picture of a glass model of the rhombicuboctahedron (see Chapter 7). The important thing to realize in discussing this period is that these ideas were very much in the air. The real issue is not who discovered a particular star polytope first. I am not sure that approach is particularly profitable. I think it is more interesting to try to understand why it was that all of this activity was going on at this particular time. Why was there so much interest in polyhedra, and why did the subject flourish during that period? The answer seems related to the emergent study of perspective and the renewed interest in the Greek classics.

Milestone 9 ◇ *Renaissance artists, architects, artisans, and scholars "discover" and "rediscover" various Platonic and Archimedean solids, star-poly-hedra, compounds, and other polyhedral objects.*

Some of the best known are Paolo Uccello, Wentzel Jamnitzer, Lorenz Stoer, Daniel Barbaro, Piero della Francesca, Luca Pacioli, Leonard da Vinci, Albrecht Dürer, Simon Stevin, François de Foix, and R. Bombelli.

The *tenth milestone* in our history is the work of Johannes Kepler, which Professor Coxeter discusses in Chapter 3. Kepler drew tilings of the plane in which he used nonconvex and nonsimple polygons. He also gave a very elaborate, case-by-case, proof that there are thirteen semiregular polyhedra. It is interesting that Pappus didn't describe the prisms and the antiprisms which, as we know, are two infinite families that are also semiregular polyhedra. But Kepler explicitly both drew and dicussed them. It is also interesting that in modern times we refer to thirteen Archimedean solids because we adopt the regularity definition based on symmetry, although almost certainly this was not the definition used by Archimedes and Pappus or Kepler. With the congruent vertex figure definition, Archimedes and Pappus missed one: the pseudo-rhombicuboctahedron. Kepler's detailed work on the semiregular polyhedra appears in Book II of *Harmonices Mundi*. Here he found only thirteen semiregular solids (plus the prisms and antiprisms). However, in an off-hand remark in the "Six-Cornered Snowflake" he refers to fourteen semiregular solids! No supporting detail is given, however.

Milestone 10 ◇ *Johannes Kepler (1571–1630)*

- *Studies tilings of the plane using convex, non-convex, non-simple polygons.*
- *"Proves" there are 13 "semiregular" polyhedra (plus two infinite families, prisms, and anti-prisms).*
- *Constructs:*
  - *Stella octangula*
  - *Two (regular) stellated polyhedra*
  - *Rhombic dodecahedron*
  - *Rhombic triacontahedron.*

Kepler also constructed the rhombic dodecahedron and the rhombic triacontahedron and two "new" "regular" (star) polyhedra. At least, today we call them regular; it is hard to tell whether Kepler really thought of them as being regular. My impression is that perhaps he did not. I also have some questions as to whether he really understood fully reciprocation and duality notions, as has been claimed. Since he was so careful, one might guess that if he understood these matters, he would have also constructed the two remaining regular star polyhedra and the duals of the Archimedean polyhedra. But that is my purely subjective view.

Unfortunately, there appears to be a pattern for many of the great discoveries about polyhedra, even for geometry in general: all too often people do very great things and then the work "goes to sleep" for long periods. Euclid's work went to sleep, the work of Archimedes went to sleep; Kepler's work, as modern in spirit as it seems to be, in fact, wasn't looked at seriously until relatively recently. Although Kepler's work is constantly cited, it does not really seem to have affected the history of polyhedra in any direct way.

Another person who made major contributions to the theory of polyhedra was René Descartes. The famous story of the lost manuscript is told in Chapter 3. In that manuscript (Milestone 11) there appears a tantalizing theorem about poly-

hedra: the sum of the defects of the vertices is 4π, where the defect of a vertex is defined to be 180° minus the sum of the face angles at the vertex. (Of course no one knows what kind of polyhedra Descartes was really taking about, but presumably he had convex three-dimensional polyhedra in mind.)

Milestone 11 ◇ *René Descartes (1596–1650)*

- *(1619–1620) (Manuscript lost, but copy by Leibniz published in 1820).*

Main Theorem:

If P is a (convex) 3-dimensional polyhedron, then the sum of the defects at the vertices is 360°.

It turns out that one can very quickly get from that theorem of Descartes to the very famous polyhedral formula of Euler, $V - E + F = 2$, and vice versa. The fact that Descartes' theorem is logically equivalent to Euler's formula has created the widespread impression that Descartes actually knew the formula, although scholars over and over again have said that this is not so. You can find in the papers of Lebesgue from the 1920s that he does not believe Descartes knew Euler's formula. You can find the same statement in Pólya, and in a very nice paper by Peter Hilton and Jean Pedersen. More recently, Federico has written a whole book on the contribution of Descartes to the theory of polyhedra, including an English translation of the manuscript and an extensive summary of this debate. In this connection it is helpful to remember remarks of Jacques Hadamard. He wrote, concerning some of his own work, that "two theorems, important to the subject, were such obvious and immediate consequences of the ideas contained therein that, years later, other authors imputed them to me, and I am obliged to confess that, as evident as they were, I had not perceived them."

Evidently there is a strong tendency on the part of many people who know of Descartes' theorem to assume that if Descartes had only gone one tiny step further, he would have discovered this or that. But if you look carefully at the work of Descartes, it is very clear that he did not think of polyhedra as *combinatorial* objects. It was not a tiny step that was needed, but a big one. That great leap forward was made in part by Euler.

I might mention that Descartes is another example of a person whose work went to sleep. (His work was literally lost.)

Aside from the material that fuels the Euler–Descartes controversy, Descartes talked about the semiregular polyhedra. He didn't enumerate all thirteen of them, only the eleven that can be obtained from the Platonic solids by truncation. Nor did any of those artists and artisans, who were obviously bright and talented people, discover all thirteen of the semiregular solids. On the other hand, Kepler explicitly referred to the fact that these objects were Archimedean solids. So we know that he had seen Pappus's work describing what Archimedes had done. This provided Kepler had a very big assist. One wonders what would have happened during this period if people had had wide access to Pappus's work. Of course, one can only speculate.

There is a long tradition of referring to 13 Archimedean polyhedra. However, how should an Archimedean polyhedron be defined? Until "modern times" the idea that Archimedes, Pappus, Kepler and others who considered the issue was to find how many different convex polyhedra there are, all of whose faces are convex regular polygons having the same pattern of faces around every vertex, but excluding the 5 Platonic Solids (convex regular polyhedra) and the two infinite families of prisms and anti-prisms having regular polygons as faces. With this definition there are 14 such convex polyhedra, not 13. The convex polyhedron that usually was not included is today sometimes called the "pseudo-rhombi-cuboctahedron." This solid may perhaps have been the one Kepler had in mind when he referred to 14 Archimedean polyhedra. It is also associated with the names of J.C.P. Miller, V.G. Ashkinuse, and the elongated square gyrobicupola (Johnson solid J37). Since Archimedes presumably did not have a "global view" of symmetry (group theory), if we continue to honor him by using the "local symmetry" definition, in the future reference should be made to 14 Archimedean polyhedra. The number of convex polyhedra with regular polygons as faces (excluding the Platonic solids and the prisms and anti-prisms) for which the symmetry group of

the polyhedron is transitive on the vertices is 13. There are 13 such solids, the ones that have been "historically" known as the Archimedean polyhedra (solids). This definition via group theory rules out the pseudo-rhombi-cuboctahedron. Unfortunately, at this point there is no easy way to fix the "mess" that Archimedes created when he did not find all 14 of the polyhedra for which we can continue to honor his memory.

I was originally going to refer to landmarks rather than milestones, and I think that Milestone 12 really deserves the title *landmark:* Euler's letter to Christian Goldbach in 1750 in which he referred to his discovery of the fact that the number of vertices of a polyhedron minus the number of edges plus the number of faces equals two. Like Euclid and Descartes, however, he did not say what kinds of polyhedral objects he had in mind, and that omission has created a long list of further controversies about the history of this subject. Suffice it to say that Euler, although he found his formula, was not successful in proving it.

Milestone 12 ◇ *Leonard Euler (1707–1783)*
- *Euler discovers that polyhedra obey:*
  Vertices + Faces – Edges = 2

The long list of very distinguished and very interesting work that was done on Euler's polyhedral formula is Milestone 13. The first proof was provided by Legendre (or, some say, I believe erroneously, Meyer Hirsh). Many people contributed to the theory of the formula by figuring out what happens for different types of polyhedra and providing different proofs. This provided the roots of modern topology; interest in Euler's polyhedral formula, and the more general idea of the Euler characteristic, brought about important developments. Second, much of the impetus for studying higher dimensional polyhedra grew out of the work on the Euler polyhedral formula.

Milestone 13 ◇ *Development of the Theory of Euler's Polyhedral Formula by:*
- *Adrian-Marie Legendre (1752–1833)*
  First proof
- *Augustin-Louis Cauchy (1789–1857)*
- *J. D. Gergonne (1771–1859)*
- *S. Lhuilier (1750–1840)*

- *J. Steiner (1796–1863)*
- *Von Staudt (1798–1867)*
- *Many others!*

The next milestone, Milestone 14, is Poinsot's 1810 discovery of the four regular stellated polyhedra. It is clear that Poinsot understood that there was a sense in which these were regular polyhedra. Poinsot's work on the star polyhedra grew out of his work on star polygons. He seems neither to have looked at Kepler's original work nor to have been aware of Kepler's discovery of two star polyhedra. (There have been allegations, however, that Poinsot plagiarized Kepler.) Other contributors to the theory of star polyhedra were A. Cauchy (1811), J. Bertrand (1858), and A. Cayley (1859).

Milestone 14 ◇ *Louis Poinsot (1777–1859)*
- *In his 1810 Mémoire, Poinsot discovers four "regular" stellated polyhedra, using both star-shaped vertices $\left(\{5, \frac{5}{2}\} \text{ and } \{3, \frac{5}{2}\}\right)$ and star-shaped faces $\left(\{\frac{5}{2}, 5\} \text{ and } \{\frac{5}{2}, 3\}\right)$—already known to Kepler).*

Milestone 15 followed about a year later, when Cauchy made major contributions to the theory of polyhedra. He gave what is the most common proof of Euler's formula using graph-theoretic ideas, and proved his famous result that polyhedra with triangular faces are rigid. He also gave a "proof" that there are no regular star polyhedra other than those found by Kepler and Poinsot.

Milestone 15 ◇ *Augustin-Louis Cauchy (1789–1857)*
- *Cauchy "proves" that polyhedra with (only) triangular faces are rigid.*
- *Gives a graph-theoretic approach to proving Euler's formula.*
- *Shows that there are 9 "regular" polyhedra.*

The first systematic account of duality of polyhedra that I have found is in the work of Catalan in 1865 (Milestone 16). In a long article he described, very explicitly, the duals of the Archimedean polyhedra. It is curious that researchers never cited this paper. In other words, this was a paper that also went to sleep. One finds Catalan's work mentioned only in historical footnotes of books written in the twentieth century.

Nobody earlier seems to have paid any attention to it.

Milestone 16 ◇ *Eugene Charles Catalan (1814–1894)*
- *Catalan gives a systematic account of the duals of the Archimedean solids.*

In the middle to later 1800s there was a tremendous flourishing of geometric activity. Max Brückner published a book (Milestone 17) in which he summarized all of what was known at the time and also gave some extensive historical notes on the subject. It has very beautiful pictures of uniform polyhedra, which served as an inspiration to people later. (The uniform polyhedra are those—not necessarily convex—that have regular polygons as faces and symmetries that are transitive on the vertices.)

Milestone 17 ◇ *Max Brückner (1860–1934)*
- *Brückner publishes an extensive summary of the known results on polygons and polyhedra, with historical notes.*

I will breeze through more recent work. I have indicated that perhaps Kepler had known the pseudo rhombicuboctahedron. This is certainly possible. This polyhedron is often referred to as Miller's solid; sometimes the Russian mathematician Ashkinuse is given credit for discovering it. But George Martin was kind enough to call my attention to a paper in 1905 of D.M.Y. Sommerville (you may know his work on n-dimensional space) in which there are Schlegel diagrams for both the rhombicuboctahedron and the pseudo one (Milestone 18). The significance of this milestone is that many proofs in this area of geometry require a delicate interplay of both theory and case-by-case analysis. Here is one of many situations where earlier work was not fully correct. It is unclear who deserves "credit" for proving there are 14 Archimedean solids.

Milestone 18 ◇ *D.M.Y. Sommerville (1879–1934)*
- *Sommerville describes the pseudo rhombicuboctahedron.*

David Hilbert (1862–1943), in his famous speech in Paris in 1900 at the International

Congress of Mathematicians, outlined problems which he felt represented important challenges to mathematics in the future. One of these problems involved cutting a polyhedron into pieces and assembling them into another polyhedron, which leads to Milestone 19.

To properly understand the problem Hilbert wanted to look at, first consider the analogous issues in the (Euclidean) plane. If one has two plane simple (non-self-intersecting) polygons S and T and one can cut one of these, say S, up into polygonal pieces (perhaps with straight line cuts) which can be assembled into T, then clearly S and T have the same area. However, what about the converse? It is not at all clear, given two simple polygons S and T with the same area, that it is possible to cut, say S, into finitely many polygonal pieces so that these pieces can be assembled into T. It is remarkable that this can always be done, and several individuals appear to have found the result independently. This result is now often known as the Wallace, Bolyai, Gerwien Theorem (for William Wallace, Wolfgang Farkar Bolyai (Janos' father) and P. Gerwien (probably Karl Gerwien, about whom little biographical information is available)). There are many variants of this result which involve such questions as whether the polygonal pieces from the first polygon S can be moved to the pieces that make up the second polygon T using particular geometric transformations. J. Sydler (1921–1988), Hugo Hadwiger (1908–1991), and P. Glur extended Dehn's work.

Hilbert was concerned with what happens when one moves from two to three dimensions with problems of this sort. He asked the question ("Hilbert's 3rd Problem") of whether it was possible to take a regular tetrahedron of volume 1 and decompose it into polyhedral parts and reassemble the pieces to form a (regular) cube of volume 1? The perhaps surprising answer is no! This remarkable fact and tools to show that it was the case were developed by the geometer Max Dehn (1878–1952). Dehn's insights and work have flowered into a theory of "equidecomposability" in three and higher dimensions. Briefly, this concerns when something can be cut up into a finite number of parts and have these parts reassembled to form another object. One perspective on Dehn's work is that "Calculus" is necessary in 3-dimensions (and higher) because one cannot base a theory of volume on ideas involving finite decomposition of the kinds that arise in the theory of equidecomposability. A somewhat related issue is that while it is possible to cut up any plane polygon into triangles using the existing vertices of the polygon as vertices of these triangles, there exist non-convex polyhedra in 3-dimensional space, which can not be decomposed, using existing vertices, into tetrahedra. An extreme version of this is that any two vertices of the polyhedron that are not joined by an edge have the property that a segment joining them includes points in the exterior of the polyhedron. Polyhedra with this property were discovered by N.J. Lennes and are now known as Lennes polyhedra.

Milestone 19 ◇ *David Hilbert (1879–1934) and Max Dehn (1878–1952)*
- *Hilbert raises the issue of equidecomposability of polyhedra.*
- *Dehn resolves Hilbert's 3rd Problem.*

By far the most important early twentieth-century contributor to the theory of polyhedra was Ernst Steinitz. Steinitz, about 1916, developed a combinatorial characterization of convex three-dimensional polyhedra (Milestone 20). This work appeared in an encyclopedia of mathematics that was published in German and in French translation. Steinitz also wrote a book on polyhedra, which was almost finished at the time of his death; it was completed by Rademacher and published in 1934. As is typical of our subject, although Steinitz's main result is extremely important, there were almost no references to it before 1963. Why does it happen so often that important work in geometry attracts so little attention?

Milestone 20 ◇ *Ernst Steinitz (1871–1928)*
- *Characterizes polyhedra combinatorially.*
- *Rademacher's completion of Stenitz's almost completed book on polyhedra is published.*

Another especially exciting avenue of work concerns Alexandrov's Theorem (Milestone 21), named for the Russian mathematician Aleksandr D. Alexandrov (1912–1999), which greatly extends the earlier work on nets that goes back to Dürer. Intuitively, what Alexandrov showed was that if one starts with a plane simple polygon and pastes its boundary together in such a manner that all of the boundary is glued or zipped up as with a zipper, that the "curvature" assembled at a point is never more than 360 degrees, and so one gets a topological sphere, then the result is a convex 3-dimensional polyhedron or a flat "double covered" polygon. For example, if one starts with a 1×1 square then one gluing could give a double covered right isosceles triangle or a $1/2 \times 1$ rectangle. Typically, one will get a 3-dimensional convex polyhedron but the whole "configuration space" that results when a square is glued as required in Alexandrov's Theorem includes a continuum of non-isometric (inequivalent) tetrahedra as well as other combinatorial types.

Milestone 21 ◇ *Aleksandr D. Alexandrov (1912–1999)*
- *Alexandrov proves (1941) an important theorem which leads to new work on when simple polygons will fold to 3-dimensional polyhedra and when 3-dimensional polyhedra can be "cut and unfolded" to simple (non-overlapping) polygons.*
- *Alexandrov's book on Convex Polyhedra appears in Russian (1950), German (1958) and English (2005).*

Next we come to Coxeter's very important work on regularity (Milestone 22). He and various others worked on the stellated icosahedra, and he developed some very important regularity concepts which allowed skew, nonplanar, and infinite polygons. His famous work on uniform polyhedra has been referred to several times. He also combined algebraic with geometrical techniques in polyhedral group theory, which led to many important results in several branches of mathematics in addition to polyhedra theory. Much of this work is summarized in his *Regular Polytopes*.

Milestone 22 ◇ *H.S.M. Coxeter (1907–2003)*
- *Coxeter (et al.) develops "regularity" concepts for polyhedra allowing skew and infinite polygons as faces.*
- *Coxeter (et al.) conjectures that there are 75 "uniform polyhedra" (+ classical examples).*
- *Coxeter pioneers work in "polyhedral" group theory.*
- *Regular Polytopes summarizes all known work, explores new material (emphasizes higher dimensions).*

A somewhat overlooked contribution to the theory of polyhedra is George Dantzig's discovery of the simplex method (Milestone 23). Dantzig's work resulted in a big explosion of attempts to study the path structure on polyhedra which was very important in the development of the combinatorial theory.

Milestone 23 ◇ *George Dantzig (1914–2005)*
- *Develops the "simplex method" for solving linear programming problem, which stimulates interest in path problems for polyhedra.*

Then there is a surprisingly neglected subject. What convex polyhedra exist all of whose faces are regular convex polygons? For example, there are eight convex polyhedra with equilateral triangles for faces. This work by several mathematicians, taken together, is Milestone 24.

Milestone 24 ◇ *Regular Faced Polyhedra*
- *Classical work.*
- *O. Rausenberger shows there are eight polyhedra with equilateral triangles for faces.*
- *N. W. Johnson conjectures there are 92 regular faced polyhedra (+ prism + antiprisms + platonic and Archimedean solids (using a count of 13)).*
- *Johnson, Grünbaum, V. A. Zalgaller et al. prove Johnson's conjecture.*

Finally, let me mention some of Grünbaum's major contributions to our subject (Milestone 25). The extremely important work of Steinitz was resurrected by Grünbaum in about 1962 when he realized that he could rephrase Steinitz's work in graph-theory terminology, making it possible to do all the combinatorial theory of polyhedra in the plane. This means that those of you who can't see things in 3-space and are interested only in the combinatorial theory can study anything you want in the plane. Grünbaum summarized what he knew on the subject of convex polytopes in 1967 in his beautiful book. Then, building on the work of Professor Coxeter, he published an article in 1977 in which he described a very general notion of regular polyhedra which allowed very general kinds of "regular" polygons as faces, not necessarily ones that can be spanned by membranes of any sort. Subsequently, Andreas Dress proved that, aside from a small omission, the list of regular polyhedra that Grünbaum gave is complete.

Milestone 25   ◇ *Branko Grünbaum (1929–)*

- *Restates Steinitz's characterization of convex polyhedra: a graph G is polyhedral if and only if G is planar and 3-connected.*
- *Publishes* Convex Polytopes, *an exhaustive account of the combinatorial theory of polytopes.*
- *Publishes a very general framework to study "regular" polyhedra, building of ideas by Coxeter.*

I think the subject will not go to sleep again as it all too often has in the past.

5

Polyhedra: Surfaces or Solids?

Arthur L. Loeb

What is a polyhedron? Since I am especially interested in the relationship between concepts and images, I decided to approach the subject from that point of view and try to relate mathematical concepts and images. A polyhedron is an image of many, many different concepts, some of them inconsistent with each other.

In Figure 5.1 you see our friend M.C. Escher contemplating the question of the apparent solid on the left and, on the right, the surface. When we talked about this print, he often said that it is very curious, it is really the reflectivity of a surface that matters. Even when material is transparent some of the light is bounced off; some of it is transmitted, but it is modified when it is transmitted. So we really cannot tell unequivocally what goes on inside. The one on the right is totally impenetrable. Everything bounces off the white surface. We cannot tell anything about the inside. Figures 5.2 and 5.3 show that Escher, who as we know was very much concerned with plane tessellations, was very much aware of the difference between tessellating a plane and tessellating a sphere; we shall return to this later.

Escher was enormously skilled as a graphic artist and his fame rose considerably in the time of conceptual art. He was truly a conceptual artist, but unlike a good many conceptual artists who had essentially become minimalists and had no more physical substance to their art, Escher had the skill to express his ideas and his concepts

visually. What Escher never did (and he said he could not) was to relate the visual concepts to equations. Nevertheless, we have here a projection of a concept in a language that is not mathematical in the sense of verbal formulas of sequences of symbols, but nevertheless is very important as a visual language.

The surface is most important. The two ways I am going to try to approach polyhedra have a certain duality in the very broadest sense (though not in the strict mathematical sense). On the one hand there is the point of view of a set of connected items of different dimensionalities, and on the other that of a set of very rigorously defined points. The first gives us the connectivity point of view, the second the symmetry point of view. We can inscribe vertices of zero dimension, edges of single dimension, faces of two dimensions, on a surface. The edges then do not have to be straight and the faces do not have to be flat; that of course means we leave the domain of defining incircles and outcircles. We simply talk about networks on a surface.

Then it matters very much how this surface is connected, whether we have a sphere or the analog to a sphere, or whether we have a toroid such as the hat that was worn in Florence many centuries ago (see Chapter 7). We call a sphere singly connected because it has no hole, a doughnut doubly connected because it has a hole. If one travels inside a doughnut, and wants to travel past its hole, one must choose one or the other of two distinct kinds of paths in order to avoid the hole. Inside a sphere one can travel between

A.L. Loeb (1923–2002)
2038 Sycamore St., Bethlehem, PA 18017, USA

M. Senechal (ed.), *Shaping Space*, DOI 10.1007/978-0-387-92714-5_5,

Figure 5.1. M.C Escher contemplating the apparent solid on the *left* and the surface on the *right. Three Spheres II.* Lithograph. M.C. Escher, April 1946. © M.C Escher Heirs c/o Cordon Art – Baarn – Holland.

Figure 5.3. *Angels and Devils.* Pencil, India ink, crayon, and guache. M.C. Escher, 1941.

Figure 5.2. *Sphere with Angels and Devils.* Stained maple. M.C. Escher, 1941.

two points along an infinity of different routes, and these routes may in principle differ one from the next one by an infinitesimal amount. Inside a doughnut there is also an infinite number of routes between two points, but they divide into two distinct groups, those going around the hole on one side or on the other. It is like driving from Northampton to Cambridge, Massachusetts: there are many ways, but one must go either north or south of the Quabbin Reservoir. Similarly, a pretzel, having three holes, is quadruply connected.

One of my students, Beth Saidel, observing that connectivity relationships are the same for all singly connected surfaces regardless of their exact shapes, decided to use the Ukrainian technique of Easter-egg painting to apply some

Figure 5.4. Tessellations in the style of Ukrainian Easter-egg painting.

tessellations which we had discussed (see Figure 5.4).

Consider a finite number of vertices V on a surface. Connect them by a number of lines called edges. To be exact, an edge is a curve (not necessarily a straight line) which joins two vertices, but does not contain any vertex except at either end. No edges cross each other: their crossing would imply a vertex at the intersection. The number of edges is called E. A region of the surface surrounded by a closed circuit of alternating edges and vertices, which does not contain either edges or vertices except on its boundary, is called a *face*. The number of faces on the surface is called F. For a surface of connectivity g,

$$V - E + F = 2 - 2g. \tag{5.1}$$

It is amazing how much practical information can be derived from this equation. For our purposes it will be convenient to translate it into an expression relating valencies. If we call r the number of edges coming into any one vertex and n the number of edges (hence also the number of vertices) surrounding any one face, then we can add up the total number of edges in two different ways. One way is to find the number of vertices, V_r, having valency r. The total number of edges coming into a vertex having valency r equals rV_r. If we then sum over all the possible values of r we would get, not the total number of edges, but twice that amount, because every edge terminates at two vertices, hence would have been counted twice. Therefore

$$2E = \sum_r rV_r, \tag{5.2}$$

and analogously

$$2E = \sum_n nF_n. \tag{5.3}$$

We can define weighted averages for both r and n:

$$r_{av} = \sum_r \frac{rV_r}{V}; \tag{5.4}$$

$$n_{av} = \sum_n \frac{nF_n}{F}. \tag{5.5}$$

Dividing Equation 5.1 by $2E$ and then substituting Equations 5.2–5.5 into Equation 5.1 produces

$$\frac{1}{r_{av}} + \frac{1}{n_{av}} = \frac{1}{2} + \frac{1}{E} - \frac{2g}{E}. \tag{5.6}$$

Both Equations 5.1 and 5.6 will prove useful. To begin with, consider the tiling of a singly connected surface ($g = 0$) with nothing but pentagons and hexagons, three tiles meeting at each vertex; in that case $r = 3$. Therefore $2E = 3V$. Moreover, $F = F_5 + F_6$. Hence Equation 5.1 becomes

$$F_5 + F_6 = 2 + \frac{E}{3}. \tag{5.7}$$

Counting edges by summing over all hexagons and pentagons, and remembering that every edge is shared by two faces:

$$5F_5 + 6F_6 = 2E. \tag{5.8}$$

Solving Equations 5.7 and 5.8 by eliminating F_6, we find that we automatically eliminate E as well and obtain $F_5 = 12$. This is a startling result: it tells us, for example, that a soccer ball must have exactly twelve pentagonal (usually black) faces. We also find that there are berries having exactly twelve pentagonal faces in the company of hexagonal faces. It has been shown that the number of hexagons can be any positive integer except 1. These results are noncommittal about the number of hexagons, but emphatically limit the number of pentagons to twelve.

Two structures are called duals to each other if to each vertex of one there corresponds a face of the other, and vice versa. The dual of the pentagon-hexagon tessellation is dealt with by interchanging V and F, and n and r. The duals therefore must have $n = 3$, hence triangular faces. If a Fuller dome were built extending all the way around a sphere instead of being anchored in the soil, it would have exactly twelve 5-valent vertices, together with a large number of hexagons that determines the size of the dome. Accordingly, the occurrence of the number 12 in connection with berries, domes, and soccer balls is not a coincidence, but is in fact a fundamental property of the space in which we live, and a constraint with which we need to be familiar if we desire to shape that space.

Structures having all vertices equivalent to each other, as well as all faces equivalent to each other, are called *regular* (see Chapter 3). For such structures r_{av} and n_{av} are integers, as of course is E. For regular structures Equation 5.6 has the solutions given in Table 5.1. No other solutions are possible; this table exhaustively enumerates all regular structures. Neither r nor n can exceed 6 except when n, respectively r, equals 2. The cases $n = 2$ (digonal faces) are very real once we accept the possibility of curved edges. There are many objects in nature that are digonal polyhedra (for instance pumpkins, which have r digonal faces meeting at the stem and at the bottom), and there are many pods which are digonal trihedra. Five of the solutions correspond to the Platonic solids. Interesting are the three solutions having infinitely many edges. If the faces are to be finite in area, these solutions can only be

Table 5.1. Solutions of Equation 5.6 for regular structures

r	n	E	Comments
2	n	n	A polygon having n sides
r	2	r	A pumpkin-like structure having r diagonal faces join at each of two points
3	3	6	Tetrahedron
3	4	12	Cube
4	3	12	Octahedron
3	5	30	Pentagonal dodecahedron
5	3	30	Icosahedron
3	6	∞	Hexagonal tiling of the plane
6	3	∞	Triangular tiling of the plane
4	4	∞	Square tiling of the plane

realized on a sphere having infinite radius. Such a sphere would be experienced as a plane, much as we experience the surface of our globe in our immediate environment as flat. Note that there is no solution having n equal to five and E infinite; the implication is that there can be no regular pentagonal tessellation of the Euclidean plane.

Note in Table 5.1 that the interchange of n and r transforms a regular structure into its dual; the symmetry of Equation 5.6 in r and n implies that if a structure represents a solution of Equation 5.6, then so will its dual.

For doubly connected surfaces ($g = 1$), we derive from Equation 5.1:

$$\frac{1}{r_{av}} + \frac{1}{n_{av}} = \frac{1}{2}. \qquad (5.9)$$

When Equations 5.6 and 5.9 are compared, it is observed that the solutions to Equation 5.9 are just those of Equation 5.6 with E equal to infinity. This means that the toroid (the Florentine hat), for example, may be tessellated just like the plane, which is not true of a sphere having a finite radius.

Besides the regular structures there are the semiregular structures and their duals which have either all faces mutually equivalent or all vertices mutually equivalent, but not both. Equation 5.6 still yields an enumerable set of solutions when either r_{av} or n_{av} is an integer, which is characteristic of these structures. Once more we find a number of solutions corresponding to infinitely many edges, which again may be interpreted as plane tessellations. It is remarkable

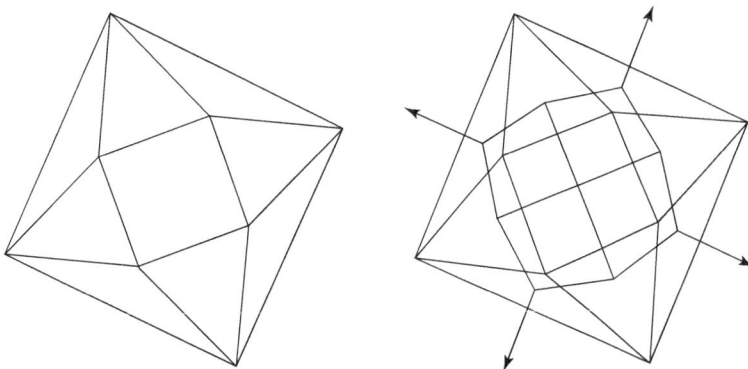

Figure 5.5. Schlegel diagram of a square antiprism (*left*) and of its dual (*right*).

that so many of the structures shown in this book correspond to the solutions of the remarkably simple equation Equation 5.6, and can be so listed and classified.

These structures may all be represented on a planar surface by their Schlegel diagrams. A polyhedron Schlegel diagram is its networks of edges and vertices drawn in a special way: if you hold the polyhedron so close to your face that one of its faces frames the entire polyhedron and you see all the other edges receding inside that frame, then you have a Schlegel diagram. Another way of looking at a Schlegel diagram is this. You can think of a truncated octahedron beautifully inscribed on a spherical blackboard and then you suddenly realize that it didn't matter where you put the vertices and edges; it was just how they were connected. You could then think of this as a peculiar kind of string bag — you could slide everything to one side of your spherical blackboard, contracting all edges into a very tiny figure. If you realize that our whole earth is really a gigantic sphere of very large radius, you could draw a gigantic truncated octahedron on the surface of the earth, and shrink it into a portion of that gigantic sphere which would just happen to be a blackboard. Then you would have a Schlegel diagram on your blackboard. Since n and r are only the valencies of the vertices and faces, it doesn't matter where these elements are located; we can say from our point of view of connectivity that a Schlegel diagram is entirely equivalent to the polyhedron itself. There is no difference

because we don't care where the vertices are. And so I will show you a number of solutions of Equation 5.6 in the form of Schlegel diagrams.

In Figure 5.5 on the left is the Schlegel diagram of a square antiprism: a square face has become extremely extended and frames the rest. All the other connections are there and you could almost solve these equations graphically just by taking a point, taking the proper valency, putting the lines in and then extending those lines until a face is created; interestingly enough, by following that procedure and not even thinking about what the polyhedron looks like, you can get your solutions directly as Schlegel diagrams. Now the question is: What about the dual? The polygon that frames the whole Schlegel diagram really represents a face corresponding to everything else in the plane — the entire universe in the plane. The reason? Remember we slid the polyhedron over to one side. But all the rest of the huge sphere is still a face, and if we now take a dual graphically, we have to put a vertex in each face (Figure 5.5). Emanating from that vertex must be the same number of edges as surround the original face, and each edge has to cross one of its companions. Then what happens when we get to the outer polygon? We have a whole cycle of faces just inside it, each of them in dualizing becoming a vertex which then has to be connected to a vertex corresponding to that outer face.

Notice that what I have done is to put arrows across the outer framing. Those arrows indicate that somewhere in the universe there is a vertex way out at the other pole of our infinitely large sphere and that is where the arrows will connect. I could have put it off center within this figure and connected everything, but it makes for a very unattractive, very ugly unsymmetrical kind of dual Schlegel diagram. I call these diagrams dual Schlegel diagrams because they are not Schlegel diagrams; they do not have everything framed by an outside face. We have in the dual Schlegel diagram a representation in which we have a vertex, which represents a real vertex, outside that frame. But it is perfectly easy to deal with those; you often can visualize much better what the dual polyhedron looks like by imagining the polyhedron flattened out, and the faces which meet at the "backside" vertex folded out.

Figure 5.6 shows a Schlegel diagram of a snub figure. It is the same one that Coxeter showed in (Chapter 3); he calculated the coordinates when the square was rotated. In this case, I happened to orient it so that the outer square is the rotated one and you see the whole tier of triangles surrounding each of those squares. So this one is the snub cube, one of the solutions of that Diophantine equation. In Figure 5.7 you see it again in the upper right, but here I put it in the company of its family having increasing numbers of edges, so corresponding to the snub cube there is a snub dodecahedron with still a finite but much larger number of edges, and finally the plane tessellation. All of these figures have in common their *chirality*, a very important property. This means they exist in forms that are distinct from their mirror images. So I could have drawn either one or I could have flipped the figure and then we would have had the other form of it. In Figure 5.8 you see that I have taken the duals of the snub cube and snub dodecahedron in the form of the dual Schlegel diagrams.

Figure 5.9 is a stellated icosahedron and again that does very well in its dual Schlegel representation; in the center you can see the structure very well, but like a polar projection it is distorted toward the outside. But if you want to build these figures, these dual Schlegel diagrams help a lot. You get your actual polyhedron from the Schlegel representation by lifting up the arrows radiating out to infinity and bringing them together.

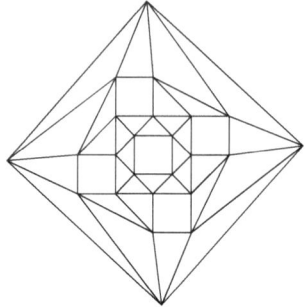

Figure 5.6. Schlegel diagram of a snub cube.

Figure 5.10 shows a pentagonal tessellation of the sphere. This is a model made by Brett Tomlinson. Next in the family is the limiting pentagonal tessellation (Figure 5.11) in which E equals infinity. Incidentally, as we saw previously, there can be no regular pentagonal tessellation of the plane. But here we have a semiregular pentagonal tessellation and we have valencies in this case of 6 and 3. You can tell fairly easily that this is also a figure that has chirality. What you have to do is make a distinction, not only between the vertices having different valencies, but also between different 3-valent vertices. You find that some of these 3-valent vertices are connected to a 6-valent as well as to two 3-valent vertices, whereas others are connected to 3-valent vertices only. Those vertices are definitely distinct. Their contexts are different even though their valencies are the same.

As we go around the pentagon, you will notice that we have five vertices and we have to make a choice, whether we are going to put one of each type of 3-valent vertices on the right-hand side or on the left-hand side. That means that we are forced to create a tessellation that has chirality, because we have these different types to distribute. That choice has to be made and depending on how we make it, we get this tessellation or its mirror image. Figure 5.12 shows two of the pentagonal tessellations about which Doris Schattschneider is an expert. You see again that we are dealing with a family of E increasing toward infinity. Polyhedra and tessellations are very closely related.

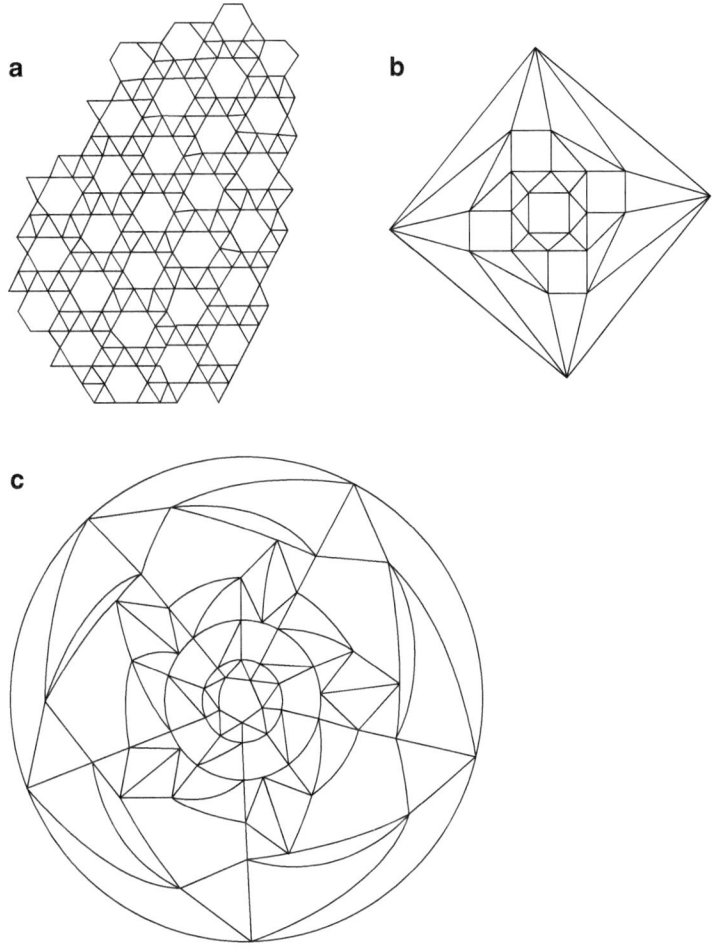

Figure 5.7. (**a**) Snub tessellation of triangles and hexagons, having a 5-valent vertices. (**b**) Snub cube. (**c**) Snub dodecahedron.

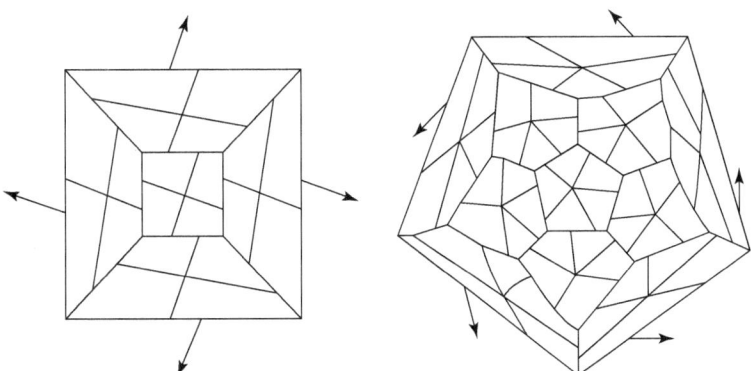

Figure 5.8. Schlegel diagrams of (*left*) a pentagonal icositetrahedron (dual of snub cube) and (*right*) a pentagonal hexacontahedron (dual of the snub dodecahedron).

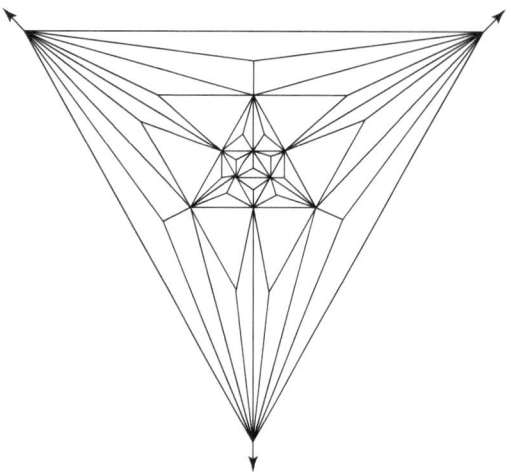

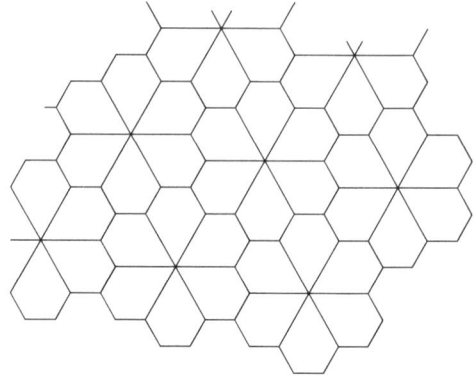

Figure 5.11. Pentagonal tessellation of the plane with $n_1 = 4, n_2 = 1, r_1 = 3, r_2 = 6$.

Figure 5.9. Schlegel diagram of a stellated icosahedron.

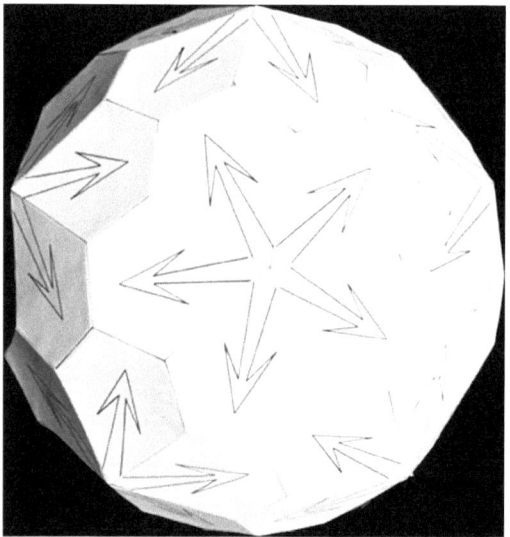

Figure 5.10. A model of a pentagonal tessellation of a sphere.

Now I am going to take the symmetry point of view. On the board in Figure 5.13 you see a number of polyhedra, all of which have exactly the same symmetry: they all have the symmetry of a cube. The cube is placed in a rather interesting orientation, having its threefold axis vertical; that is done on purpose. Each polyhedron has been put there with a threefold axis vertical. We now want to think about the question: If they all have the same symmetry, then how are we going to distinguish among them?

One way of distinguishing is by the process of truncation. When we talk about this we can think of ordinary knives slicing pieces of cheese. But I am going to look at the coordinates of the vertices. It is these configurations made by the atoms and ions in the crystals, that give us the so called coordination polyhedra.

Suppose that we have a point whose coordinates are x, y, and z. If this point is part of a structure having threefold rotational symmetry, then there must be two other points whose coordinates are cyclic permutations of x, y, and z; and which are related to the point (xyz) by threefold rotational symmetry. The three points form the vertices of an equilateral triangle whose coordinates are, respectively: xyz, yzx, and zxy.

Cubic symmetry moreover implies mirrors diagonally through the cartesian axes, hence an additional triplet whose coordinates are zyx, xzy, and yxz. Reflection of these six points in each of the coordinate planes produces 48 points whose coordinates are those of the above six combined with the eight possible combinations of plus and minus. These 48 points constitute the vertices of a greater rhombicuboctahedron (Figure 5.14).

There are special circumstances under which the symmetry elements of the cube do not generate a full complement of 48 distinct vertices. This happens when x, y, or z have special values which cause these vertices to lie precisely on a symmetry element. For instance, a point lying on one of the threefold axes would have

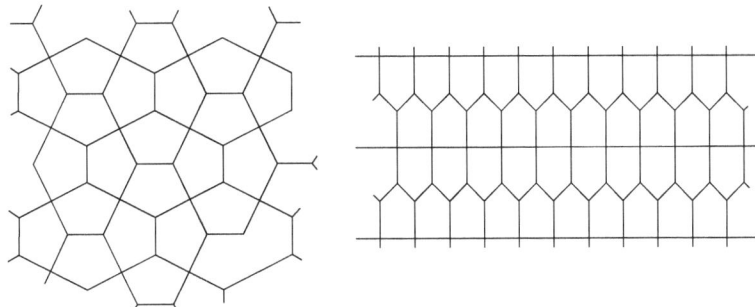

Figure 5.12. Two pentagonal tessellations of the plane, each with $n_1 = 3, n_2 = 1, r_1 = 3, r_2 = 4$.

Figure 5.13. A collection of wooden polyhedra, each with the symmetry of a cube. Each polyhedron has been oriented with a threefold axis vertical.

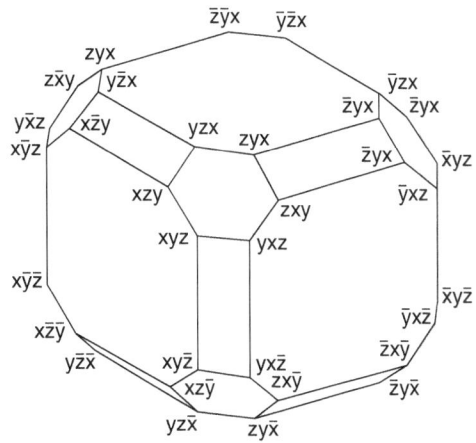

Figure 5.14. The greater rhombicuboctahedron.

$x = y = z$, with the result that the six points whose coordinates were all the permutations of x, y, and z have fused into a single point, which when reflected into the cartesian planes, produces merely the eight vertices of a cube. All special cases are listed in Table 5.2 together with the names of the polyhedra whose vertices are defined by the resulting special combinations of coordinates. These polyhedra are shown in Figures 5.14–5.20. Here we call the initial point on which the cubic symmetry elements act to generate the entire point complex the "generating point."

The computer can very quickly perform all the permutations inherent in the cubic symmetry, regardless of the specific values of x, y, and z,

Table 5.2. Special cases in which the symmetry elements of a cube do not produce 48 distinct vertices

Coordinates of generating point	Special condition	Number of vertices	Polyhedron generated	Figure
xyz	None	48	Greater rhombicuboctahedron	5.14
$xyz, y > x$	$x = y$	24	Lesser rhombicuboctahedron	5.15
$xy0$	$z = 0$	24	Truncated octahedron	5.16
$xxy, y < x$	$y = z$	24	Truncated cube	5.17
$xx0$	$x = y, z = 0$	12	Cuboctahedron	5.18
xxx	$x = y = z$	8	Cube	5.19
$x00$	$y = z = 0$	6	Octahedron	5.19
000	$x = y = z = 0$	1	A single point	

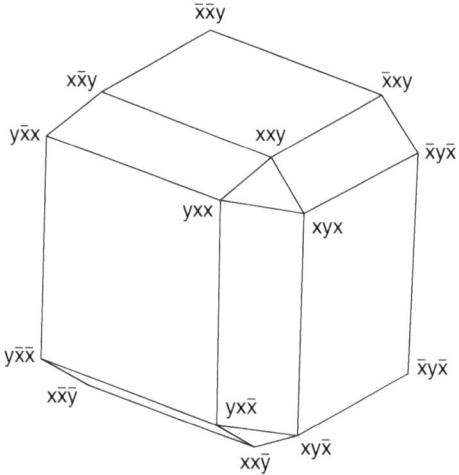

Figure 5.15. Lesser rhombicuboctahedron.

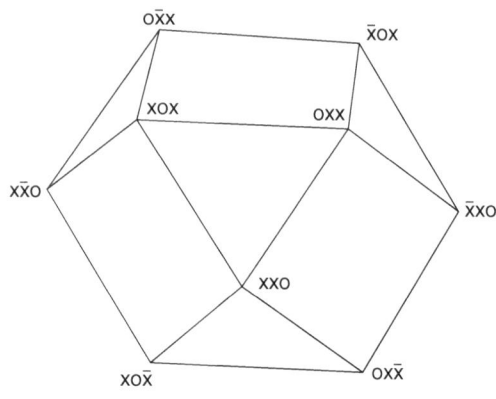

Figure 5.18. Cuboctahedron.

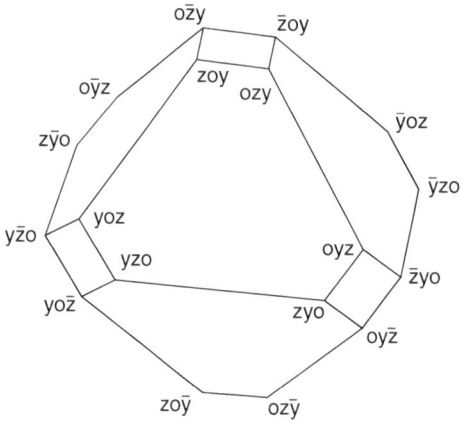

Figure 5.16. Truncated octahedron.

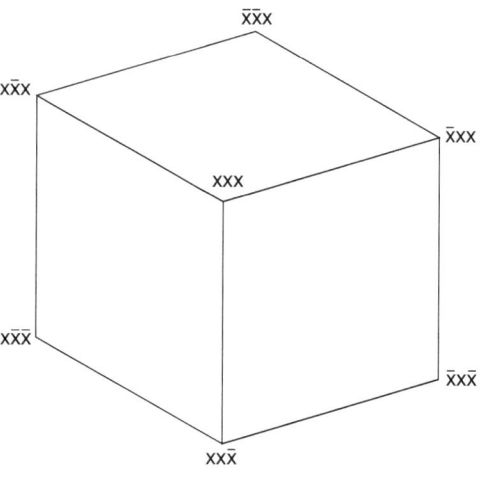

Figure 5.19. Cube.

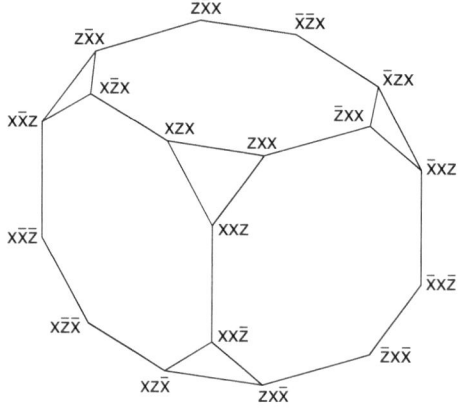

Figure 5.17. Truncated cube.

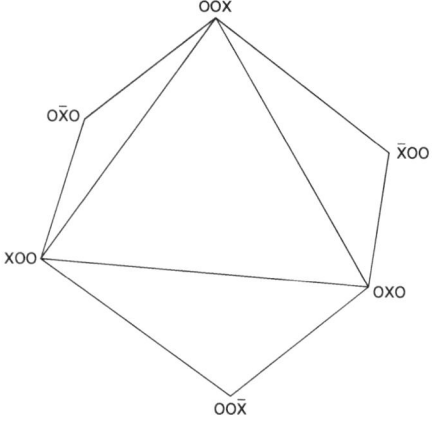

Figure 5.20. Octahedron.

and then determine how many distinct points are generated. Table 5.1 translates that number into the name of the appropriate polyhedron; the computer recognizes the polyhedron on the basis of the number of distinct vertices, a task at which it is much more adept than that of slicing cheese.

Table 5.2 shows three different polyhedra having 24 vertices. It is easy enough for the computer to distinguish them: if one of the coordinates is zero, then the polyhedron is a truncated octahedron, otherwise it is either a lesser rhombicuboctahedron or a truncated cube, depending on the relative magnitudes of u and w. This latter distinction appeared to me rather subtle for two apparently so different polyhedra and hence prompted me to compare the two. As a result, I came upon a previously unrecognized relationship between the two forms. This relationship is based on the circuits traced by the edges of either polyhedron on a (spherical) surface on which it may be projected. The lesser rhombicuboctahedron has triangular, rectangular, and square faces, and truncated cube triangular and octagonal faces. If, however, we look at a square face on the former, we note that its four vertices are also vertices of four triangles whose bases combine with four additional edges to form an octahedron. If we take the eight triangular faces on either form and flip them upside down, changing the relationship between u and w, then we interchange the squares and octagons, and transform one form into each other. A model of this transformation has been built, confirming the close relationship between the two forms (Figures 5.21 and 5.22).

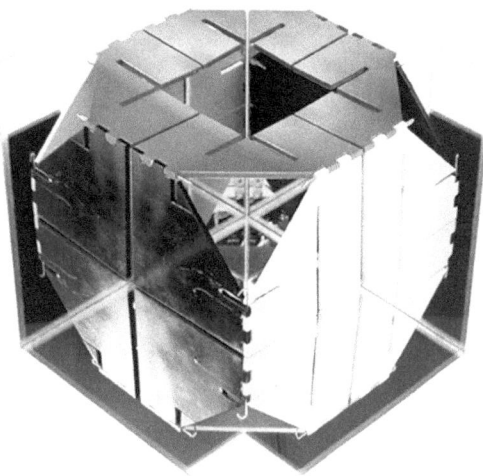

Figure 5.21. Model of truncated cube.

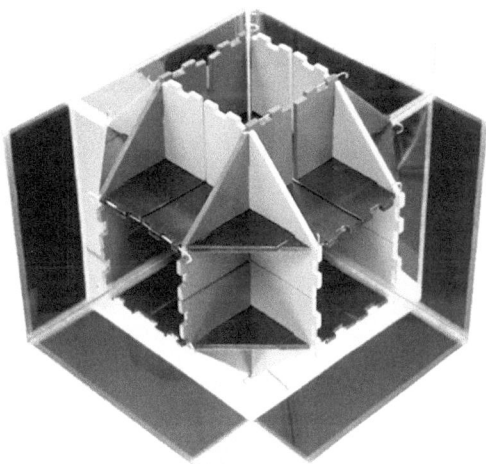

Figure 5.22. Model of lesser rhombicuboctahedron.

6

Dürer's Problem

Joseph O'Rourke

In 1525 the German painter and thinker Albrect Dürer published his masterwork on geometry, whose title translates as "On Teaching Measurement with a Compass and Straightedge."

The fourth part of this work concentrates on polyhedra: the Platonic solids, the Archimedean solids, and several polyhedra "discovered" by Dürer himself. (This work was hailed in Chapter 4 as Milestone 8.) His interest in polyhedra was evident at least a decade earlier, when he used an apparently original polyhedron in his famous engraving *Melencolia I*. In his book he presented each polyhedron by drawing a *net* for it: an unfolding of the surface to a planar layout. The net makes the geometry of the faces and the number of each type of face immediately clear to the eye in a way that a 3D drawing, which necessarily hides half the polyhedron, does not. Moreover, a net almost demands to be cut out and folded to form the 3D polyhedron.

Examples of Dürer's nets are shown in Figures 6.1 and 6.2. The first is a net of the *snub cube*, which consists of six squares and 32 equilateral triangles. The second is a net of the truncated icosahedron, consisting of 12 regular pentagons and 20 regular hexagons, the spherical version of which we know as a soccer ball.

In the half-millennium since Dürer, nets have become a standard presentation method for describing polyhedra. For example, Figure 6.3 shows a modern display of nets for the so-called Archimedean solids.

But no one has proved that a net exists for every convex polyhedron. It is this long-unsolved problem that is the focus of this chapter.

Convex Polyhedra and Nets

We will be concerned only with polyhedra without holes, so we exclude polyhedral tori and other such objects that could be hung on a thread through a hole. Edges will be straight line segments. Edges meet at vertices, the sharp corners on the surface.

The most famous polyhedra are the five "regular" or "Platonic" solids shown in Chapter 1, known for at least 2,500 years. Despite the name "solid," we will view a polyhedron as the thin surface enclosing the volume, rather than the solid itself.

The Platonic solids are all convex polyhedra, which means essentially that they have no dents. The notion of convexity applies both to 2D and to 3D, and we will need both.

A convex polygon is a closed figure in 2D composed of straight edges, with the property that if you walked around the boundary, counterclockwise from above, you would make only left turns at each vertex. A right turn would constitute a dent. Another way to view this is that the internal angle at every vertex of a convex polygon is less than 180°. A vertex at which the internal

J.O'Rourke
Deptartment of Computer Science, Smith College,
Northampton, MA 01063, USA
e-mail: orourke@cs.smith.edu; http://cs.smith.edu/~orourke/

M. Senechal (ed.), *Shaping Space*, DOI 10.1007/978-0-387-92714-5_6,

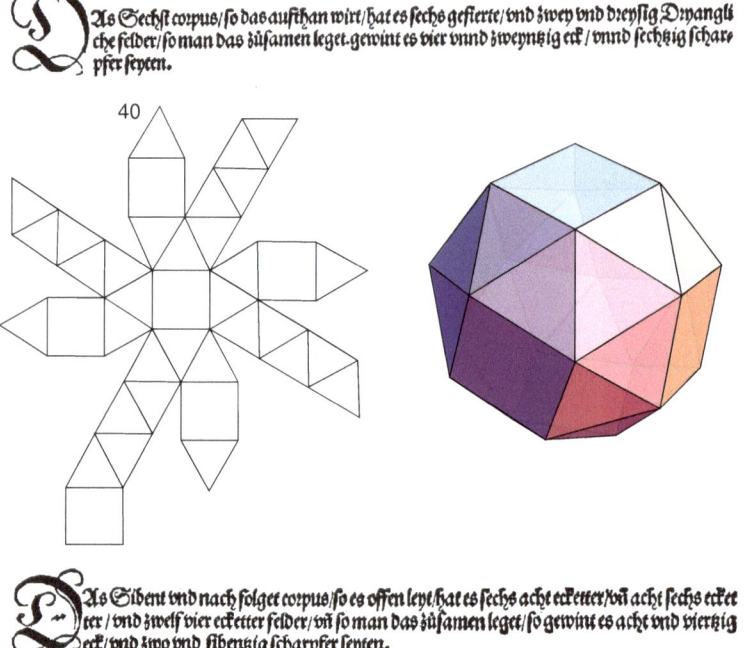

Figure 6.1. Dürer's net for a snub cube.

angle exceeds 180° is called a "reflex vertex" (a term more memorable than "concave vertex").

Just as a convex polygon has no reflex vertices, a convex polyhedron has no reflex edges: the internal dihedral angle between the two faces meeting at each edge of a convex polyhedron is less than 180°. For the cube, this dihedral angle is 90° at each convex edge.

Now we finally define what constitutes a net. A *net* is an unfolding of the surface of a polyhedron produced by cutting the polyhedron along some of its edges and flattening it to a single, non-overlapping piece in the plane. The key aspects of this definition are:

1. The net is planar.
2. It is a single piece.
3. It is the result of cutting polyhedron edges: it is an *edge unfolding*.
4. It is non-self-overlapping: non-cut points do not unfold on top of one another.

These four properties mean that one can cut out a net drawn on paper with scissors, crease along edges, and fold to the 3D polyhedron. Dürer's unfoldings are nets of course; our

definition is intended to capture the standards he set down by his example. Note that his net for the truncated icosahedron (Figure 6.2) is just barely non-overlapping, with some pentagons almost touching some hexagons. It makes sense to permit touching of boundaries, which could still permit the net to be cut out with scissors. What is forbidden is more substantive overlap, when points not on the boundary of the figure overlap in the plane.

The Open Problem

Now we can phrase the unsolved question, which we take the liberty of calling "Dürer's Problem" even though there is no evidence that he recognized it as a claim that needed proof:

> **Dürer's Problem:** Does every convex polyhedron have a net?

Despite almost 500 years of many people drawing nets for convex polyhedra, no one has

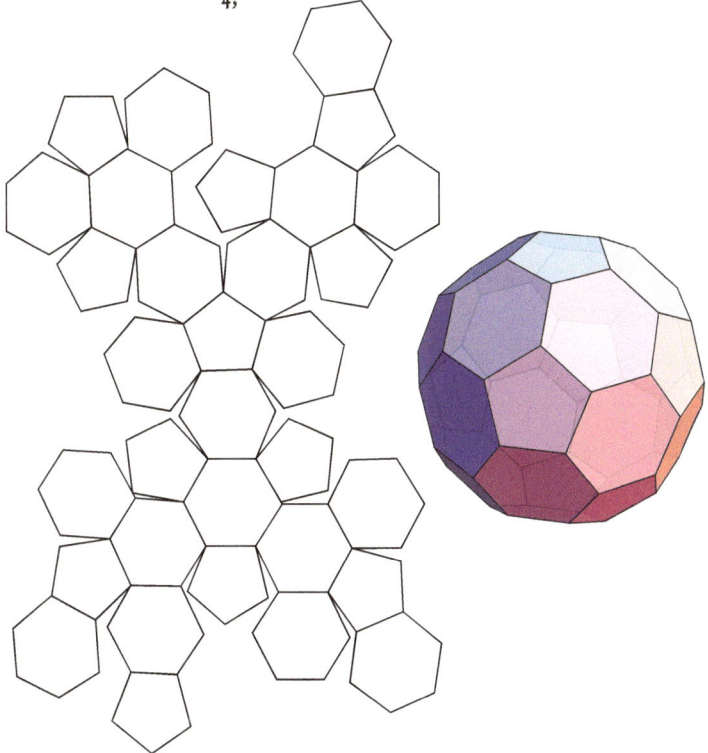

Figure 6.2. Dürer's net for a truncated icosahedron.

come up against an example that has no net. On the other hand, there is no proof that every convex polyhedron does have a net, despite years of effort since the problem was first formalized as a sharp mathematical question by Shephard in 1975.

You might think it is obvious that a convex polyhedron has a net, because when the surface is cut and flattened, it "spreads out," and so there should be plenty of room in the plane. This is certainly true for the regular and semi-regular polyhedra that have been the focus of attention for hundreds of years. But that intuition weakens when considering more complicated and irregular convex polyhedra, such as the egg-shaped object in Figure 6.4.

Or maybe you don't see how overlap can occur at all. Then look at Figure 6.5, in which (a) shows a cube with one vertex truncated, and (b)

an unfolding with overlap. On the other hand, (c) shows that it is easy to avoid this overlap by repositioning one face.

You might think that surely every tetrahedron—that is, every convex polyhedron with exactly four vertices—has a net. This is true (and has been proved), but even here some care must be exercised. Consider the thin, nearly flat tetrahedron shown in Figure 6.6a. One choice of cutting leads to overlap (b), although again it is easy to find other cutting choices (c) that do produce nets for this tetrahedron.

Nonconvex Polyhedra

In the late 1990s several groups of researchers independently discovered *nonconvex* polyhedra that have no net. Figure 6.7 shows an especially

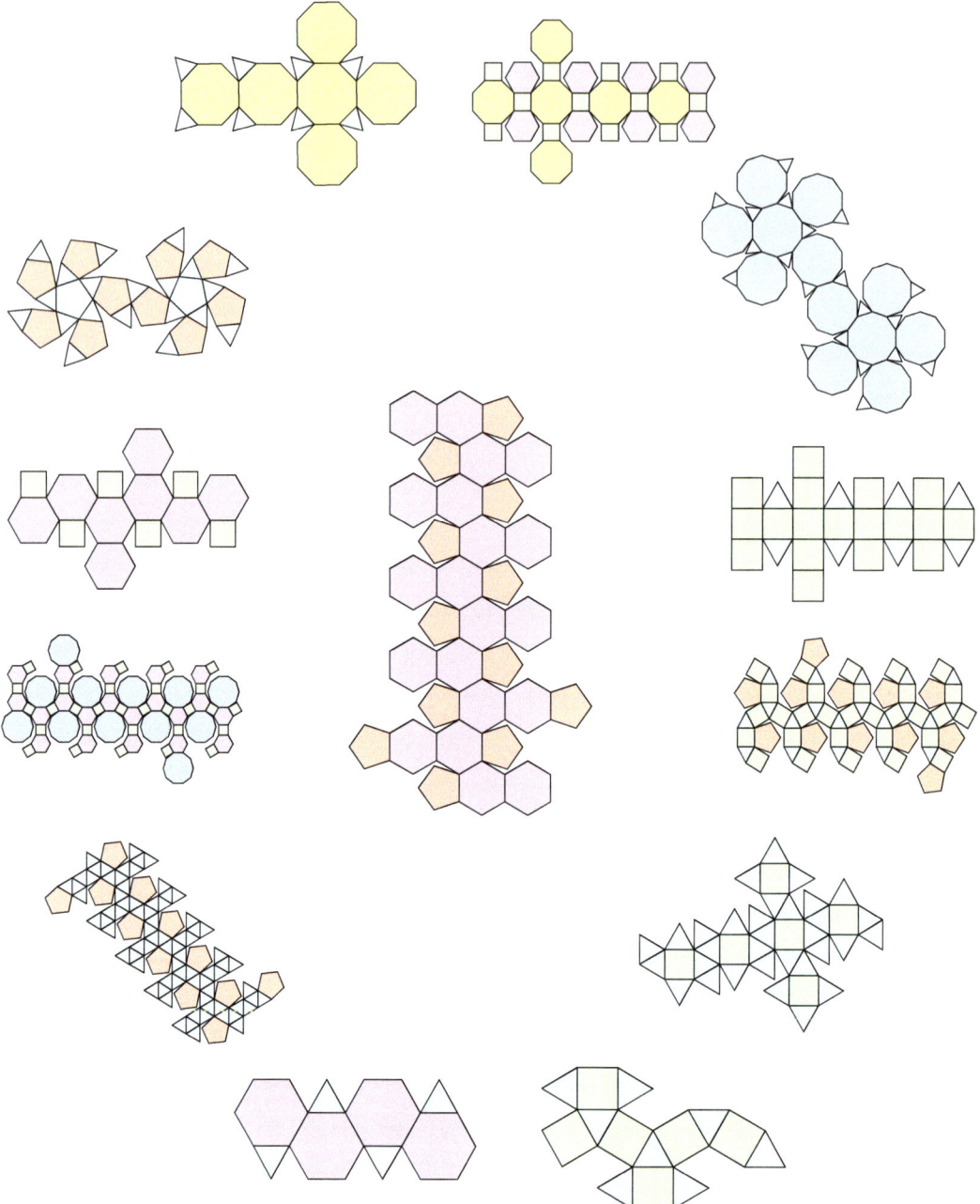

Figure 6.3. Unfoldings of the 13 Archimedean solids. The snub cube (Figure 6.1) is at the 5 o'clock position, and the truncated icosahedron (Figure 6.2) is in the center.

elegant example, a "spiked tetrahedron." The proof that this polyhedron has no net must show that every possible edge unfolding leads to overlap. (And there are many possible edge unfoldings.)

Because any proof that a polyhedron has no net must grapple with "every possible edge un-folding," we turn next to a useful proposition that captures some necessary properties of edge unfoldings.

Spanning Cut Tree

If we cut along a given edge of a polyhedron as part of an unfolding, we call that edge a *cut edge*. The following proposition states necessary conditions on the collection of cut edges using technical jargon; the terms are explained below.

Proposition 6.1. *The cut edges of an unfolding that results in a net for a polyhedron form a spanning tree of the skeleton of the polyhedron.*

The *skeleton* of a polyhedron is the network of edges and vertices on the surface—effectively a wireframe view of the polyhedron without the faces. A *tree* is a collection of edges that never loops back to form a closed loop, or *cycle*. It is called a tree because, drawn in the right way, it resembles a tree (or a bush; see Figure 6.8a):

Figure 6.4. A convex polyhedron whose 100 vertices are randomly sprinkled on the surface of an ellipsoid.

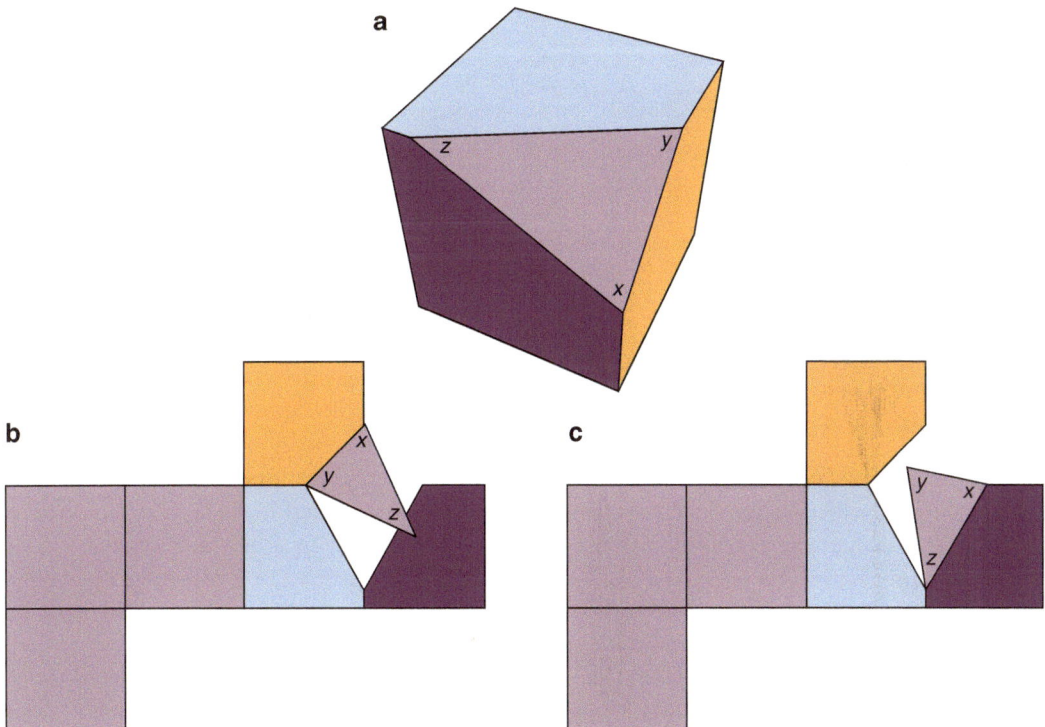

Figure 6.5. (**a**) Cube with corner truncated. (**b**) Overlapping unfolding. (**c**) A net: non-overlapping.

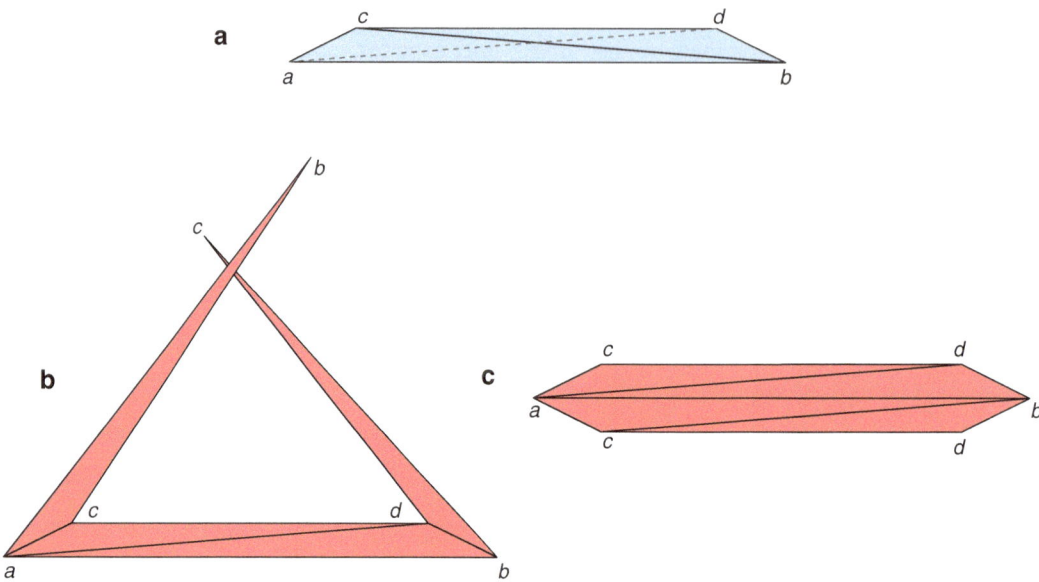

Figure 6.6. The four faces are colored *blue* on the outside and *red* on the inside. (**a**) A nearly flat tetrahedron. Edge ad is in back. (**b**) Overlapping unfolding from cutting (a, b, c, d). (**c**) A net obtained by cutting (a, c, d, b).

Figure 6.7. A nonconvex polyhedron of 36 triangle faces that has no net.

the branches cannot dovetail and connect, for that would form a cycle of edges. Finally, a *spanning tree* is a tree that touches every vertex. Figure 6.8b shows that the cut edges that Dürer used to obtain his net of the snub cube (Figure 6.1) indeed form a spanning tree of its 24 vertices.

Now we can prove the proposition. If the cut edges do not form a tree, then they contain a cycle. A cycle of cuts on the surface separates

the surface inside the cycle from the rest of the surface. Thus we would be left with at least two separate pieces, the surface inside the cycle and the surface outside. This violates condition (2) of the definition of a net. So the cut edges must form a tree.

Suppose the tree is not spanning. That means that there is some vertex v of the polyhedron not touched by any cut edge. But this means that v retains its 3D structure, and so cannot be flattened. This violates condition (1) of the definition of a net: for the net to be planar, the cuts must span the vertices.

Thus we have established the proposition: the cut edges must form a spanning tree to have any hope of reaching a net. So the spanning tree condition is necessary. But notice that we did not employ the crucial 4th condition: the net should be non-overlapping. And indeed the spanning tree condition is far from sufficient to guarantee a net. The cuts used to produce our overlapping examples (Figures 6.5 and 6.6) form spanning trees.

To this day no one has discovered necessary and sufficient conditions for a collection of cut edges to unfold to a net. Consequently, any

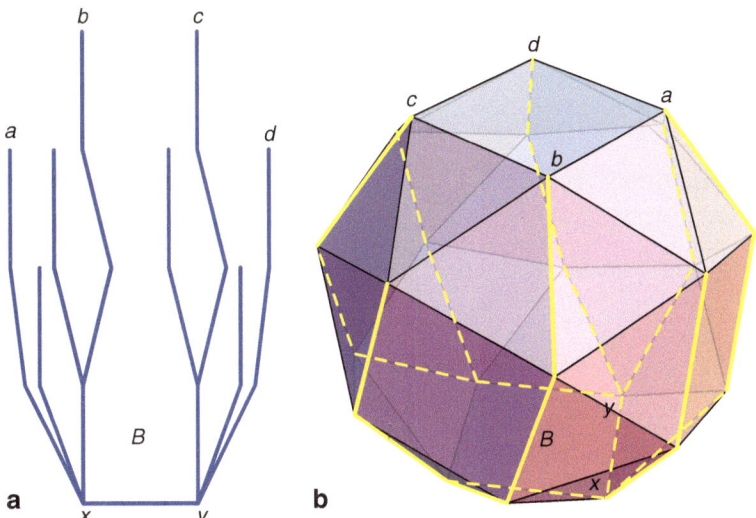

Figure 6.8. (**a**) A tree. (**b**) Dürer's spanning cut tree on the snub cube. The tree in (**a**) is structurally the same as the tree in (**b**). The base face B and several vertices are labeled in both.

potential counterexample to the conjecture that every convex polyhedron has a net must thwart every possible spanning tree. And there are a lot of spanning trees! The number is exponential in the square root of the number of faces. For the egg-shaped polyhedron in Figure 6.4, which has $F = 196$ faces, there are more than 500,000 distinct spanning trees.

Some Polyhedra with Nets

There is no grand theorem as yet in the investigation of Dürer's Problem, just scattered results of a few classes of convex polyhedra that are known to have nets. We will describe these classes without proving the results, and end with a challenge to settle another natural class.

We need the notion of the *convex hull*, which is easiest to understand in 2D. Suppose we mark a set of points in the plane by pounding in nails at each point, leaving much of the nail above the plane. The convex hull of the points is the convex polygon determined by the shape of a stretched rubber band that encompasses all the points. In 3D, we have to imagine a set of points fixed in space. The convex hull of the points is the the convex polyhedron determined by wrapping a set

of points in 3D as tightly as possible with plastic wrap. I constructed the polyhedron in Figure 6.4 by computing the convex hull of random points on an ellipsoid.

Now with this notion, we can define the classes of polyhedra for which nets are known to exist. Let B be a convex polygon in a plane (B for base). Make a second copy of B vertically above B; call that copy A (A for above). The convex hull of the vertices of A and B is a *right prism*, "right" because all the lateral faces are rectangles with right angles. A *prism* (or *oblique prism*) does not insist that A be directly above B; rather it can be shifted (*translated*) with respect to B. The lateral faces of a prism are parallelograms. See Figure 6.9a, b. Both of these classes are subclasses of a more general shape, a *prismoid*, whose lateral faces are trapezoids. There is a proof that every prismoid has a net, as illustrated in Figure 6.9c: All lateral edges are cut, and all but one edge of A is cut. The only delicate part of the proof is deciding which edge of A not to cut; not all choices lead to non-overlap.

A *pyramid* is the convex hull of a base convex polygon B and a single point a (the apex) above B. The natural *petal unfolding* determined by cutting every edge incident to a leads to a net; see Figure 6.10a.

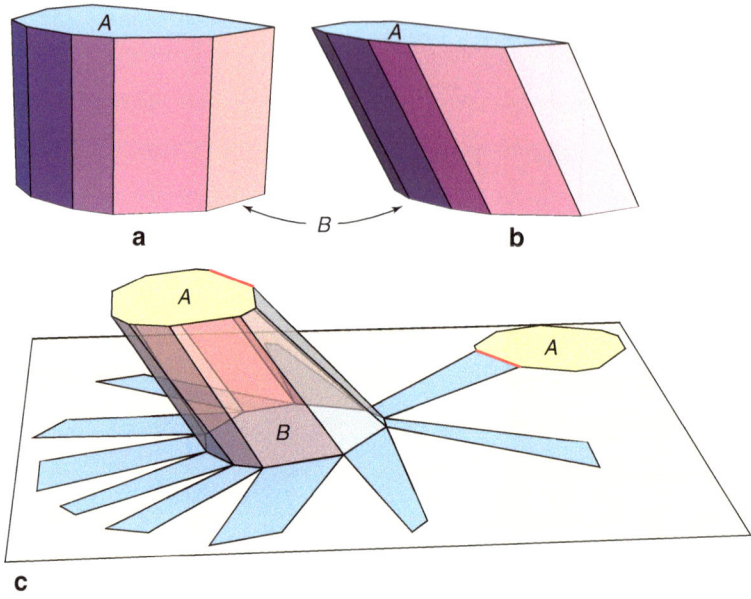

Figure 6.9. (**a**) A right prism. (**b**) An oblique prism. (**c**) Unfolding of a prismoid to a net. The *red* edge of A is not cut.

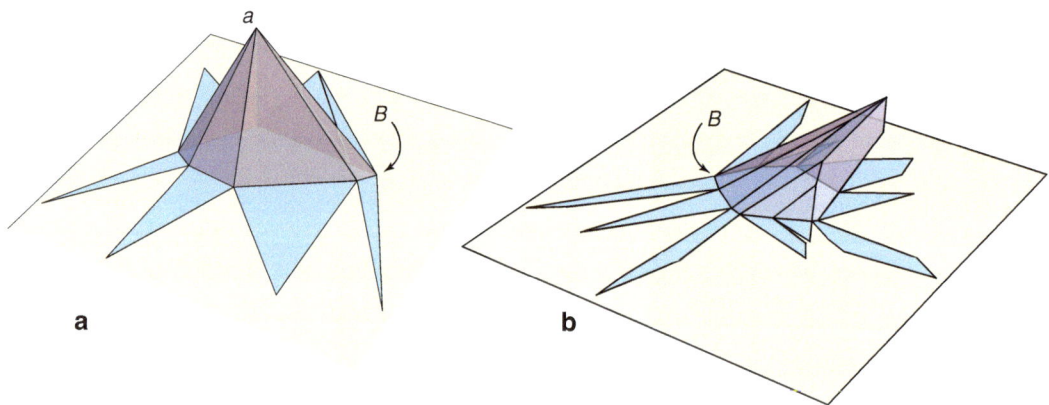

Figure 6.10. (**a**) A net for a pyramid. (**b**) A net for a dome.

A *dome* is a generalization of a pyramid with the property that every face shares an edge with the base convex polygon B. Again the petal unfolding leads to a net, a theorem for which there are now three different proofs.

One more class, "higher-order deltahedra," was proved to always have a net by Daniel Bezdek as part of his award-winning 9th grade Canadian Science Fair project. These are polyhedra whose surface is composed entirely of equilateral triangles, including faces (e.g.,

hexagons) partitioned into several coplanar triangles (see Chapter 2). These few classes are essentially the only infinite classes of convex polyhedra for which it is established that there is always a net.

One natural class of polyhedra for which there is as yet no proof that nets always exist is the prismatoids. A *prismatoid* is the convex hull of two arbitrary convex polygons A and B lying in parallel planes. Thus it is very similar to a prism or a prismoid (and its name is similarly

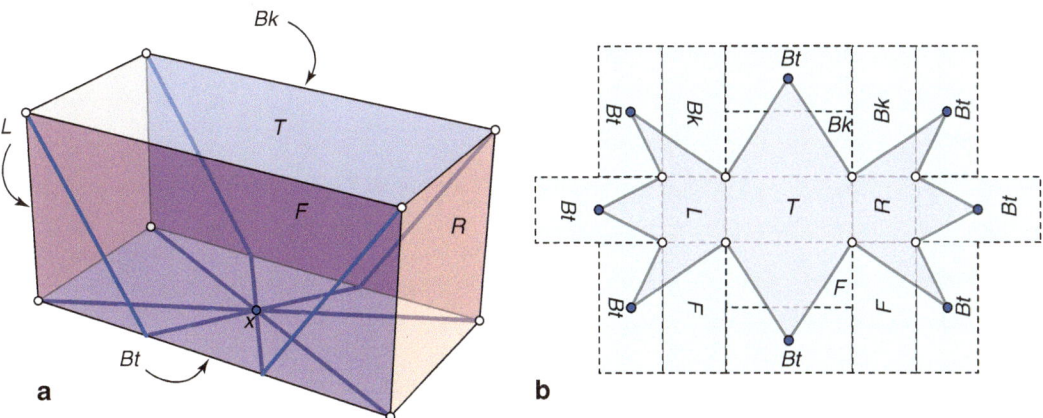

Figure 6.11. (**a**) $2 \times 1 \times 1$ box. Box faces are labeled: *Bt, F, T, R, L, Bk* for *Bottom, Front, Top, Left, Right,* and *Back* respectively. (**b**) Star unfolding with respect to x.

confusing!), but A and B may be very different. This means that, in general, the lateral faces are not quadrilaterals, but rather triangles. I invite the enterprising reader to tackle this class.

General Unfoldings

Although there are no broad theorems on nets, if the restriction to cutting along edges of the polyhedron (Condition (3) in our definition of a net) is relaxed, there are some nice theorems, one of which we describe here. In contrast to edge unfoldings, which demand the cuts be along edges, a *general unfolding* permits arbitrary cuts to produce the unfolding—the cuts can run through the interior of polyhedron faces. The goal is otherwise the same: find a collection of cuts that unfold the surface to a single, planar, non-overlapping piece. Let us call this, for lack of a better term, a *general net*. There are now several different proofs that every convex polyhedron has a general net. All of them depend on the notion of a shortest path on the surface between two points a and b. As its name suggests, a *shortest path* is the minimum-length route to travel from a to b on the surface, i.e., the optimal path for an ant to walk between the two points.

Shortest paths have many mathematical properties, two of which we need here. First, a shortest path never passes through a polyhedron vertex: it is always shorter to go around a vertex than through it. Second, when a shortest path crosses a polyhedron edge, it does so in such a way that the planar unfolding of the two faces sharing the edge straightens the path: a shortest path unfolds straight across every edge.

The *star unfolding* always produces a general net. This concept was introduced by Alexandrov in 1948 (and so sometimes called an "Alexandrov unfolding") but only proved to be non-overlapping in the 1990s.

Pick any "generic" point x on the surface of the polyhedron not at a vertex, generic in a sense soon to be clarified. Now draw the shortest path from x to each vertex of the polyhedron in turn. There are usually many choices of x for which the shortest path to each vertex is unique. This is the sense in which x must be generic. Figure 6.11a illustrates these paths for a rectangular box, with x in the middle of the bottom face. Notice that the collection of these paths satisfies the Proposition: they form a tree that spans the vertices. The star unfolding is produced by cutting all the shortest paths and unfolding, as shown in (b) of the figure. Note that each shortest path unfolds to a straight line segment, as it must by the second property of shortest paths mentioned above.

Although non-overlap is almost obvious in this symmetric example, it is less obvious, although now proved, for more generic convex

Figure 6.12. The star unfolding of a polyhedron of $n = 18$ vertices.

polyhedra. Figure 6.12 shows a more typical star unfolding.

If this unfolding is cut out, and the cuts taped back together, it forms a convex polyhedron of 18 vertices. (Moreover, it uniquely folds to this polyhedron, by a theorem of Alexandrov.)

The star unfolding has now been generalized to cutting shortest paths to a closed curve Q on the surface rather than to a point x, yielding the same result when Q shrinks down to x.

Nonconvex General Unfoldings

We have seen that it remains unresolved whether every convex polyhedron has an edge unfolding—a net (Dürer's Problem), but all these polyhedra do have general nets (via, e.g., the star unfolding). We have also seen (in Figure 6.7) that not every nonconvex polyhedron has a net. So it is natural to ask whether every nonconvex polyhedron has a general net. This enticing problem remains unsolved today. The widest class for which this is established is *orthogonal polyhedra*, polyhedra all of whose edges are parallel to orthogonal Cartesian axes. There is an algorithm that unfolds any orthogonal polyhedron (of genus zero—i.e., without holes) to a single, non-overlapping piece.

Finally, although it would seem that the reverse of unfolding—folding polygons to form convex polyhedra—could hold no new mysteries, that is far from the case. Unsolved problems abound in this topic.

Part II

Polyhedra in Nature and Art

7

Exploring the Polyhedron Kingdom

Marjorie Senechal

Having paid our respects to the rulers of the Poly-hedron Kingdom and their extended families, we're ready for a walking tour.

An Architectural Walking Tour

The first polyhedral buildings we see on our tour are perhaps the most famous of all: the pyramids of Egypt, built about 2500 B.C.E. (Figure 7.1). Yes, a pyramid is a polyhedron; one of its faces is a polygonal base (of $3, 4, 5, \ldots, n$ sides) and the others are congruent isosceles triangles joined to the base along its edges, meeting above it in a single point. The bases of the Egyptian pyramids are squares.

As we walk along, we see buildings based on prisms. In fact, most buildings are prisms, since the rectangular "boxes" that constitute much familiar architecture are prisms with rectangular bases. Even boxes can become interesting polyhedral structures when juxtaposed in imaginative ways. A-frames are triangular prisms resting on a rectangular side. A notorious pentagonal prism is located near Washington, D. C. (Figure 7.2) It is rare to see a prismatic building with more than eight flat sides, but if we allow the meaning of "prism" to include polyhedra with curved sides, then we find that they are quite common.

The Coca-Cola Building (Figure 7.3), designed by Erwin Hauer for the 1964 New York World Fair, is a prism with many curved sides. The detail of the outer grill show in Figure 7.4 shows that the design can be considered a slice through a packing of truncated octahedra. (Interesting polyhedral structures are often designed for world fairs and moved later to their permanent sites in the Kingdom).

Other interesting polyhedral buildings are almost spherical in form. In Figure 7.5 we see one of Buckminster Fuller's geodesic domes, the United States Pavilion at Expo '67 in Montreal. Its faces are triangles, grouped into hexagons and pentagons (see Chapters 3 and 9). The geodesic dome has been the inspiration for countless buildings, large and small. The house shown in Figure 7.6 was constructed by a family in Hadley, Massachussetts. Many other interesting domes are described in do-it-yourself publications (see, for instance, Figure 7.7). The faces of some dome structures are deliberately arranged in asymmetric ways. For example, the dome grid

M. Senechal
Department of Mathematics and Statistics,
Smith College, Northampton, MA 01063, USA
e-mail: senechal@smith.edu, http://math.smith.edu/~
senechal, http://www.marjoriesenechal.com

Figure 7.1. The pyramids of Mycerinus, Chefren, and Cheops at Giza.

M. Senechal (ed.), *Shaping Space*, DOI 10.1007/978-0-387-92714-5_7,
© Marjorie Senechal 2013

Figure 7.2. The Pentagon is an enormous pentagonal prism.

Figure 7.3. Coca-Cola Building at the 1964 World Fair.

Figure 7.4. Detail of the Coca-Cola Building at the 1964 World Fair, showing a slice through a tight packing of opaque acrylic truncated octahedra.

(shown in Figure 7.8) of the Five-College Radio Astronomy Observatory at Quabbin Reservoir, Massachusetts, is deliberately random to prevent interference patterns with the incoming signal.

Once your eyes are opened, you will find many interesting examples of polyhedral architecture in your own neighborhood. Figure 7.9, shows are some "geometric" student residences at Hampshire College in Amherst, Massachusetts.

The rulers of the Polyhedron Kingdom have instituted a Polyhedral Hall of Fame to honor human beings who use polyhedra in especially unexpected and delightful ways. The first person to be elected to the Hall was the Israeli architect Zvi Hecker, cited for his multipolyhedral synagogue in Negev desert (Figure 7.10) and

Figure 7.5. The U.S. Pavilion at the Expo '67 World Fair Montreal.

his dodecahedral housing complex in Ramot, Israel (Figure 7.11).

Architecture reminds us that the most important part of some polyhedral structures is the network of edges and vertices. If we agree that such networks themselves constitute polyhedra, then many bridges are polyhedra, and so are common (and uncommon) jungle gyms (such as those shown in Figures 7.12 and 7.13). In fact there is no end to the polyhedral structures around us.

The Nature Preserve

The nature preserve is a vast region of the Polyhedron Kingdom, whose known extent keeps growing larger as it becomes possible to study structures on increasingly smaller scales. On this brief visit we will only have time to glance casually at some natural polyhedra which can be seen with the naked eye or with a simple microscope.

Our first stop is at a mine, where polyhedral crystals of many different kinds can be found. Look carefully in Figure 7.14 at crystals of the familiar mineral quartz. Quartz crystals are essentially prisms with terminating facets, which are arranged in interesting ways. In Figure 7.15 we see a leucite crystal in the shape of an *icositetrahedron*; it has 24 trapezoidal faces. In other

Figure 7.6. Geodesic dome house under construction in Hadley, Massachusetts.

Figure 7.7. Baer's fused triple rhombicosidodecahedra at Drop City, Colorado.

Figure 7.8. The Five-College Radio Astronomy Observatory in Massachusetts.

Figure 7.9. Modular residences, Hampshire College, Amherst, Mass.

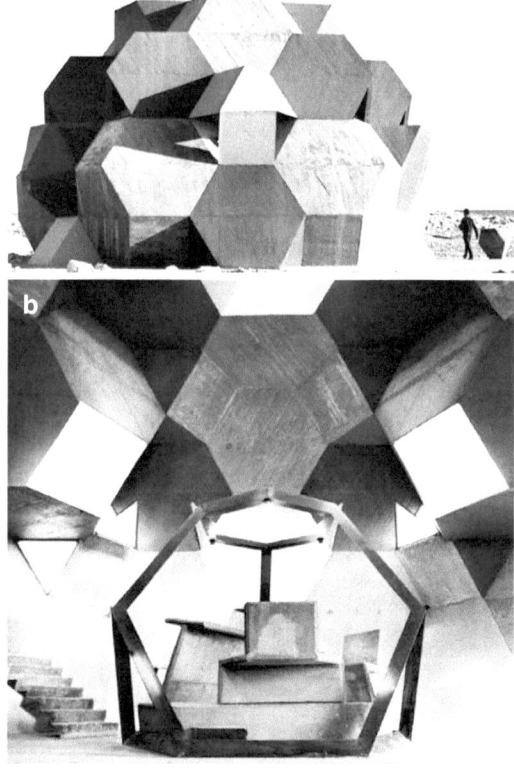

Figure 7.10. Synagogue in the Negev Desert, Israel, 1969–1970. (**a**) Exterior view. (**b**) Interior view.

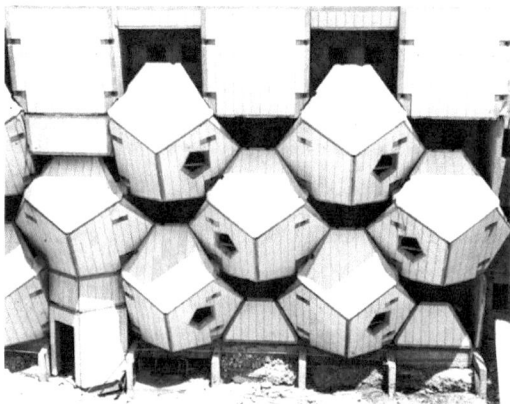

Figure 7.11. Housing complex in Ramot, Israel, 1972–1980.

crystals, the faces are truncated, as in the crystal of benitoite shown in Figure 7.16. Some kinds of crystals come in many forms. Sixteen drawings of gold from the famous twenty-volume *Atlas der Krystallformen* are shown in Figure 7.17.

Figure 7.12. Jungle gym in Cambridge, U.K.

Figure 7.14. Quartz crystals.

Figure 7.13. Jungle gym at the Nonotuck Community Child-Care Center adjacent to the Smith College campus.

Figure 7.15. Icositetrahedral leucite crystal.

The polyhedra that occur as plants and animals are usually less standard in form than polyhedral crystals, but are no less intriguing. The purple sea urchin of Peru combines features reminiscent of both star polyhedra and geodesic domes. The pufferfish is a polyhedron of uncommon charm (see Figure 7.18). Radiolaria are single-celled sea creatures whose skeletons have very interesting polyhedral forms. The sketches in Figure 7.19 were made by Ernst Haeckel on his trip abroad H.M.S. *Challenger* with Charles Darwin. Some insects—for example, the bees—build polyhedra for their own purposes. Among the culinary delights of the Nature Preserve are its many honeycombs. As we see in Figure 7.20, a comb is an aggregate of half-open polyhedra.

Aggregated polyhedra are also found in plants. How would you describe the examples shown in Figure 7.21? Aggregates of polyhedra will be discussed in more detail later on this tour and in Chapters 10 and 22.

This concludes our tour of the Nature Preserve. A much deeper discussion of polyhedra in nature is found in Chapters 9 and 12.

At the edge of the Nature Preserve we come to the Gallery of Polyhedral Art.

Figure 7.18. Inflated spiny pufferfish.

Figure 7.16. Benitoite crystal.

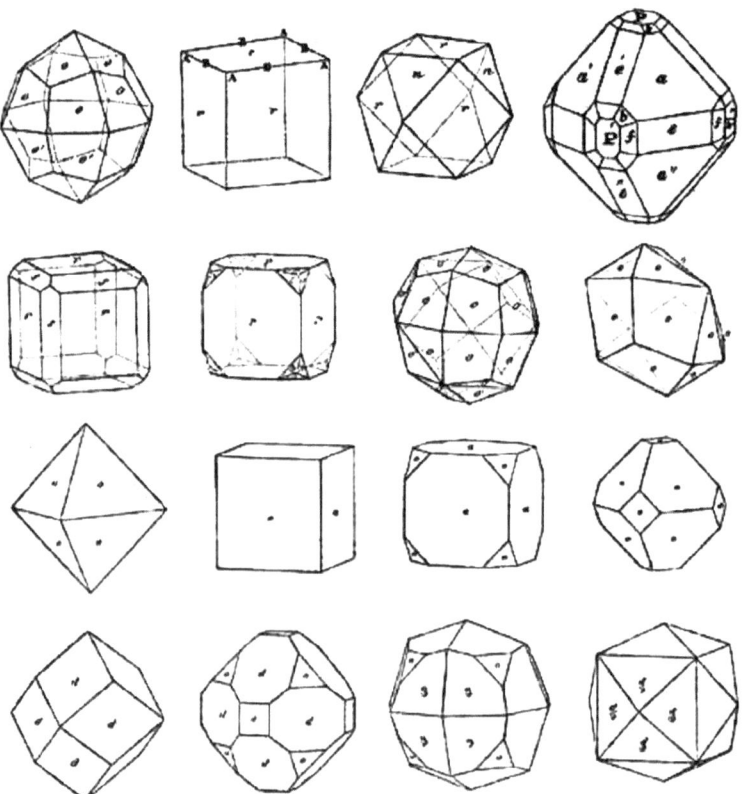

Figure 7.17. Drawings of crystals of gold.

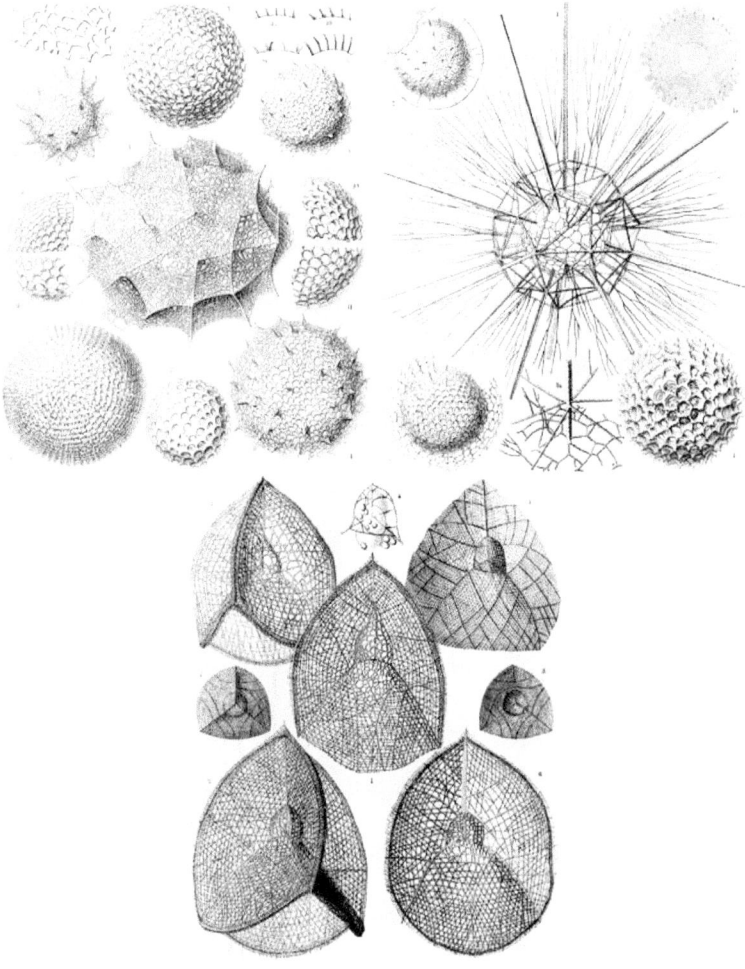

Figure 7.19. Radiolaria.

Figure 7.20. A honeybee comb.

a   b   c

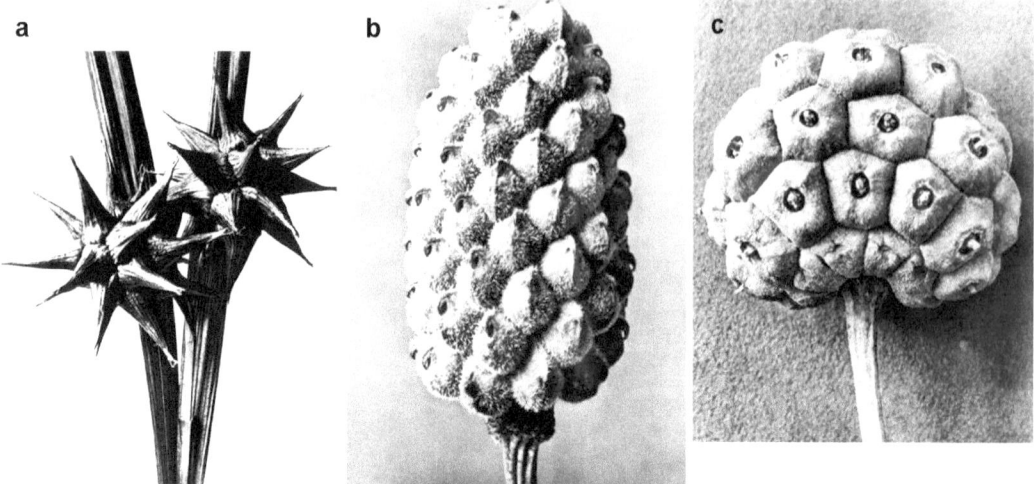

Figure 7.21. (**a**) *Carex grayi.* (**b**) *Adonis pernalis.* (**c**) *Cornus kousa.*

The Gallery of Polyhedral Art

Polyhedral art can be found throughout the world. In honor of your visit to the Kingdom, a small but exquisite collection of sculpture, paintings, and graphics in which polyhedra are an important theme has been assembled. The Renaissance and Modern exhibits are especially strong. Renaissance artists seem to have been very fond of the regular polyhedra, both because of their association with Plato and because they offered opportunities for the study of perspective (see Chapter 4). We saw one of Jamnitzer's engravings in Figure 1.16; Leonardo da Vinci (1452–1519) also drew many polyhedra. Polyhedra often appear in Renaissance paintings; the gallery is proud to display Jacopo de Barbari's portrait of Fra Luca Pacioli (author of *Divina Proportione*), we see in Figure 7.22. The mazzocchio, a doughnut-shaped polyhedral hat popular in fourteenth-century Florence, appears in many paintings by Paolo Uccello; details of two of his paintings are reproduced in Figure 7.23. The mazzocchio was revived for the Shaping Space (Figure 7.24).

Modern polyhedral art includes three striking paintings by Salvador Dali: *The Sacrament of the Last Supper* (Figure 7.25), *Cosmic Contemplation* (Figure 7.26), and *Corpus Hypercubicus* at the Metropolitan Museum of Art (where you can also study works by such renowned polyhedrists

Figure 7.22. Jacopo de Barbari, Portrait of Fra Luca Pacioli and His Student Guidobaldo, Duke of Urbino.

as Picasso, Bracque, and de la Fresnaye.). At first glance the polyhedron in *Cosmic Contemplation* appears to be a pentagonal dodecahedron (cf. Plato), but then we notice that it has one hexagonal face. Such a structure is impossible (this can be deduced from Euler's formula)! How do you think Dali envisioned its other side? A well-known impossible structure appears in the painting by Mary Bauermeister reproduced in Figure 7.27. The gallery's exhibit of cubist painting is very good; it includes important works by Josef Albers (Figure 7.28) and M.C. Escher (Figures 7.29 and 7.30).

Figure 7.23. *Top*: Paolo Uccello, *The Rout of San Romano* (1456–60, tempera on panel. Louvre, Paris.), detail. Museé de Louvre, Paris. *Bottom*: Paolo Uccello, *After the Flood* (Frescoed lunette in terraverde. Green Cloister of Santa Maria Novella, Florence), detail.

The crowded sculpture court contains a wide variety of noted works. The largest sculptures in the court are Isamu Noguchi's *Red Rhombohedron* (Figure 7.31) and Charles Perry's *Eclipse* (Figure 7.32). There is also *Polyhedral Fancy* by Arthur L. Loeb (Figure 7.33) and *Tetrahedron* by Lee Burns

(Figure 7.34). The gallery is also proud to display Hugo Verheyen's sculpture with movable parts (Figure 7.35) and Max Bill's *Construction with 30 Equal Elements* (Figure 7.36).

It is perhaps here in the structure court that we first become acquainted with the boundaries of the Polyhedron Kingdom. As we move away from the center of the Kingdom, the population variation becomes greater and greater, until we cannot really say what is a polyhedron and what is not. Is *Eclipse* a polyhedron? If not, why not? Figures 7.37 and 7.38 are two sculptures by Erwin Hauer. Is either of them a polyhedron? What about Alan Holden's *Ten Tangled Triangles* (Figure 7.39)?

Before leaving the gallery, take a close look (in Figure 7.40) at the *Truncated 600-Cell* by Harriet Brisson. The 600-cell is a name of the four-dimensional polyhedron whose 600 "faces" are three-dimensional regular tetrahedra! The viewer entering the large tetrahedron is surrounded by mirror images approximating the experience of the fourth dimension extending to infinity. It is fascinating to think about the ways in which four-dimensional polyhedra can be represented in our three-dimensional world (see Chapters 18 and 20).

A Note on Polyhedral Society

Polyhedra communicate with one another in a variety of subtle ways (see Figure 7.41). Indeed, the sociology of polyhedra is extremely complicated, as polyhedra tend to be related to one another through many different kinship structures. Some of them are related by geometry; for example, some can be inscribed inside one another, as in Figure 7.33. Others can be grouped into families whose members are related by truncation, that is, by successively slicing off larger and larger corners and edges (see Figure 7.42). (As we have seen, this is the way that Archimedean polyhedra are related to their regular forebears.)

Crystals of the same kind are often related by truncation, and the discovery of this fact by

Figure 7.24. Busts of scientists Florence Sabin, Smith College Class of 1893 (*left*), and Dorothy Mott Reed, Smith College Class of 1895 (*right*), wearing the Shaping Space version of a fourteenth-century mazzocchio (see Figure 7.54). The busts, by Joy Buba, are in the Young Science Library of Smith College.

Figure 7.25. 1963.10.115 *The Sacrament of the Last Supper*. Salvador Dali.

Figure 7.26. Salvador Dali, *Cosmic Contemplation*. Watercolor and ink, 1951.

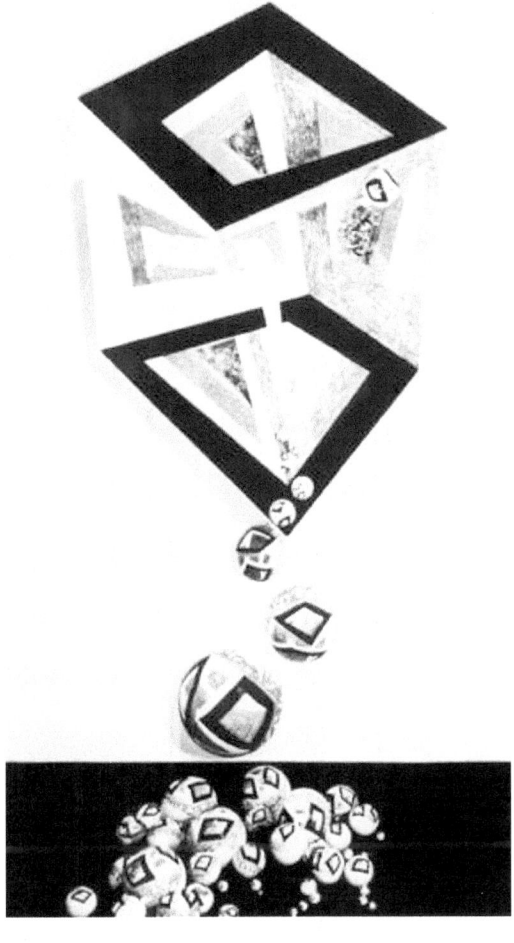

Figure 7.29. M.C. Escher, *Waterfall*.

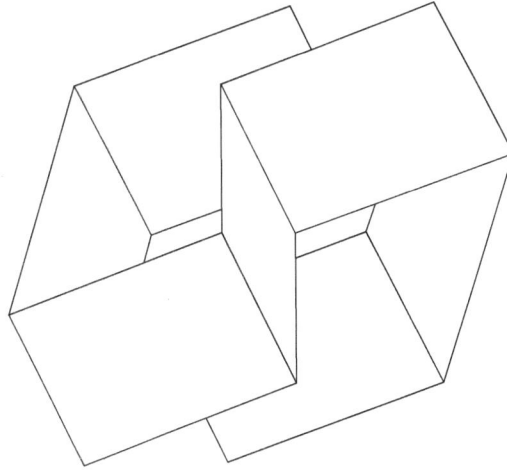

Figure 7.27. *Fall-Out* from the series *Unsculptable* by Mary Bauermeister.

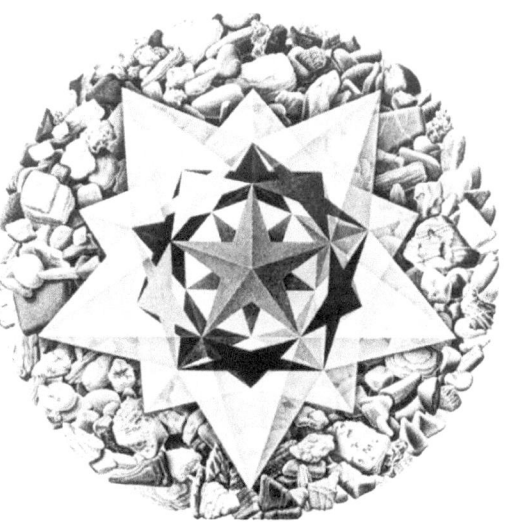

Figure 7.30. M.C. Escher, *Order and Chaos*.

Figure 7.28. Josef Albers, *Structural Constellation*.

J.B.L. Romé de Lisle in 1783 was a milestone in our understanding of crystal structure. Some such relationships are recorded in Figure 7.43.

A major eighteenth-century discovery—see Chapter 4—was that of Leonhard Euler (1707–

Figure 7.31. *Isamu Noguchi, Red Rhombohedron,* in the plaza of the Marine Midland Bank Building, New York.

Figure 7.32. Charles O. Perry, *Eclipse.* The helical explosion of every face rotating from a dodecahedron through the icosidodecahedron to the small rhombicosidodecahedron. Hyatt Regency Hotel, San Francisco.

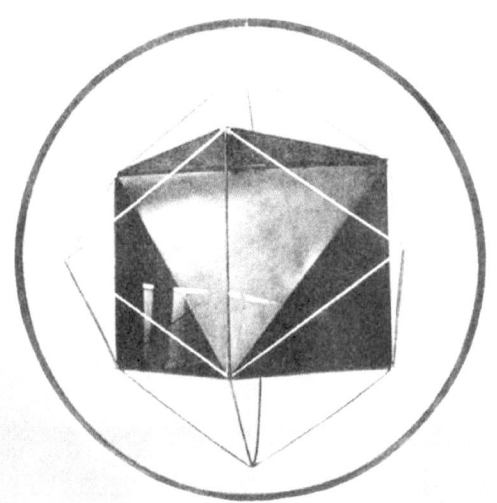

Figure 7.33. *Top*: Arthur L. Loeb standing next to his sculpture *Polyhedral Fancy* in Burton Hall, Smith College. *Bottom*: *Polyhedral Fancy*, a part of the permanent collection of Smith College, is a copper tetrahedron with a Plexiglas ® cube within an octahedral framework within a brass cross-section of a sphere.

1783), who found a simple equation that has great theoretical importance: for every *convex* polyhedron, the number of faces (F), edges (E), and vertices (V) are related by the equation $V - E + F = 2$. This suggests another way of classifying polyhedra into families: classification according to the triple of numbers (V, E, F) (see Figure 7.44). Yet another important relationship among polyhedra is duality, which is rather intimate, and about which there is still a great deal to be learned (see Chapter 15).

In addition to belonging to such families, polyhedra often form voluntary associations to provide important services to nature and to society. Like human and animal associations, these associations require a great deal of conformity but can be very effective in achieving

Figure 7.36. Max Bill, *Construction with 30 Equal Elements*.

Figure 7.34. A. Lee Burns, *Tetrahedron*. A polished brass tetrahedral "soap bubble," inspired by a soap bubble in a tetrahedral frame.

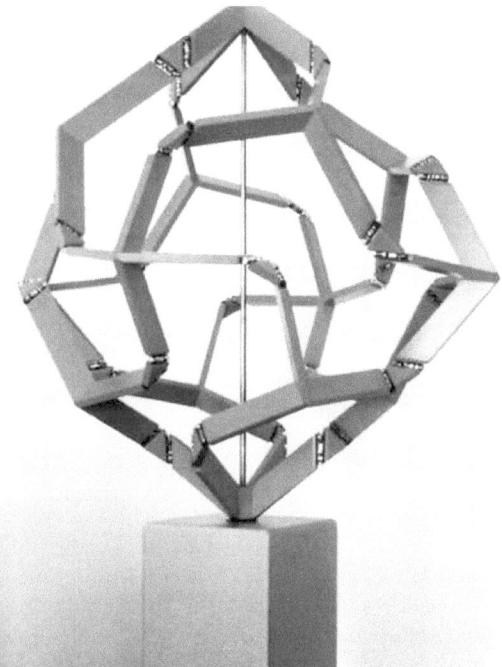

Figure 7.37. Erwin Hauer, *Rhombidodeca*. An excerpt from an infinite, continuous and periodic surface, WPI. The inner labyrinth is expressed as a solid volume. Produced of organic composite materials, 28 × 26 × 26 inches.

Figure 7.35. *IRODO*, an expandable polyhedral sculpture based on the impandable rhombic dodecahedron; by Hugo Verheyan.

their goals. Bubbles in a froth have polyhedral forms which, although appearing to be quite varied, have a very restrictive property: in each bubble, exactly three faces must meet at every vertex. You can see some explorations into the nature of soap films in Figures 1.7, 7.45 and 7.46. Froths are important models for many biological structures (see Chapter 12).

Figure 7.39. Alan Holden, *Ten Tangled Triangles,* Smith College, sent by the artist as his surrogate representative to the Shaping Space Conference.

Figure 7.40. Harriet E. Brisson sitting in the *Truncated 600-Cell,* a four dimensional form made by Harriet E. Brisson and Curtis LaFollete, 1984.

Figure 7.38. Erwin Hauer, *Obelisk*, also an excerpt from WPI, but along the diagonal bisectors of the constituent cubes. The outer labyrinth is maximized in volume and appears as the largest perforations through the sculpture. The shallow exterior spaces are what remains of the inner labyrinth. Produced in cast stone, 1 inch thick, the sculpture measures $9 \times 2 \times 2$ feet.

Crystal architecture is another cooperative polyhedra endeavor (see Chapter 10). The atoms in a crystal come together in more or less regular arrays, like building blocks, to form the crystals that we see with our eyes. With an electron microscope, we can "see" the arrays themselves

(Figure 7.47). The Russian crystallographer E. S. Federov showed 100 years ago that there are exactly five polyhedral building blocks. That is, there are five combinatorial types of convex polyhedra whose copies fill space completely when they are stacked face to face in parallel position; they are shown in Figure 7.48. Many other examples of polyhedral cooperation are found in human-constructed architecture. We have already seen some examples on our walking

Figure 7.41. *Ginger and Fred* by Robinson Fredenthal.

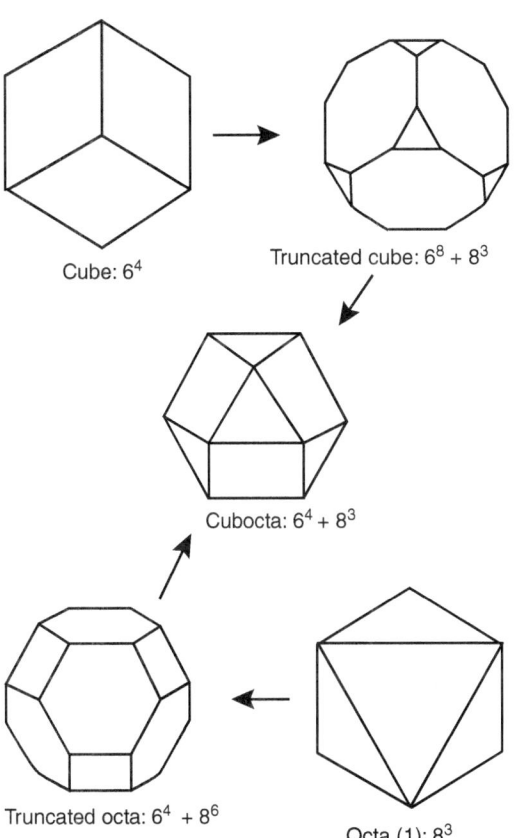

Cube: 6^4

Truncated cube: $6^8 + 8^3$

Cubocta: $6^4 + 8^3$

Truncated octa: $6^4 + 8^6$

Octa (1): 8^3

Figure 7.42. From cube to octahedron.

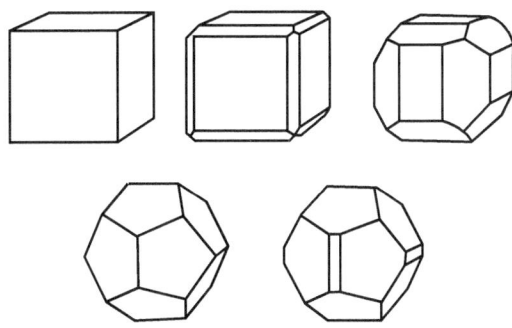

Figure 7.43. Drawings of pyrite.

polyhedra of the family in Figure 7.49 related to one another?) This will continue as the theory of polyhedra expands to include the study of function as well as form (see Chapters 12 and 14).

A Polyhedral Artisan Fair

Polyhedral artisan fairs are held from time to time in the Kingdom. Here you can browse among the many delightful items that the more artistic natives of the Polyhedron Kingdom have created for your enjoyment. As you wander among the many displays, you will find such things as:

Wooden puzzles: See Figure 7.50.
Polyhedra kits: Kits for building star polyhedra (and other polyhedra, too) are sold in many stores. In Figure 7.51 we see M.C. Escher

tour. And of course the bees' cells stack together to make the honeycomb in Figure 7.20.

New relations among polyhedra are being found all the time (for example, how are the

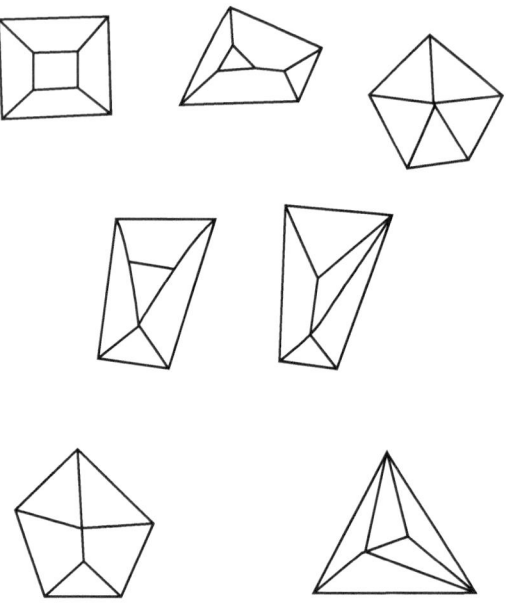

Figure 7.44. There are seven polyhedra which have six faces. Classified according to number of vertices, edges, and faces—the triple of integers ($V, E = V + F - 2, 6$)—they belong to four families.

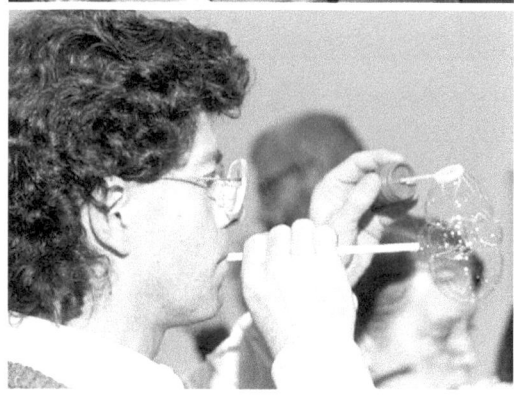

Figure 7.45. *Top*: A. Lee Burns leading the Gala Soap Bubble Workshop at the Shaping Space Conference. *Bottom*: Godfried Toussaint creating a polyhedral bubble.

contemplating a polyhedron constructed from parts provided in such a kit. Uttara Coorlawala and Matthew Solit are shown (in Figure 7.52) building a Rhombics structure.

Dome kits: Figure 7.53 shows a dome built from instructions in *Domebook 2*.

Minerals:

Polyhedral jewelry:

The mazzocchio: The Shaping Space adaptation is shown in Figures 7.24 and 7.54

An unusual lamp shade: Figure 7.55 was inspired by the rhythm and repetition of a Persian mystical poem.

Vases and other pottery: See Figure 7.56.

"Total photos":   A total photo is reproduced in Figure 7.57.

Unusual toys: Deltahedra are polyhedra whose faces are equilateral triangles but which are not regular because the numbers of triangles at the vertices can vary. One of them seems to have been the inspiration for the crawl-through toys shown in Figure 7.58. You can build this deltahedron and all the other convex ones by following instructions in Chapter 2.

Figure 7.46. Soap bubbles in a froth.

But if you do not have time to linger at the fair, do not be disappointed; you will find many delightful polyhedra for sale in shops everywhere.

Figure 7.47. Electron micrographs of crystals, showing arrays of individual molecules. Protein from southern bean mosaic virus (magnification 30,000) (*left*). Protein from tobacco necrosis virus (magnification 73,000) (*right*).

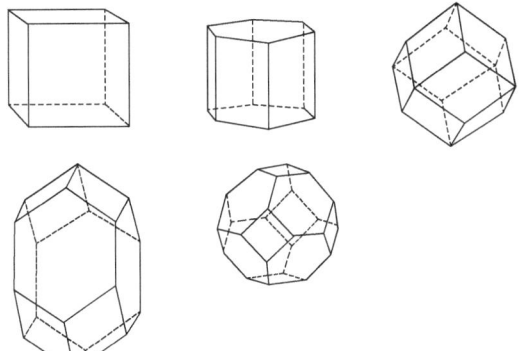

Figure 7.48. The five kinds of polyhedra which fill space in parallel position.

Figure 7.49. Twelve deltahedra-regular polyhedra.

Figure 7.50. Wooden Puzzles.

Figure 7.53. A plywood dome.

Figure 7.51. M.C. Escher contemplating a homemade polyhedron.

Figure 7.54. The fourteenth-century mazzocchio was adapted for the Shaping Space Conference by Helen Connolly, who prepared a do-it-yourself kit.

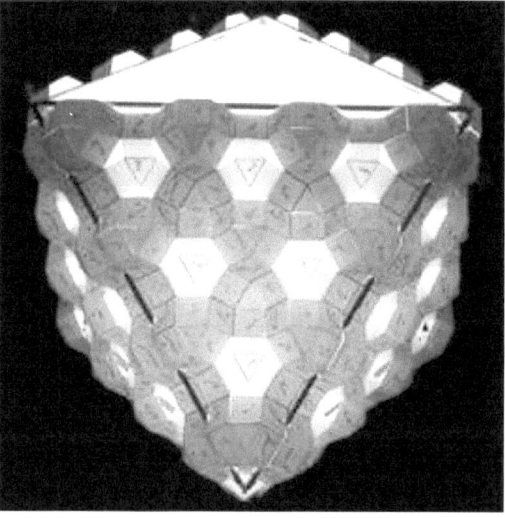

Figure 7.52. Uttara Coorlawala and Matthew Solit building a tetrahedron with Rhombics parts.

Figure 7.55. Lamp shade inspired by the hexagonal structure of a poem by Rumi (1207–1273), teacher of Islamic Sufiism.

Figure 7.57. Total photo. Unfolded dodecahedron total photo of the Chicago Art Center.

Figure 7.56. Corner-posed cubical vase by unknown Kyoto potter, 1964: collection of A. Taeko Brooks.

An easy-to-build crawl-through toy

Cut 11 triangles out of ⅜-inch exterior-grade plywood. You can get 11 of them from 4x8 sheet of plywood. Use grade A-A.

Triangles are equilateral 2 feet on edge. Round edges slightly with block plane and a rasp.

Nine of the triangles have 1¼-inch holes on two sides; 2 have 1¼-inch holes on three sides. They are laced together with cord.

To drill holes in triangles, clamp all 11 of them together and drill holes on two sides first, spacing them on 5-inch centers about 1½-inch from the edges.

Clamp 2 triangles together and finish drilling holes on the third side. Sand, seal, and finish with exterior trim paint.

To lace triangles together, fold A to A, B to B. Pull taut.

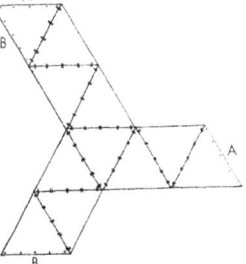

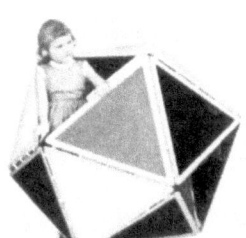

Versatile tunnel toy

Here's a larger crawl-through toy that uses 16 equilateral triangles 2 feet on edge in its construction. You cut out and finish component parts the same as you would the smaller toy pictured above.

By clamping the pieces together and drilling the holes at one time, lacings match better, making it easier to assemble and fold away.

Lace triangles together, as drawing shows, and fold A to A, B to B, and C to C. Pull laces taut.

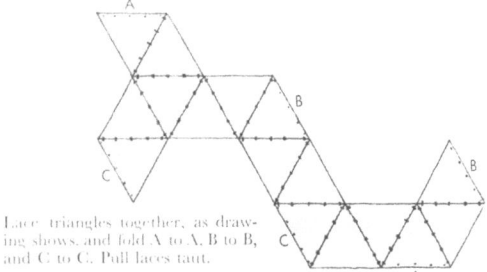

Figure 7.58. Polywood crawl-through toys, with construction plans.

8

Spatial Perception and Creativity

Janos Baracs

I come from Montreal, where I belong to a group called the Structural Topology Research Group. "Structural topology" is an often criticized term, but we are stuck with it.

In Chapter 15, Branko Grünbaum and Geoffrey Shephard talk about misconceptions that may arise when amateurs get mixed up in mathematics; they named the result "mathematical folklore." I like this label and I admit that I am a folklorist.

Now I will prove this point for you: I will discuss spatial perception and creativity. This term suggests some competence in the fields of psychology, philosophy, logic, and so forth, and I have none. But my experience has led to some success in understanding the creative process in morphology. We study this through a sequence of actions that are translated in geometric terms.

Let me start by defining the field with which we are dealing. When we talk about spatial perception we can talk about the physical, social, and other types of spaces. We narrow our interest to the geometrical space, in the structural and formal sense. By structural perception, I mean the combinatorial study of the topological, projective, affine, and metrical properties of configurations, of spatial models. In formal perception, we are interested in quantitative properties, such as ratios, proportions, measures, and coordinates. This part of the field is very interesting and

involves working with sculptors and architects; in this chapter, however, we consider only structural perception.

Figure 8.1 is a diagram describing the three major, distinct phases that should occur when we start with the exposure to a spatial model and end with the perception of the spatial model. These phrases are the creation of an *image*, the *imagery*, and the *imagination*. Studying the first phase required very little. I asked an ophthalmologist to help me find out whether there are people who have some deficiencies in stereoscopic vision. He gave me a very simple tool, a little booklet with polarizing glasses. With this instrument, I was able to code everybody's stereoscopic vision. I gave the test to students, colleagues, everybody I could find. I found that an extremely small percentage of people have deficiencies in stereoscopic vision (R. Buckminster Fuller and A.L. Loeb were in this extremely small group). So the problem is not that some of us have difficulties in creating images.

The next two phases are *imagery* and *imagination*. Imagery is a phase of comprehension, of understanding space. The last phase, imagination, is the process of intervention, which is the creative process or, in our profession, the design.

For many years I have been teaching courses like descriptive geometry and structural topology, and I have also worked with architects and sculptors. Eventually—I think it is a sign of age—one starts to analyze the mental process. The diagram in Figure 8.1 is a result of such an analysis. I divided the second phase, imagery, into three

J. Baracs
21 Springfield, Westmount, QC H3Y 2K9, Canada
e-mail: baracs@sympatico.ca

M. Senechal (ed.), *Shaping Space*, DOI 10.1007/978-0-387-92714-5_8,

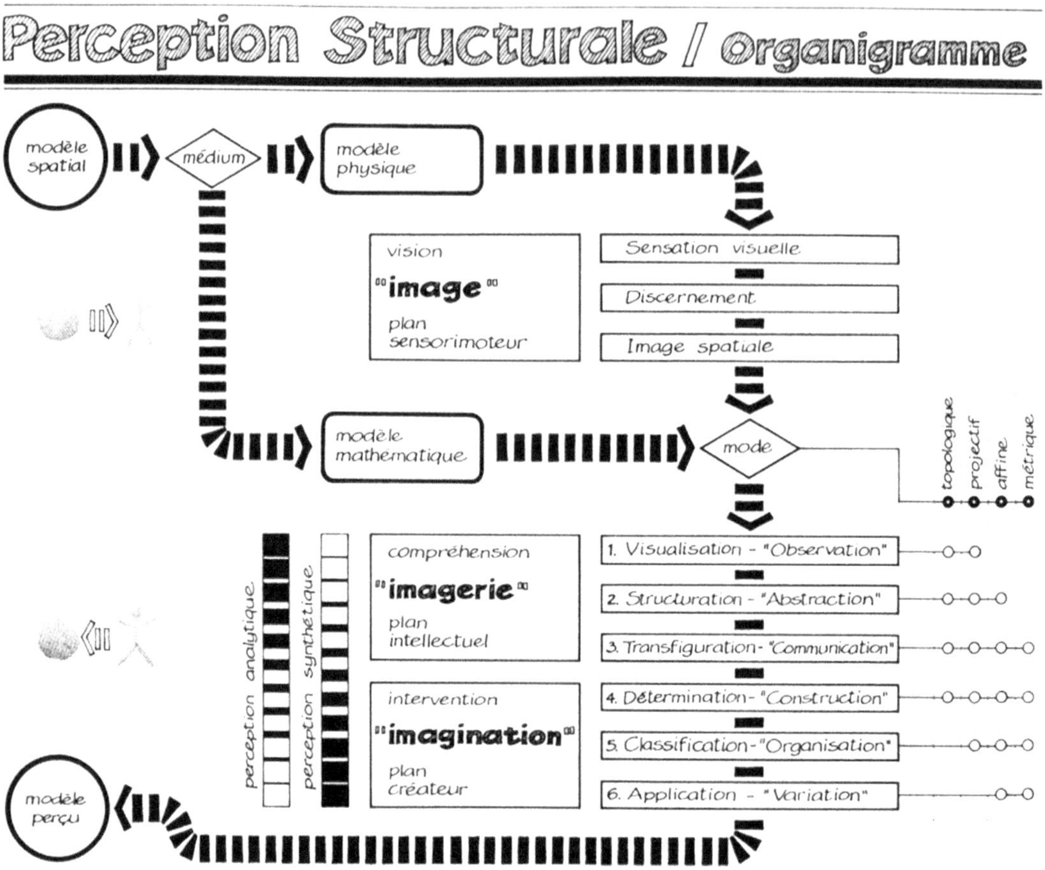

Perception Structurale / Organigramme

modèle spatial ⫸ ⟨médium⟩ ⫸ modèle physique

vision
"image"
plan sensorimoteur

Sensation visuelle
Discernement
Image spatiale

modèle mathématique ⫸ ⟨mode⟩

topologique · projectif · affine · métrique

perception analytique — perception synthétique

compréhension
"imagerie"
plan intellectuel

intervention
"imagination"
plan créateur

modèle perçu

1. Visualisation - "Observation"   o–o
2. Structuration - "Abstraction"   o–o–o
3. Transfiguration - "Communication"   o–o–o–o
4. Détermination - "Construction"   o–o–o–o
5. Classification - "Organisation"   o–o
6. Application - "Variation"   o–o

Figure 8.1. Three phases: the creation of an image, the imagery, and the imagination.

actions, and the last phase, imagination, also into three actions. I will go through a simple form in space and show at each step what I am proposing. We have six clearly defined geometrical actions. They are *visualization, structuration, transfiguration, determination, classification,* and *application*. Next to the actions I have written those terms we call skills or aptitudes. The actions to be performed are linked to these aptitudes. What we have been trying to do in the last two or three years is to devise exercises in order to introduce people, young and old, to these skills.

The model shown in Figure 8.2 may not be as attractive as others at this conference, but it does something that other models do not do. I don't want to present the shape in a frozen, rigid form with particular metric properties; I want you to view it as I slide the vertices as a movable object

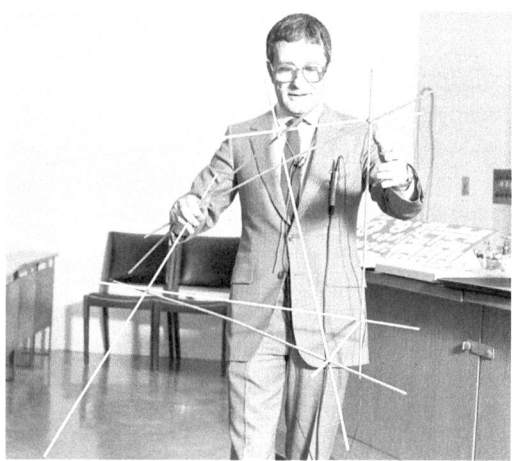

Figure 8.2. Janos Baracs demonstrating a model of a polyhedron with movable vertices.

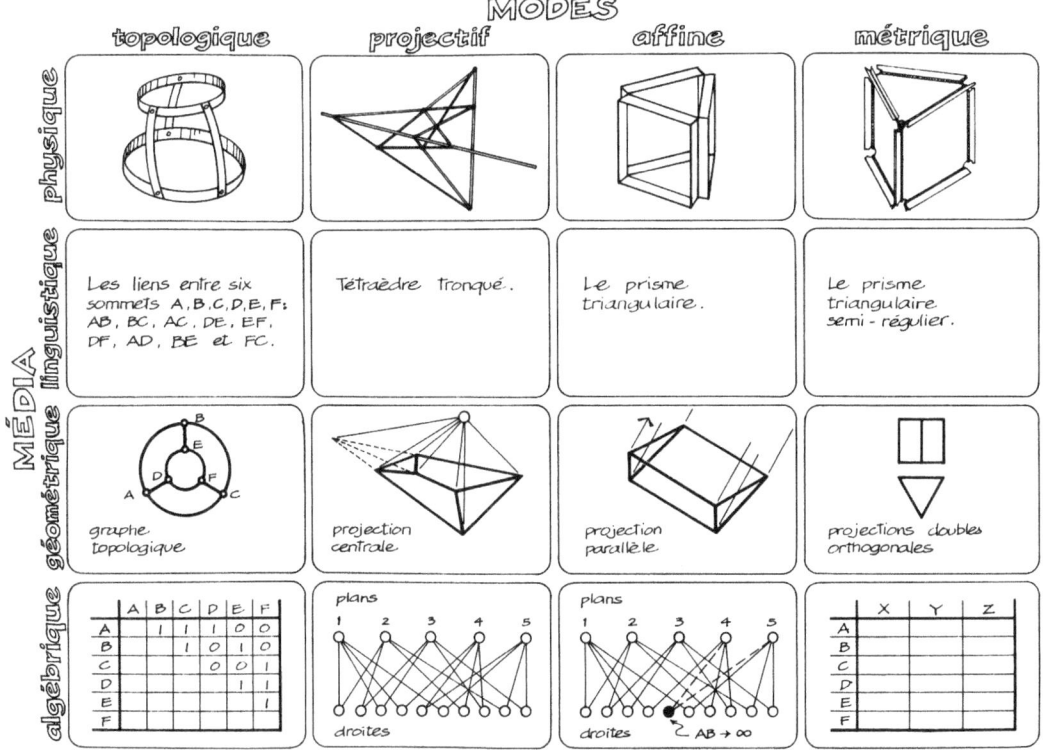

Figure 8.3. Matrix of representations of a spatial model.

which I can continuously transform. I can change lengths and angles at will, I can change symmetries, and I can study many different properties.

If you can create an imagery that is movable, transformable, and which you can manipulate, that is the best start for imagination and creation. This is the beginning of the voyage, an excursion in space. The model is a particular combinatorial structure composed of six vertices, nine edges, and five faces. I will subject it to different motions.

Before describing the six actions, I should clarify the meaning of "spatial model." Figure 8.3 is a matrix of representations of a spatial model. We may use a topological model, a projective model, an affine model, or a metric model. Each model exhibits only those properties that are conserved during the proper transformations in a particular phase. These are the models

of representation. The media of representation may be physical (like the model in Figure 8.2), linguistic (a verbal or written description), geometric (mapping or different types of projections) and finally algebraic (matrices or lattices). It is worth mentioning my suprise when I noticed in my experiments the link between linguistic abilities and the aptitude of spatial perception; students with limited verbal skills also proved to be handicapped in creating geometric imagery! We shall return to this representation matrix when we discuss Action 3: transfiguration.

And now let us go on with the description of the six actions listed in Figure 8.1.

Action 1. The first action is *visualization* (Figure 8.4). There are two distinct steps because in architecture you are either "outside" or "inside." If you are outside you have to walk around the

1. Visualisation (Observation)

1a. Intégration des images partielles

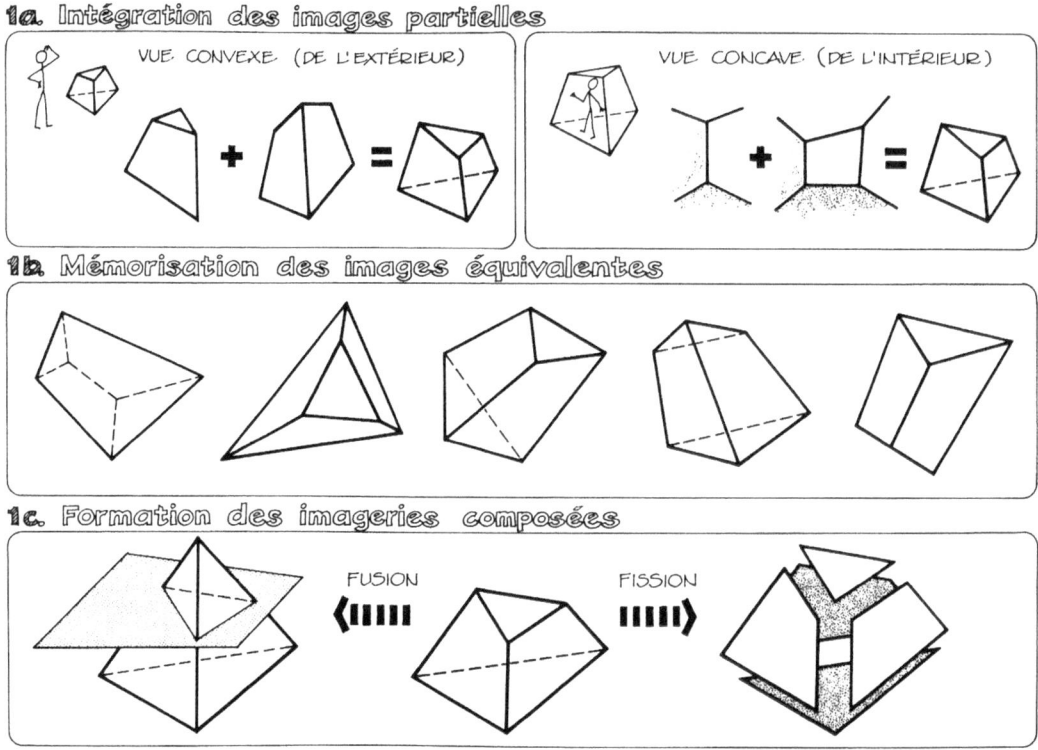

1b. Mémorisation des images équivalentes

1c. Formation des imageries composées

FUSION   FISSION

Figure 8.4. Action 1: Visualization.

object to receive a complete image, while if you are inside you have to turn around, unless you have 360° vision. In both cases, you have to integrate partial images. There are various simple tests and exercises to show that this integration is not a simple process. In a cubic space the process is well exercised, but we may not be living in cubes for the rest of our lives.

The next step is to memorize images. If we cannot store a mental image of a seen object, then I think we are stuck. This can be tested easily. (People with excellent memories sometimes fail, while others who have very poor memories can be excellent here.) In the third step we are looking for composite images. For instance in Kepler's drawing, the icosahedron was shown as a pentagonal antiprism with two pentagonal pyramids; this is a composed image to help you memorize the structure. This completes the first action.

Action 2. The second action is *structuration* (Figure 8.5). Here we want to study the topological, projective, and affine structures of the object. Remember that we do not measure in this phase, we do not care about angles or distances, we do not study symmetries.

As a first step, we recognize and classify incidences. In the second step, we integrate these incidences in a combinatorial structure, in the topological, projective, and affine modes. The third step is a synthesis of the two completed actions: visualization and structuration. We should now possess a geometric imagery of the spatial model in the topological, projective, and affine modes.

Action 3. Our next action is *transfiguration* (Figure 8.6), an apprenticeship to communication. In the first step we are using the representa-

2. Structuration (Abstraction)

2a. Description des incidences

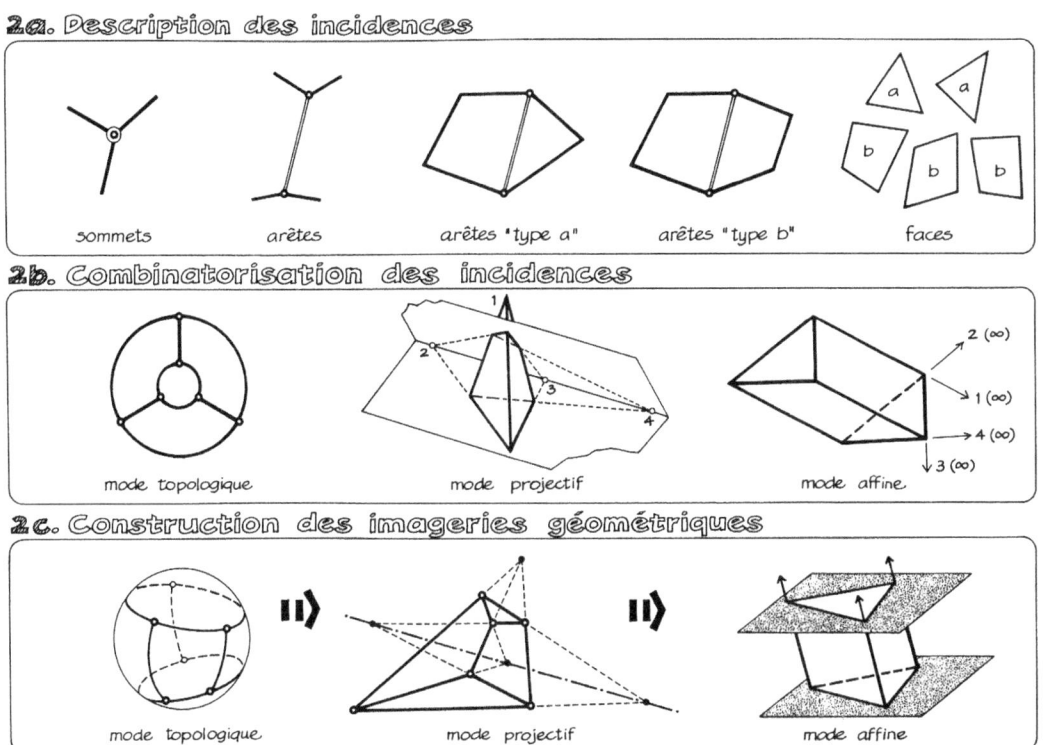

2b. Combinatorisation des incidences

2c. Construction des imageries géométriques

Figure 8.5. Action 2: Structuration.

tion matrix of Figure 8.3. A spatial model in a given mode and medium will be transferred into another (or the same) mode and medium. For instance we may ask to prepare the perspective drawing (projective mode, geometric medium) of a triangular prism (affine mode, linguistic medium). Or we may start with an incidence matrix (topological mode, algebraic medium) and decide to produce the graph of the polyhedron (topological mode, geometric medium). Essentially, we decide on a method or representation of a given model, taking into account the purpose of the representation and the type of properties we wish to exhibit.

The second step is codification (standard or created for a particular task) with a legend that allows our representation to be read by others. The third and final step is the prepared representation. This passage from our mode and medium to another mode and medium is more than mere communication; it is an essential process in order to perceive the model itself.

Action 4. Action 4 is *determination* (Figure 8.7). The problem is rooted in combinatorial geometry and linear algebra. The example presented here is for metric determination; the same action, however, can be equally applied for projective or affine properties.

The first step is to enumerate the (metric) invariants of the spatial model. As you can see, the seven types of invariants (distances and angles) result in 148 pieces of information in the case of the truncated tetrahedron.

Now we need to know the least number of invariants that uniquely determine the polyhedron in space. If we exclude the Euclidean motions (six degress of freedom), we need exactly nine

3. Transfiguration (Communication)

3a. Décision sur la représentation

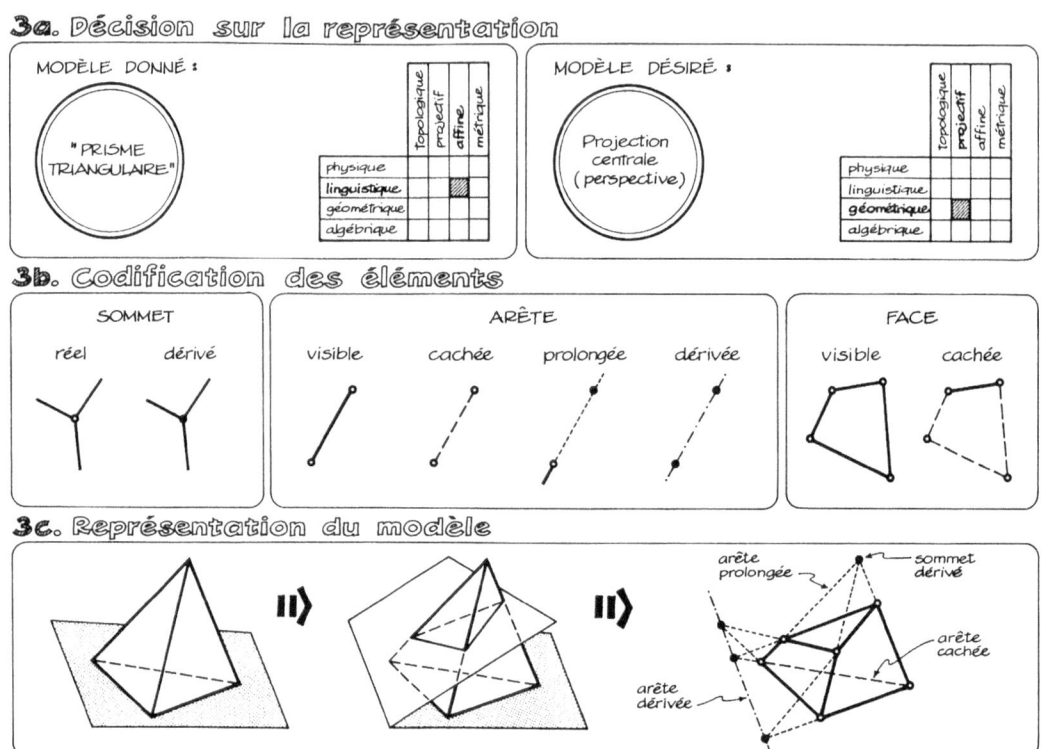

Figure 8.6. Action 3: Transfiguration.

invariants (this number, C_d, is equal to the number of edges in the case of a spherical polyhedron). The number of necessary invariants is reduced if we impose a symmetry group on the polyhedron. For instance, we have the choice of one invariant only if we wish the truncated tetrahedron to be realized as a semiregular triangular prism.

Going back to our general position, the choice of nine invariants out of 148 elements is a terrific number (7.32×10^{13}). It is our duty in the third step to select a combination whose elements are linearly independent. This selection was done intuitively (Figure 8.7, section 4c) in our example. Thus determination is an integral part of the perception process and also an important practical tool to design and to realize (to construct).

Action 5. Classification is the fifth action. We now have reached a point in our actions where

we are able to classify the available options of our model into multiple groupings (Figure 8.8). In the first step we gave as examples topological, projective, affine, and metric groupings. The derivation of the combinatorial types (affine in this example) is shown in the second step, where the option of parallelism is explored.

The third step is symmetrization or regularization. Shown in the figure is the semiregular triangular prism and all its symmetry subgroups. Faces, which have to be regular polygons for each group, are also incidicated.

Action 6. The last action—*application*—is the least understood and the most mysterious of all. It is the action of conception, creation, or (in my profession) design. The three steps shown in Figure 8.9 represent three levels of application in an increasing order of complexity.

4. Détermination (Construction)

4a. Enumération des invariants

EXEMPLE: invariants métriques – Les 148 paramètres

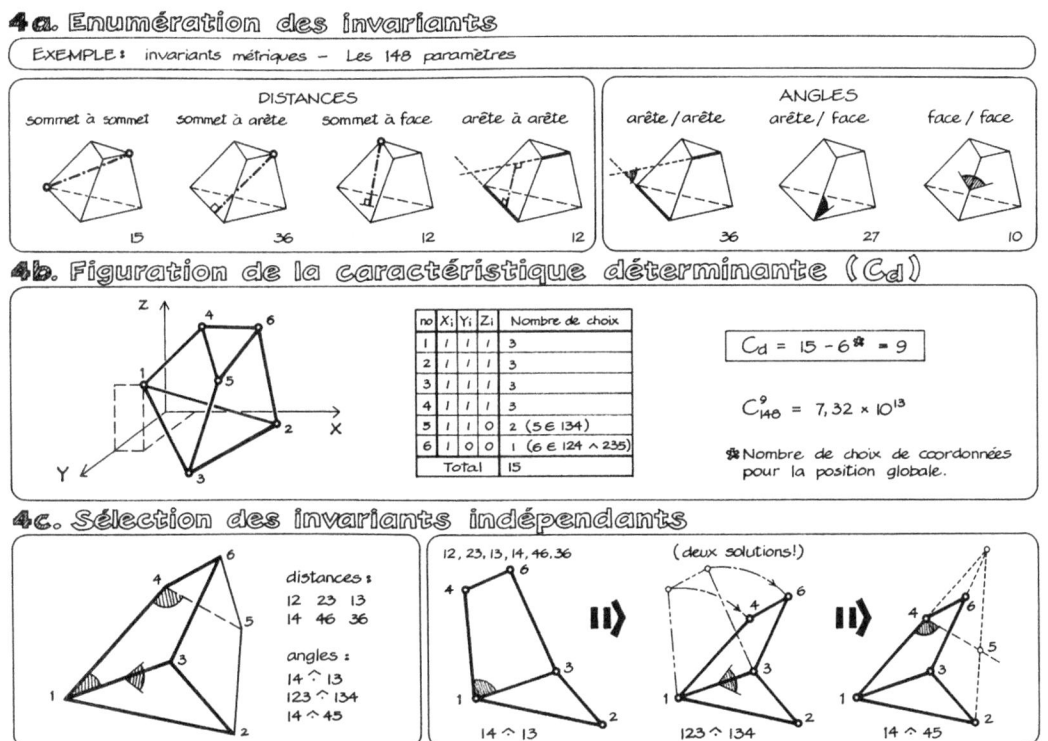

Figure 8.7. Action 4: Determination.

The first level is the resolution of a given problem, which requires a certain degree of imagination. At the second level, the creative process is more advanced. Teachers will concur with me that finding a good problem is usually harder and more rewarding than solving one.

We named the last step "creative manipulation." To demonstrate it, we use again the triangular prism, which in this example was "manipulated" into a creative toy, the serpent, by Ernö Rubik.

We propose these six actions as a logical, sequential mental process to shape space. The first three actions represent the analytic perception of a spatial model, resulting in an imagery. The last three actions, the synthetic perception of the model, provokes the imagination, ending in a creative application. During the six actions, we applied topological, projective, affine, and metric transformations in a gradual fashion to a given spatial model.

The actual process of design in my profession is somewhat different: it does not begin usually with a given spatial model. It starts with a program that describes functions, criteria, and so forth. So the first step is to generate the spatial model itself, which fulfills the functions and satisfies the criteria. This model should be quite general, free of details, possessing only some essential, intrinsic geometric properties. It follows that I am proposing a topological model, to be found by enumerating the available options. We refine this model through projective, affine, and metric transformations, ending up with the desired product.

The process is illustrated in Figure 8.10. To make this presentation brief, the example

5. Classification (Organisation)

5a. Inclusion multiple

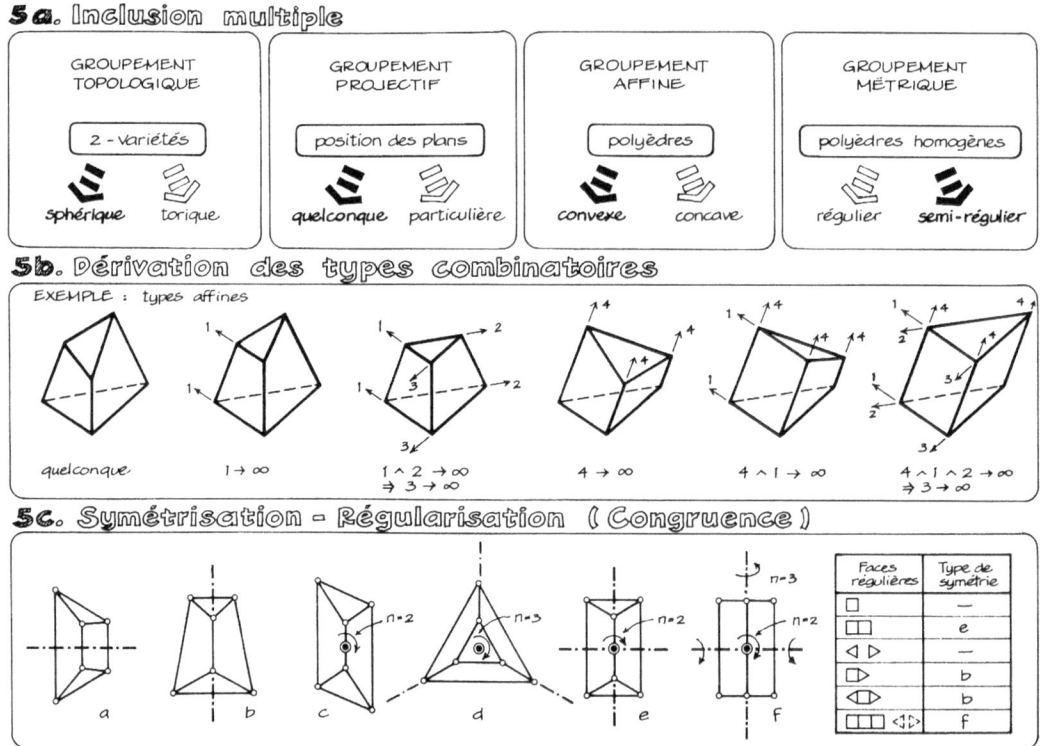

GROUPEMENT TOPOLOGIQUE	GROUPEMENT PROJECTIF	GROUPEMENT AFFINE	GROUPEMENT MÉTRIQUE
2 - variétés	position des plans	polyèdres	polyèdres homogènes
sphérique torique	quelconque particulière	convexe concave	régulier semi-régulier

5b. Dérivation des types combinatoires

EXEMPLE : types affines

quelconque $1 \to \infty$ $1 \wedge 2 \to \infty$
$\Rightarrow 3 \to \infty$ $4 \to \infty$ $4 \wedge 1 \to \infty$ $4 \wedge 1 \wedge 2 \to \infty$
$\Rightarrow 3 \to \infty$

5c. Symétrisation - Régularisation (Congruence)

a   b   c   d   e   f

Faces régulières	Type de symétrie
□	—
□□	e
◁ ▷	—
□▷	b
◁▷	b
□□□ ◁▷	f

Figure 8.8. Action 5: Classification.

is simplified. The selected topological model is a closed curve with six labeled vertices (a, b, c, d, e, and f): no lengths, no angles, no parallelism, not even straight lines are specified in this original choice. As a result, we have a large family of figures (all the plane and spatial hexagons) by selecting a few properties, which are, however, the most instrinsic ones of this family. Let us now imagine that our model is a rubber band marked with the six vertices. While stretching this rubber band, adjacent vertices will remain adjacent and the band will stay as a closed curve. This type of transformation is called a *continuous mapping*, while the invariant properties are *adjacency* and *continuity*; we are in the realm of topology. With so few properties to scrutinize, it is suprising how rich the content of topology remains.

We now enter the second phase in shaping our hexagon. While the properties established in the topological phase are kept unchanged, we decide that the curved edges shall be straight lines, all in one plane, and we choose some particular incidences among them. Let the lines af, be, and cd meet in the point s, arranged in such a way that the common point p of the lines ab and ef, the common point q of the lines bc and de, and the common point r of the lines ac and df are on one line. Many other choices and their combinations are possible, each set of choices representing a distinct hexagon with a visual impact on its shape. If we now project this figure from a point onto another plane, the projected figure will preserve all the chosen properties. We may state that in projective geometry, we study properties of a figure which are invariant

6. Application (Variation)

6a. Résolution des problèmes

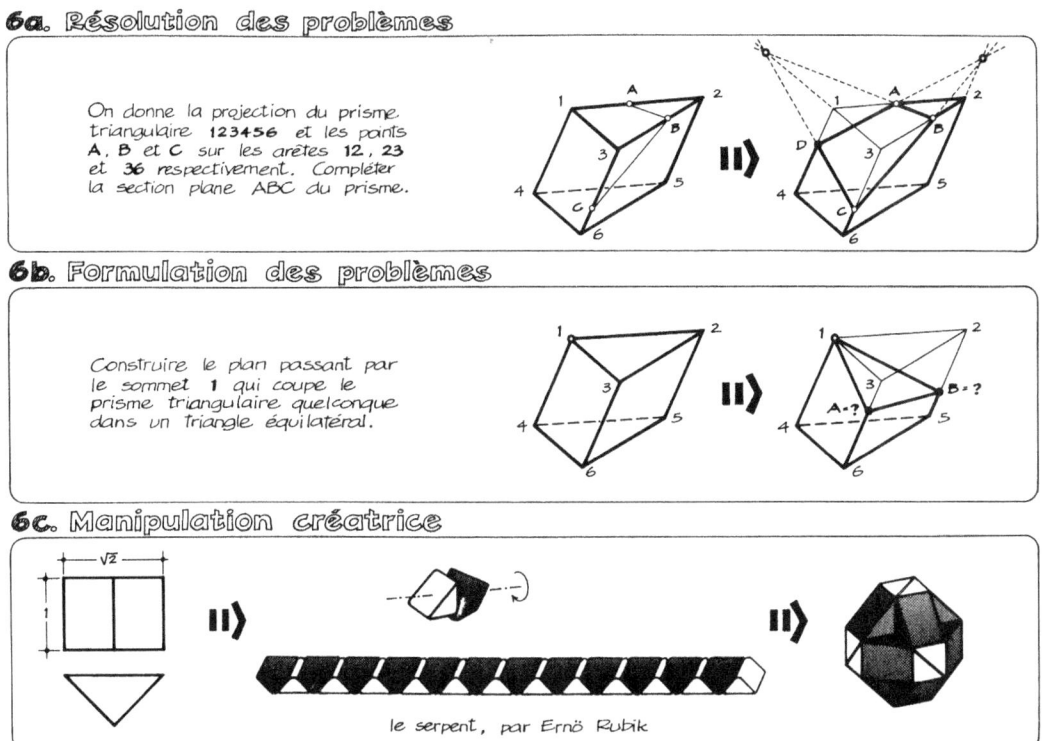

On donne la projection du prisme triangulaire **123456** et les points **A**, **B** et **C** sur les arêtes **12**, **23** et **36** respectivement. Compléter la section plane **ABC** du prisme.

6b. Formulation des problèmes

Construire le plan passant par le sommet **1** qui coupe le prisme triangulaire quelconque dans un triangle équilatéral.

6c. Manipulation créatrice

le serpent, par Ernö Rubik

Figure 8.9. Action 6: Application.

under central projections. These new projective properties are incidences and flatness (straight lines and plane surfaces).

We continue to refine the shape of the hexagon by choosing new properties in the third phase, but again all previous choices—at this time topological and projective properties—are kept invariant. The option in this phase is the selection of certain lines to be parallel. Another way to phrase it is to say that certain chosen points are moved to infinity along their incident lines. For instance, we choose the points r and s to be moved to infinity. The point r is common to three lines (ac, pq, and df), which are now parallel, and similarly the lines be, af, and cd become parallel because their common point s is at infinity. If we now project this figure with parallel rays onto any other plane, the projected figure will preserve all the properties chosen so far. In affine geometry, we study properties of a figure that are invariant under parallel projection. The new affine properties are parallelism and convexity.

We have gone through three new geometries by now, but the most familiar geometric properties, such as distances and angles, have not yet been mentioned. These choices are left for the fourth and last phase of our program. Let us choose the line be to be perpendicular to the line pq. This decision will force the lines cd and af to be also perpendicular to the line pq, since these lines were parallel in the affine phase. Just as before, we do not alter choices made in a previous phase. The last choice we make now concerns distances; we want the line pq to bisect the segments be, cd, and af. An important new property has emerged in this final shape

Conception

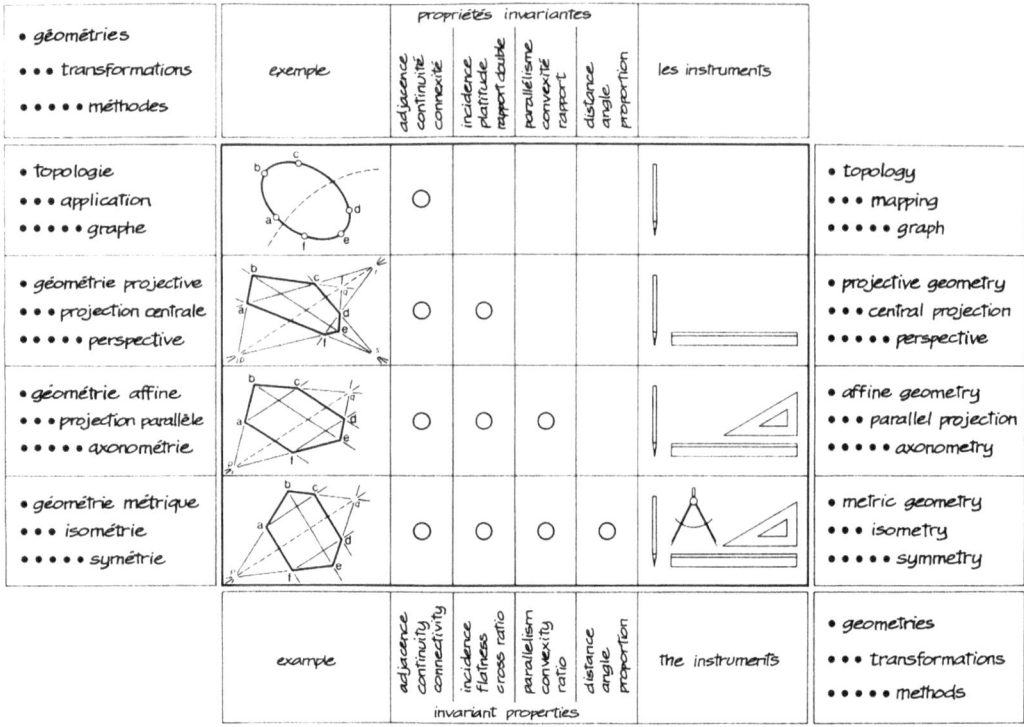

Figure 8.10. An illustration of the actual process of design.

of the hexagon: if you consider the line pq as a "mirror-line," the vertices f, e, and d are the mirror-views of the vertices a, b, and c, respectively or, simply, the hexagon possesses bilateral symmetry. The reflection of a figure in a line preserves distances and angles. Such an operation is called an isometric transformation. If all the distances of a figure are preserved, so are all the other geometric properties. In metric geometry, we study properties of figures, which are invariant under isometric transformations. The new metric properties are distances and angles.

We have completed a sequential approach to shape a simple plane figure. Within each of the four geometries sketched above, certain properties of a figure may be determined quite simply by counting, for instance, the number of vertices, edges, incident lines in a point, the number of lines parallel to each other, the number of equal distances and angles. These properties are the subject matter of the combinatorial theory, which has been introduced on the mathematical scene in the last fifty years.

You might say now, yes it is a nice procedure, but why not start simply at the other end and just draw a hexagon that has a mirror symmetry? This could be done for a simple example like a plane polygon. But if I had chosen a polyhedron, even one with not more than six or seven faces, you would not recognize immediately, for instance, how to manipulate the planes of the faces in order to arrange them according to certain symmetry groups; it would be an extremely difficult task. However if you go through this process, you will become aware of the available options, you will be able to control your form, you will be in charge of it. Usually, the forms are in charge of us. We have to reverse the process. We should not have to use catalogs and say, "I want this form, I want that form" as if we were shopping in a supermarket! We should be in full, complete control; we should be able to shape space, as the title of this book says.

As we progressed from the topological figure to the metric figure of the hexagon, the invariant properties have been increasing while including all the previous properties. We have been progressing from the most general towards the most specific. This concept is also supported by illustrating the necessary and sufficient drafting tools to produce the drawings in Figure 8.10. Only a pencil is needed to draw a topological figure of the hexagon; a straight edge is required in projective geometry. To draw parallel lines in an affine figure, a new instrument is necsesary, a straightedge gadget that slides on the first one. The symbol of a right triangle is used in the illustration, being the standard tool in the drafting practice (but remember that perpendicularity is irrelevant in affine geometry). Finally we have to add a compass to our repertoire of instruments to draw a metrically equivalent (isometric) figure.

The methods (practical applications in our profession) listed in Figure 8.10—graph theory, perspective, axonometry, and symmetry operations—presently are used mostly to analyze and communicate preconceived forms. We are convinced that geometries can do much more for us; we propose to apply them in the synthetic process of conceiving forms.

I have presented the six actions to perceive a spatial model and the gradual transformations to design a product. At first glance this methodology may seem to you as one of those fashionable intellectual exercies based on personal beliefs and prejudices. That is not so! As I mentioned earlier, these methods evolved during these decades of professional practice and teaching, which gave me ample opportunity to test their didactic and practical use. I hope the following illustrations will convey the message: I do try to practice what I preach!

Figure 8.11 is a picture of my office. Each year I keep some souvenirs and make room for the new projects. These models are important because

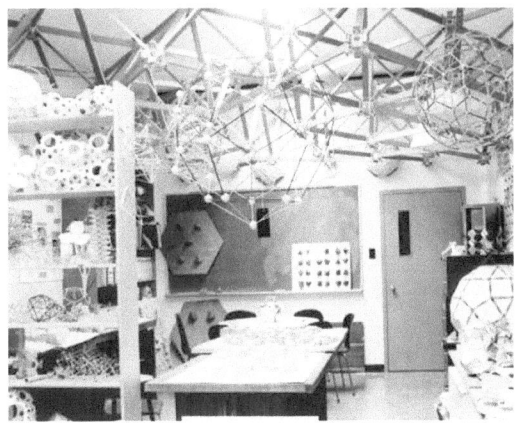

Figure 8.11. "...I do and I understand".

I believe in the old Chinese saying: "I hear, I forget; I see, I remember; I do and I understand." But, there is a drawback. Meaningful models are painstakingly slow to build, to take apart and to transform. To solve this problem, I designed five kits that allow fast assembly, easy transformation, and high precision.

Figure 8.12 shows a kit named Poly-Form. It is a topological and projective kit; its purpose is to demonstrate through simple manipulations the links between the concepts of polyhedral graph, adjacency matrix, embeddings, projective conditions, and projective realizations. I tried Poly-Form with 8- to 12-year-old children with a result suprising to me. At first I was wary of explaining it to them; I found it hard to avoid the fancy terms. Then I was chagrined when I realized how fast they grasped the concepts and how happily they went on to explore. College students (ages 18 to 20) sometimes are slower than those kids and certainly are more inhibited about exploration! These and other experiments with younger children and college students led to the following proposals for teaching geometry:

1. Spatial geometry should be introduced at an early age (10 to 12 years).
2. The subject matter should be polyhedra.
3. The starting notions should be topological and projective, to be followed later with affine and metric properties.

Figure 8.13 shows Poly-Kit No. 1. The diecut cardboard polygons are to be attached with rub-

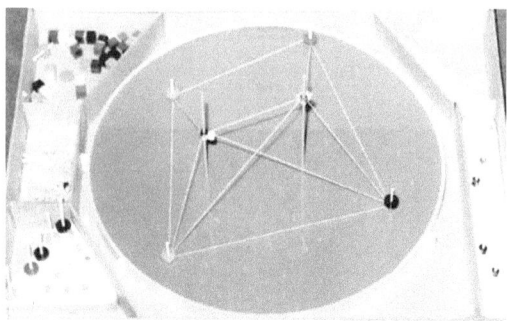

Figure 8.12. Poly-Form: From graphs to projective polyhedra.

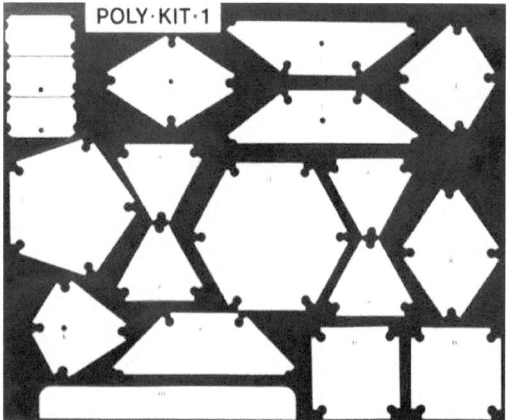

Figure 8.13. Poly-Kit No. 1: Metric polyhedra.

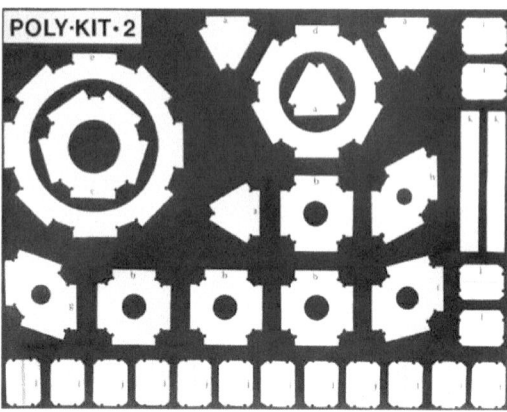

Figure 8.14. Poly-Kit No. 2: Space-filling polyhedra.

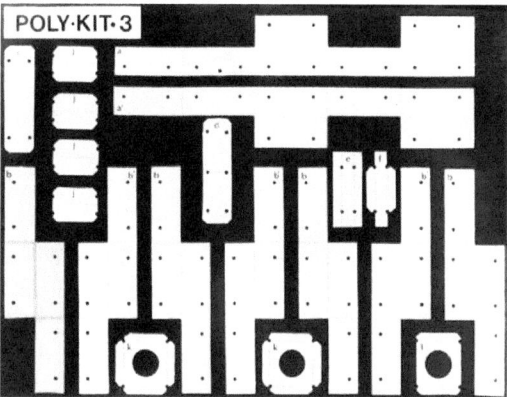

Figure 8.15. Poly-Kit No. 3: Affine polyhedra (zonohedra).

ber bands to form the five regular polyhedra, the thirteen semiregular polyhedra, six of the family of prisms and antiprisms, four of the semiregular duals, the five parallelohedra, and all other polyhedra with regular faces.

Poly-Kit No. 2 (Figure 8.14) is a space-filling kit. Here the special connection between the die-cut cardboard polygons makes it possible to attach three or more faces along an edge. The circular holes in the polygons allow the user to inspect the incidence structure of the juxtaposition. Poly-Kit No. 2 allows a fast assembly of the space-filling by the 5 parallelohedra and others where the component polyhedron is composed of regular faces.

Figure 8.15 shows the die-cut cardboard bands of Poly-Kit No. 3. This kit was conceived for building a fascinating infinite family of polyhedra

called zonohedra. (A *zonohedron* is a convex polyhedron; all of its faces are parallelograms). The repetitive use of one type of band is sufficient to build any zonohedron. This is a truly affine kit, because the models that can be built allow the affine motions of the zonohedron to be demonstrated; the model retains its central symmetry during the deformations. Another type of band in the same kit may be used to build all polyhedra that can be realized with equal edge lengths (equilateral polyhedra).

Poly-Kit No. 4 (Figure 8.16) also serves two purposes. It can be used to demonstrate the geometric rigidity of regular grids composed of bars or bars and tension members. The same kit can also be used to build polyhedra projected onto a sphere.

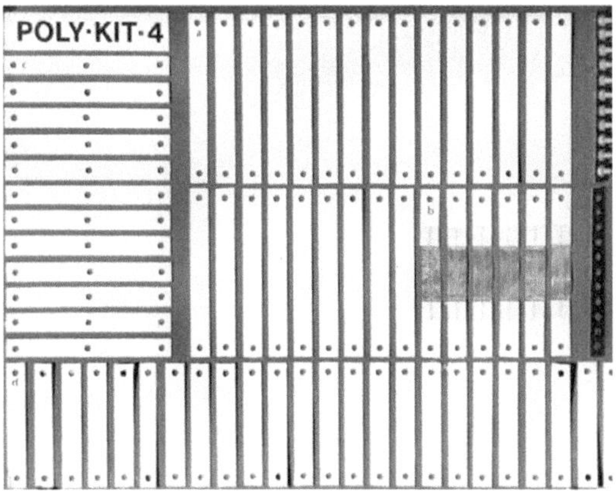

Figure 8.16. Poly-Kit No. 4: Geometric rigidity of braced grids.

Figure 8.18. Joint detail of "octet" spaceframe.

Figure 8.17. "Octet" spaceframe (student project).

Figure 8.19. Articulated ring-dome (student project).

My students completed their apprenticeship using the different kits, then moved on to build large-scale models. In this phase they were confronted with structural and technological considerations. Figures 8.17 and 8.18 show an "octet" spaceframe and its detail built with wooden bars and plastic joints, injected for this particular project.

In Figure 8.19 we see a dome built with tubular aluminum rings. The geometry is based on the inscribed circles of the faces of a dualized semiregular polyhedron. The joints are all of one type and despite their articulations (they are "hinged" to each other), the dome when attached to the ground becomes rigid.

The last few figures are examples from my professional practice. I selected five projects where the geometric content is evident, and I

had the good fortune to work with truly talented and adventurous architects and sculptors. My duties included proposing a geometric concept, calculating the stresses, and devising modes of fabrication and erection while respecting the ever-present budgetary constraints.

In Figure 8.20 we see the main theme building *Man the Explorer* of Expo '67 in Montreal. It is a giant, integrated steel spaceframe, based on the juxtaposition of truncated tetrahedra and tetrahedra. Figure 8.21 is also a theme building of Expo '67 (*Man in the Community.*) Here concentric hexagonal plywood-box rings are superposed; the reduction in size results in a logarithmic silhouette. Figure 8.22 is a prefabricated concrete

Figure 8.20. *Man the Explorer.* Theme building, Expo '67, Montreal. Architects: Affleck, Desbarats, Dimakopolous, Lebensold, and Size. Structural engineers: Eskenazi, Baracs, de Stein, and associates.

Figure 8.21. *Man in the Community.* Theme building, Expo '67, Montreal. Architects: Erickson and Massey. Structural engineer: J.J. Baracs.

Figure 8.22. *Plaza Côtes des Neiges.* Shopping center, Montreal. Architects: Mayers and Girvan. Structural engineers: Baracs and Gunther.

Figure 8.23. Sculpture, station Namur, Montreal. Sculptor: Pierre Granche. Structural engineer: J.J. Baracs.

Figure 8.24. Sculpture, Palais de Justice, Québec. Sculptor: Louis Archambault. Structural engineer: J.J. Baracs.

spaceframe with the tetrahedron-octahedron geometry, built for a shopping center in Montreal.

The last two projects are large-scale sculptures. The first one (Figure 8.23) is a set of juxtaposed general zonohedra built with identical hexagonal aluminum frames, articulated at their joints. The integrated lighting system adds to the visual impact for this Montreal subway station. The last figure shows a sculpture in the enclosed mall of a large public building in Québec City (Figure 8.24). It is interesting to note that each of these last two sculptures, despite their different appearances, is based on one of the 230 space groups: three mutually perpendicular screw motions with half-turns. In the first case, the hexagonal frames, in the second case the extruded H beams are subjected to the same symmetry operations. Even the orientations of the bolts and nuts are consistent with the motions.

The German mathematician Felix Klein gave an overall view of geometries in his famous address at the University of Erlangen in 1872. He was the first to propose transformations as criteria to distinguish geometries. This concept was applied by the Swiss psychologist Piaget. He demonstrated that growing children perceive space in a sequential fashion. They go through a topological stage until the age of 6, then progress through a projective stage (at ages 6–10); the perception of affine and metric properties begins around the age of 10.

The results of my own experience and research in the field of the synthetic process of creating form led me to a similar approach: in order to exploit fully the stunning richness of three-dimensional space, the initial method for conceiving form is a sequential series of combinatorial choices taken at the different levels of geometries in the order of topology, projective, affine, and metric geometry. But, despite the clearly established hierarchy of the geometries, creating form cannot be simplified to a linear thinking process. Certain early choices in the topological and projective level are dependent on affine or metric criteria, imposing simultaneous considerations. This linking of seemingly unconnected concepts (bisociation) is the theme of Arthur Koestler's famous book, *The Act of Creation*.

This approach, where the structure of forms is studied rather than the forms themselves, stimulates the imagination and appears to be more conducive to creative design. I used the term "creative design"; it may seem that the adjective "creative" is redundant, but I do not think so. Take a look at our cities, buildings, and objects: they are the results of "design," but in most cases, with little sign of creativity. Many interesting books and essays have been written on the theme of creativity, and any definition is obviously subject to debate. I like what the mathematician-philosopher D.R. Hofstadter wrote in a recent article, "Making variations on a theme is really the crux of creativity." This statement confirms our geometric view of morphology: creating form is not an invention, it is a process of transformations. The same article began with a quotation of G.B. Shaw: "You see things, and you say 'why?' But I dream things that never were; and I say 'why not'?" It certainly takes a poet to express so well the contrast of minds: the analytical versus the synthetic, the critic versus the design.

9

Goldberg Polyhedra

George Hart

The regular polyhedra—see Chapter 1—are famous for their history, applications, beauty, and mathematical properties. Though not yet famous, the *Goldberg Polyhedra* too are notable in all these ways.

A *Goldberg polyhedron* is a polyhedron with three properties:

- all its faces are pentagons or hexagons,
- all its vertices are trivalent vertices, i.e., three faces meet at each vertex, and
- the polyhedron has the rotational symmetry of the icosahedron.

The regular dodecahedron (Figure 9.1) is the only regular polyhedron that meets these conditions. A more typical example of a Goldberg polyhedron is shown in Figure 9.2.

The two simplest Goldberg polyhedra can be traced back at least 2000 years to classical Greek mathematics, but as a group these shapes have been identified and studied only since the twentieth century. The credit for the geometric insight and formal definition of this family goes to the mathematician Michael Goldberg, who published the definitive paper in the 1930s. As with many mathematical ideas, these started out as the imaginings of a pure mathematician exploring an abstract puzzle that interested him, never guessing that these polyhedra would be found in subatomic particle detectors, loudspeaker design, virus macromolecules, and carbon chemistry. Other applications include architecture, spherical game boards, golf ball

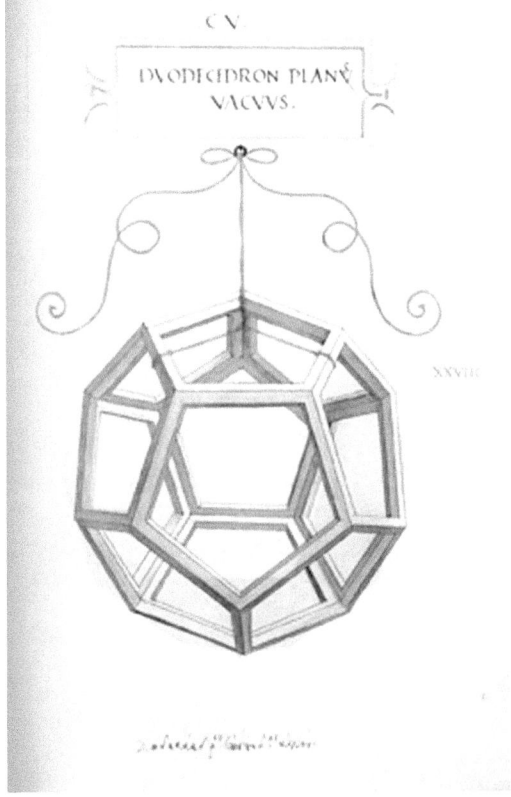

Figure 9.1. Leonardo's drawing of the regular dodecahedron The dodecahedron is the limiting GP case with zero hexagons.

G. Hart
Chief of Content, Museum of Mathematics,
New York, NY, USA
e-mail: george@georgehart.com, http://georgehart.com

M. Senechal (ed.), *Shaping Space*, DOI 10.1007/978-0-387-92714-5_9,
© Marjorie Senechal 2013

dimple patterns, pave jewelry, cartography, and abstract artwork.

Goldberg polyhedra are similar to spherical patterns of packed living cells, though more regular. Their organic quality resonates with our sensibility for natural structure. They remind us of various microscopic organisms, plants' seed pods, and skin pattern textures. In my sculpture, I seek to make forms that are simultaneously mathematical and organic. So I find these polyhedra to be a natural foundation for design.

Goldberg Polyhedra

I will refer to Goldberg polyhedra generically as GP, and identify specific examples with a pair of whole numbers, GP(a,b). The numbers a and b indicate a type of pentagon-to-pentagon "60-degree knights move." For example, the regular dodecahedron is GP(1,0) and the Goldberg polyhedron in Figure 9.2 is GP(3,2).

Like all simple polyhedra, Goldberg polyhedra are subject to the constraints of Euler's formula, $F - E + V = 2$. Let's see what the formula tells us. Since the faces of a Goldberg polyhedron

can only be pentagons and hexagons, we can write the number F as a sum, $F = F_5 + F_6$, where F_5 is the number of pentagons and F_6 is the number of hexagons. Then (figure out why) $E = (5F_5 + 6F_6)/2$. And since all vertices are trivalent, $V = (5F_5 + 6F_6)/3$. Putting these expressions for F, E, and V into Euler's formula, we discover that the whole thing simplifies to $F_5 = 12$. That is, Euler's formula says every Goldberg polyhedron has exactly 12 pentagonal faces. On the other hand, it says nothing about the number of hexagonal faces.

In fact, one can construct polyhedra with 12 pentagonal faces and any number of hexagonal faces other than one. But they will not, in general, be Goldberg polyhedra because of the third condition, icosahedral symmetry. This condition tells us something about how the hexagons surround the pentagons. In particular, it means every pentagon must be surrounded in exactly the same way.

The best way to understand Goldberg polyhedra is to study the simplest possible examples. These have 12, 32, 42, 72, 92, 122, ... faces. What is the pattern in the sequence of numbers? Why does the number of faces always end in a "2" and why are there none with 22, 52, 62, or 82 faces? But before considering such questions, you should make some GPs.

32 Faces, GP(1,1), The Truncated Icosahedron

Our name for the second in the series, the well-known 32-faced form shown in Figure 9.3, is GP(1,1). To mathematicians, the shape is the "truncated icosahedron," as Johannes Kepler named it in the 1600s. But the form was well known to earlier artists and mathematicians in the Renaissance, and Archimedes had already written about it before 200 BC. It is the familiar soccer ball (football in Europe). The GP soccer ball was invented in the 1950s in Denmark; the familiar black and white version became official starting with the 1970 world cup. The shape has also received considerable attention since the Nobel prize winning discovery in 1985 that

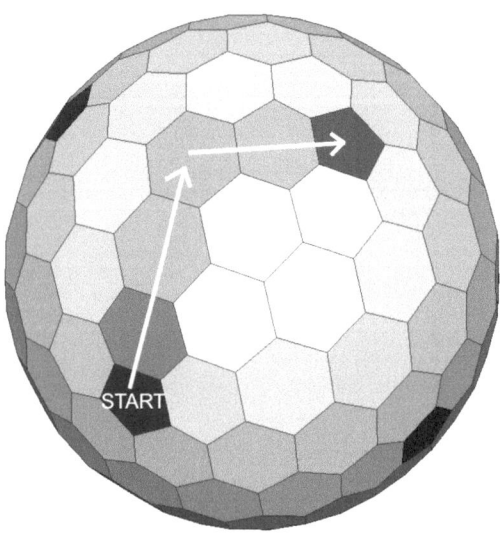

Figure 9.2. A typical Goldberg polyhedron. Its name, "GP(3,2)," describes the pentagon-to-pentagon walk, shown highlighted to indicate the meaning of the 3 and the 2.

carbon atoms naturally join into structures which chemists refer to as "C60." The 60 refers to the number of 3-way joints, which are the positions where carbon atoms sit.

Twelve of the faces are pentagons and the remaining twenty are hexagons. GP(1,1) is not one polyhedron; it is a family of them. Consider the icosahedron, shown at the left in Figure 9.4.

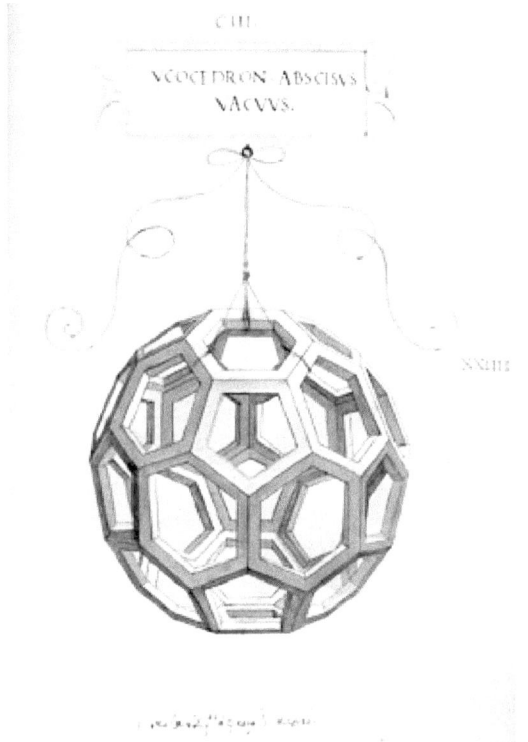

Figure 9.3. Leonardo's drawing of the truncated icosahedron. Our name for it is GP(1,1).

To truncate means to cut off the vertices. Since five triangles meet at each vertex, truncation reveals a pentagon under each cut. It also leaves a hexagon from each original triangle. Each vertex must be truncated in exactly the same way to preserve icosahedral symmetry. We can, however, vary the depth of truncation. This gives a family of GP(1,1) forms with different shapes for the hexagons. The particular truncation depth which results in regular hexagons is very attractive, so I use it as a representative form.

42 Faces, GP(2,0), The Truncated Rhombic Triacontahedron

When most people first see GP(2,0), they assume that it is the familiar soccer ball shape, but GP(2,0) has 42 faces, not 32. As far as I know, this shape has not been used in athletics. Notice also that GP(2,0) has some vertices where three hexagons (and no pentagons) meet, but in GP(1,1) every vertex has exactly one pentagon. So a GP(2,0) sports ball would have some all white vertices, while the familiar soccer ball has black touching every vertex. A third difference is that the twenty hexagons in GP(1,1) are regular, while GP(2,2)'s thirty hexagons are not. Figure 9.5 shows a model of GP(2,0) constructed from paper polygons and tape.

GP(2,0) is a truncation of the rhombic triacontahedron, a polyhedron with thirty rhombic faces. By truncating all the 5-fold vertices, each rhombus is reduced to a hexagon. Again, we have a choice of how deep to cut. A natural choice is the depth which results in equilateral

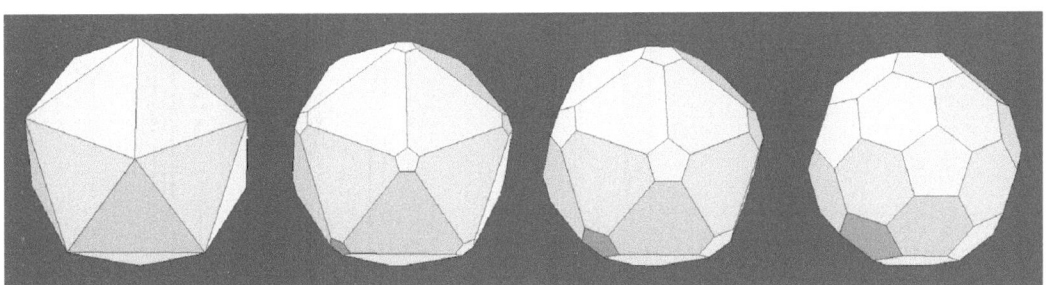

Figure 9.4. Icosahedron, and truncation process to generate various forms of GP(1,1).

hexagons, but they will not be equiangular. We do not truncate the vertices where the obtuse angles meet (in groups of three). These obtuse angles are about 116.5 degrees. Thus the GP(2,0) hexagon differs from a regular hexagon, which has all 120-degree angles.

72 Faces, GP(2,1), The Truncated Pentagonal Hexecontahedron

GP(2,1) is the simplest GP which is chiral, meaning it has distinct left-hand and right-hand forms,

Figure 9.5. Five inch paper model of GP(2,0) colored dark and light like a super-soccer ball.

shown in Figure 9.6. The lack of mirror symmetry makes this form more interesting visually.

To derive GP(2,1), we can start with the pentagonal hexecontahedron (Figure 9.7). This is the well-known Catalan polyhedron (to those who well-know it). It is the dual of the Archimedean snub dodecahedron and has sixty congruent 5-sided faces. Truncating its twelve 5-fold vertices as shown, we create twelve pentagons and sixty hexagons.

At this step, we leave the realm of shapes that have good classical descriptions and move into territory best described as "GP(a,b)." Knowledge of this family provides a storehouse of useful forms that should be in the mental inventory of any geometric designer. For example, physicists designing the Spin Spectrometer at the Oak Ridge Heavy Ion Research Facility needed a way to arrange a sensor array spherically around a target to capture a 3D distribution of particles emitted in all directions. They could afford about six dozen sensors, each covering a roughly equal, roughly circular area. The solution they chose is the geometry of pentagonal and hexagonal particle detectors spherically arranged with the structure of GP(2,1). In another application, one manufacturer has recently produced a wiffle ball of this shape, claiming its aerodynamic properties make it take a curved path when thrown.

GP(2,1) is the foundation for a computer image "Puzzle" which I made in 1995, shown in Figure 9.8. As I wrote at the time, the first part

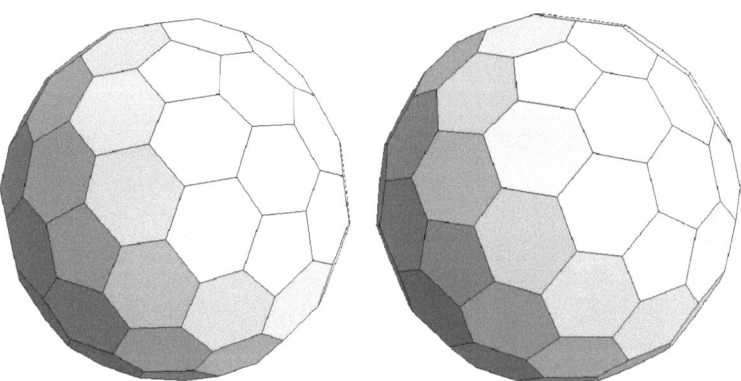

Figure 9.6. GP(2,1) consists of twelve pentagons and sixty congruent hexagons. It comes in two enantiomorphic forms.

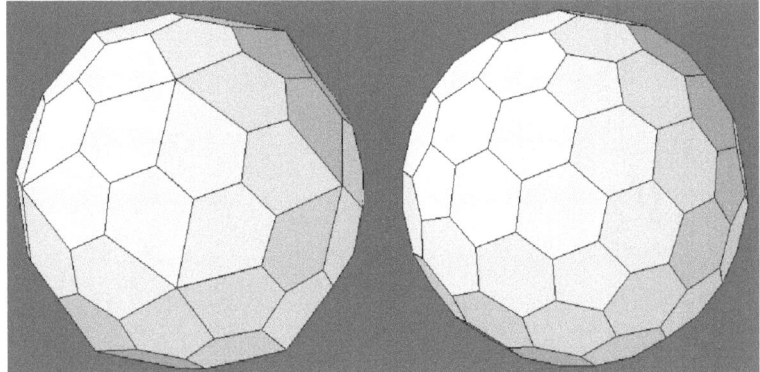

Figure 9.7. GP(2,1), on the *right*, can be derived by truncating the 5-fold vertices of the Pentagonal Hexecontahedron on the *left*.

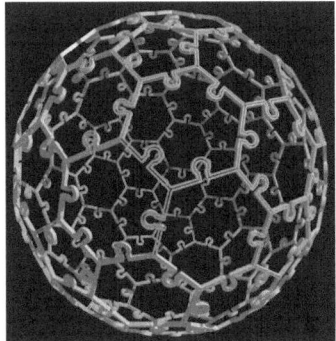

Figure 9.8. "Puzzle" digital image, 1995.

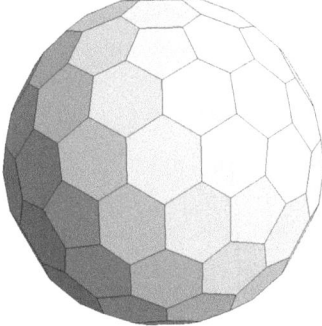

Figure 9.9. GP(3,0) consists of twelve pentagons and eighty hexagons. It is reflexible.

of the puzzle is to figure out the underlying polyhedron. Later I made GP(2,1) as the hollow wood sculpture, carefully bevelling 72 laser-cut wood polygons, giving them the appropriate dihedral angles, then epoxying them edge-to-edge.

92 Faces, GP(3,0), Dual to a Three-Frequency Icosahedron

The next example, GP(3,0) in Figure 9.9, is dual to a 3-frequency icosahedral sphere. Figure 9.10 shows the derivation of a 3-frequency icosahedral sphere. (Duality is explained in Chapter 15.)

This is a good place to mention differences between GPs and "Fullerenes," the spherical allotropes of carbon also called "Buckminsterfullerenes" or "Bucky balls." Fullerenes can contain rings of five and six carbon atoms

assembled like GPs, but also allow a much wider range of structures, including rings of size seven or more. Fullerenes typically contain non-planar rings, because the carbon atoms minimize energy functions based on angles and distances, and usually have no incentive to fall into planes. Furthermore, Fullerenes are not necessarily icosahedral and most have much less symmetry.

The discoverers of C_{60} chose a name honoring Buckminster Fuller because they did not realize the truncated icosahedron was already a well-known mathematical structure. They alluded to geodesic domes for their lightness, strength, and the internal cavity, but I think Fuller would disavow the name. He was specifically interested only in triangulated structures. "If we want to have a structure, we have to have triangles." He understood that triangulated structures remain

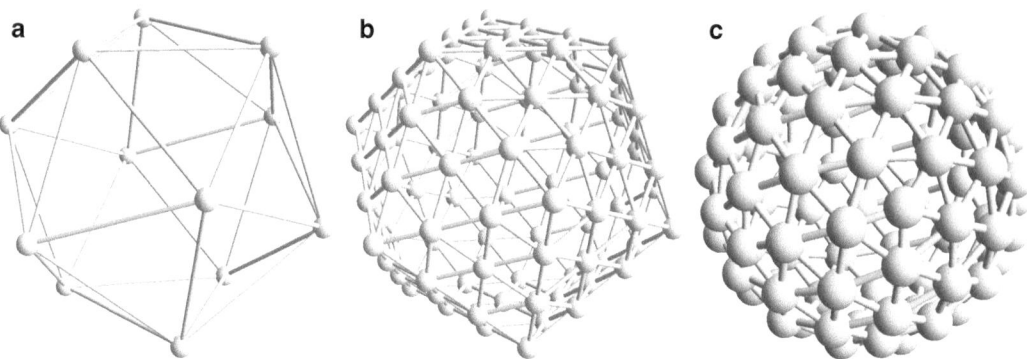

Figure 9.10. Derivation of the 3-frequency icosahedron: (**a**) underlying icosahedron, (**b**) edges divided into thirds, (**c**) projection to sphere.

rigid due to the lengths of the struts even if the angles at the joints are not fixed. So C_{60}, like all GP having open hexagons and pentagons, is not even a "structure" by Fuller's definition.

122 Faces, GP(2,2)

The next simplest member of the family is GP(2,2), shown in Figure 9.11. This is a good example to introduce the notion of "paths." In Figure 9.2, we saw a kind of "knight's move" with a steps forward and b to the side, hence the GP(a,b) notation. Now imagine walking on a giant GP globe, taking one step per face, but always trying to go straight. When you step into a hexagon from a neighboring face, there is an edge on the opposite side of the hexagon, shared with another face. If you step into a pentagon, there is no opposite edge to cross, so the path ends.

Figure 9.11 shows that there are two types of paths in GP(2,2). If you step off a pentagon and start walking straight, you end up at another pentagon after six steps. A second type of path goes once around the world, like an equator, with 18 steps in its cycle. Each hexagon is the crossroads for three paths. When holding a physical GP model, it is fascinating to follow the various types of gently meandering paths as you spin the ball in your hands, trying to make sense of where you end up from a given starting pentagon. Look back to GP(2,1) for an example where every path

Figure 9.11. GP(2,2) consists of twelve pentagons and 110 hexagons. A pentagon-to-pentagon path and an all-hexagon round-the-world path are highlighted.

takes the traveler halfway around the world, from a pentagon to its antipodal pentagon. In GP(2,2), every face can be reached starting from at least one pentagon, but look at GP(3,0) for the simplest example where some of the hexagons can not be reached in a straight path if one always starts at a pentagon. GP(3,0) also illustrates parallel equatorial paths—you and a friend can walk together around the world side-by-side holding hands, each on your own 15 step cycle.

132 Faces, GP(3,1)

GP(3,1) is borderline between simple and intricate. Its chirality makes it an interesting subject for a sculpture. Figure 9.12 shows two versions I made of wood, using a laser-cutter to fabricate the 132 planar parts. To fit precisely, the parts must be made with high accuracy. Cutting the 120 irregular hexagons might drive one insane if using traditional tools such as a band saw. But a computer-controlled laser-cutter is ideal for producing irregular parts from flat material. My primary goal was the open-faced one, because Leonardo's solid-edge style lucidly presents an unfamiliar form with front and back simultaneously visible. The parts of the smaller, solid-faced one are the holes removed from the larger, open-faced one. For this to work out, one

must be careful to design the openings to be proportional to the face shapes.

492 Faces, GP(5,3) and GP(7,0)

Consider a GP(a,b) with 492 faces. As explained below, Euler's formula tells us that the polyhedron has 1470 edges and 980 vertices. However, this data does not determine the polyhedron. There are *two*, topologically different, forms of GP with 492 faces, GP(7,0) and GP(5,3). Each has only one type of pentagon-to-pentagon path, shown in Figure 9.13. In GP(7,0) these paths are of length seven, but in GP(5,3) the paths are of length 49 and go more than once around the world.

2562 Faces, GP(16,0)

Figure 9.14 shows GP(16,0), which has 2562 faces. It can be found as one of the layers in Buckminster Fuller's two-layer Expo67 dome in Montreal.

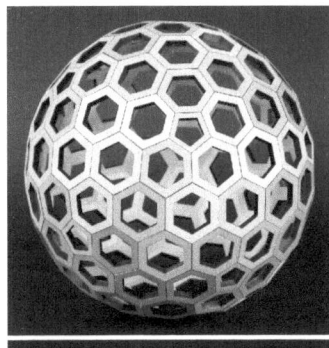
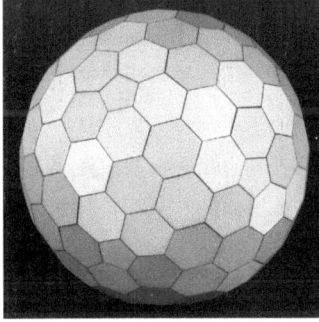

Figure 9.12. Wood sculptures, 8 inch and 6 inch diameters, of GP(3,1).

A General Approach

Goldberg's key insight is indicated in Figure 9.15. Using the vertices of triangular graph paper, one can generate a large equilateral triangle, ABC, by moving *a* steps to the right and then *b* steps to the upper right. Any integer pair (a,b) takes

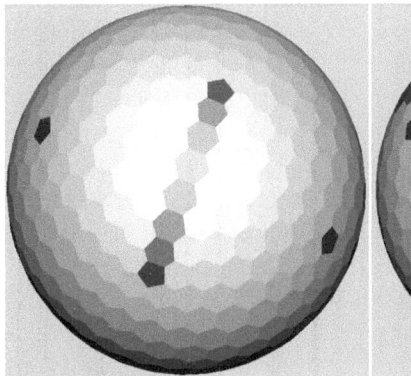

Figure 9.13. GP(7,0) and GP(5,3) each have 492 faces but their pentagon-to-pentagon paths have very different structures.

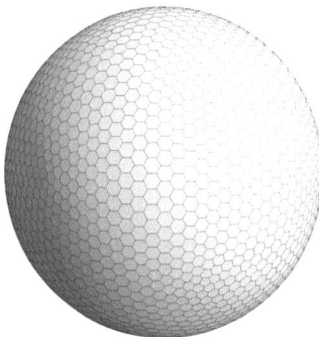

Figure 9.14. GP(16,0) has 12 pentagons and 2,550 hexagons. It underlies the Expo67 dome in Montreal. Find the pentagons in this chicken wire!.

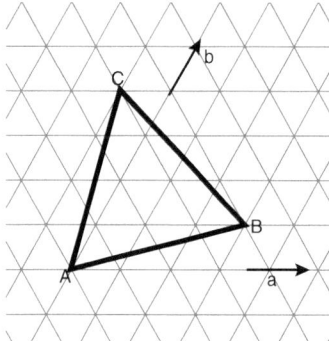

Figure 9.15. An equilateral triangle, ABC, drawn on triangular graph paper. The oblique (a,b) coordinate system indicates how to move from A to B. Here (a,b) = (3,1).

us from vertex A to some vertex B. Then that same motion, but rotated counterclockwise 120 degrees, takes us from B to C, and rotated once again takes us from C back to A. The three movements are congruent, so the large triangle, ABC, is equilateral yet has all three vertices on the triangular lattice (the edges of ABC need not be parallel to the graph paper axes). If we take twenty copies of ABC and assemble them as an icosahedron, we get a continuous pattern of the small triangles. The result is like the k-frequency icosahedron of Figure 9.10, but allowing for a rotation between the triangular tessellation and the icosahedron faces. Small triangles at the edges of ABC can "fold" along an icosahedron edge and be part of two icosahedron faces. Extending the terminology "k-frequency" icosahedron, this can be called an "(a,b)-frequency icosahedron."

The special case of $b = 0$ gives the simplest, parallel type subdivision. Another special case, where $a = b$, makes the edges of ABC perpendicular to the edges of the small triangles and the result is again reflexible. The in-between values, where $0 < b < a$, give the chiral patterns.

Generating (x,y,z) coordinates for these vertices is a straightforward exercise in coordinate geometry. A linear transformation mapping ABC to an icosahedron face will map the small triangle grid points to our desired points in the icosahedral face planes. Each vertex is then projected to a unit sphere to give a triangulated geodesic sphere as in Figure 9.16. Many references on geodesic domes give the basic geometric ideas. For architectural purposes, the points may be redistributed slightly to reduce the number of different edge lengths, as that simplifies the physical construction if cutting metal struts.

Geometric Realizations

There are many ways to choose particular coordinates for the pentagon and hexagon vertices. Here are four:

1. *Vertex Elimination.* In a triangulated geodesic sphere, delete one third of the vertices (and their incident edges) to create a hexagon pattern in the triangular grid. Starting at a 5-fold vertex, take i steps to the right and j steps to the upper right. Then omit the vertices where i–j is a multiple of 3. This leaves an open pentagon surrounded by open hexagons. If the neighboring 5-fold vertex is also omitted, i.e., if a–b is a multiple of 3, then this pattern matches with itself all over the sphere to give a grid of hexagons and pentagons. However, in general they are not planar, so this method is suitable for making a structure of edges, but it is not suitable for architectural purposes where one wants to cut flat sheets, e.g., plywood, to use as faces.

2. *3D Reciprocal Construction.* Given the triangulated geodesic sphere, construct a plane tangent to the sphere at each vertex. Each plane defines two half-spaces, one of which

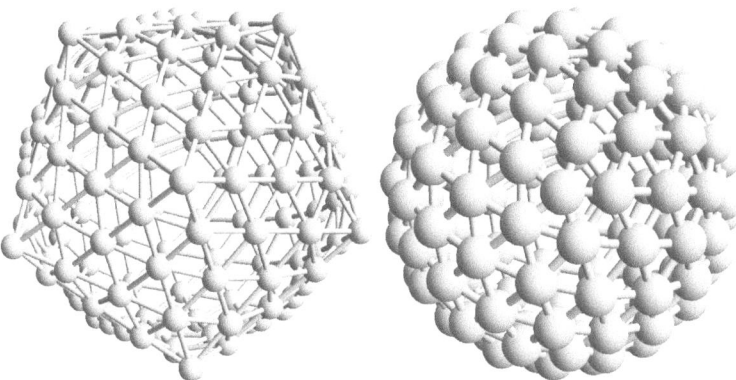

Figure 9.16. Vertices of (3,1)-icosahedron in twenty icosahedral face planes and after projecting to unit sphere.

includes the sphere. Consider the intersection of all the sphere-containing half-spaces. Its boundary faces will be planar hexagons and pentagons. Each vertex is easily located at the intersection of three planes by solving three simultaneous linear equations. (This method is known as taking the reciprocal in the sphere of the convex hull of the original vertices.)

3. *Centroid Tangency.* Goldberg was interested in polyhedra that maximize the volume-to-surface-area ratio. In the optimal solution each face will be tangent to the unit sphere at its centroid. We would like to solve for the set of tangency points which have this property, but it gives a complex set of nonlinear equations to solve simultaneously, so we do not hope to solve it exactly in general. But one can find a numerical solution by a simple iterative algorithm. Start with an approximate solution for the tangency points, e.g., use Method 1. Find the face planes, calculate the centroid of each face, and project each centroid to the sphere to get a new set of points that can again be used to generate tangent planes. The optimal solution is a fixed point of this iteration. A simple computer program shows that iterating this process many times leads numerically to an approximation of the fixed point.

4. *Canonical Form.* The canonical form is one in which in which all the edges are tangent to the unit sphere and the center of gravity of these tangency points is the origin. It is of mathematical interest in part because the dual

can also be made so its edges are tangent to the same sphere at the same points. There are various algorithms to calculate this canonical form, one of which is a simple fixed-point iteration analogous to that in Method 2.

Note that Methods 2 and 3 give polyhedra circumscribed around a sphere. Method 4 gives one which is "midscribed," meaning the edges are tangent to a sphere. Method 1 attempts to give an inscribed result, with vertices on the sphere, but that is not generally attainable with planar faces. The dodecahedron and truncated icosahedron are inscribable, but whether or how the more complex GPs are inscribable while preserving symmetry is an open problem. The simple Method 2 is suitable for most sculptural design purposes that motivate me.

Counting Components

How many faces, edges, and vertices does GP(a,b) have? To answer this question, consider its dual, the triangulated sphere (a,b). Moving one unit in the a direction is equivalent to moving (1,0) in XY coordinates and moving one unit in the b direction is (1/2, $\sqrt{(3)/2}$) in XY coordinates. So the Pythagorean theorem and a bit of algebra show that the total movement from A to B has length squared $d := a^2 + ab + b^2$. An area argument shows there will be d small triangles for each icosahedron face. As an icosahedron has 20 faces, there are $20d$ small triangles in a complete sphere. Imagine cutting those $20d$ small triangles out of paper to assemble them. There are three edges each, so we would have $60d$ small triangle edges in a pile of loose paper triangles. After taping the edges together in pairs to make a geodesic sphere, there would be $30d$ edges in the polyhedron. Knowing F and E, we can solve for V in Euler's theorem. There are $10d + 2$ vertices in the triangulated geodesic sphere.

Since GP(a,b) is the dual to the triangulated sphere, the values of V and F are swapped, and the number of edges is unchanged. So the parts counts for GP(a,b) are:

Number of Faces: $10d + 2$
Number of Vertices: $20d$
Number of Edges: $30d$

As a and b are integers, so is d, and the formula $10d + 2$ explains why the number of faces is always 2 more than a multiple of 10. As to the mysterious sequence 12, 32, 42, 72, 92, 122...these are all the values obtainable when using a positive value for a and a non-negative value for b. The smallest case where there are two sets of (a,b) values that give the same d are (5,3) and (7,0) which both give 492 faces. Goldberg gives a construction showing that there are larger examples with not just two, but any desired number of different forms resulting in the same number of faces.

Paths

Straight paths of hexagons correspond to the edges of the small triangles in the underlying lattice. Because they are highly structured, yet we can only see a small part of them at once as they disappear over the horizon, they provide a pleasant sense of mystery.

Analysis of Figure 9.15 shows that the length of a pentagon-to-pentagon path in GP(a,0) is a and in GP(a,a) is $3a$. In the chiral cases, the length is d in GP(a,b) if a and b are relatively prime. More generally, one must divide out the greatest common divisor, so the path length is d/GCD(a,b) in the chiral cases. The length of the equatorial paths is $5a$ in GP(a,0) and $9a$ in GP(a,a). In the chiral cases where a and b are relatively prime, there are no equatorial paths. But the chiral cases with GCD(a,b) > 1 are mysterious; "equatorial" paths can wrap around several times, crossing over themselves repeatedly. I can write a program to build the structure and count the answer, but not a simple analysis or a simple formula for the length(s).

The question of where a finite path ends up does not seem to have an easy answer. Starting at a pentagon, there are three possibilities: the five paths may end at the five near neighboring pentagons, they may end at the five medium distance pentagons, or they may all end at the one opposite pentagon. No path can end at the pentagon is starts from. (Exercise: why not?) When $b = 0$, paths end at a near neighbor and if $a = b$, paths end at a further neighbor. But the chiral cases can be subtle. Paths across hexagons correspond to lines of the small triangle lattice. It is easy to follow them and see where they meet a vertex of the large triangle lattice. For example, if $a = 2b$ or if $2a = 7b$, the path can be traced to end at the opposite pentagon. The difficulty is in understanding how long paths wrap multiple times around the icosahedron when it is unfolded. I leave it as an open problem to give a simple method of determining for which GP(a,b) the paths end at the opposite vertex.

A related open question is to give a formula for determining how many hexagons cannot be reached on any path that starts at a pentagon. Here again, the reflexible cases are straightforward. In the chiral case, the answer is easy when a and b are relatively prime, as then every hexagon can be reached. But the remaining cases are another open problem.

Models and Artwork

Paper models are a very cost-effective means for gaining 3D understanding in a hands-on manner. The model in Figure 9.5 was expediently assembled using tape on the inside. For a more elegant, tape-free model, one can cut out the faces with tabs around the edges and then fold back the tabs and glue them internally. In the "one-tab" method, at each edge just one of the two faces has a tab, which is glued to the inside of the other face. In the "two-tab" method, all edges of all faces have a tab and they are glued together in pairs to make a rib that lies under each edge. I find this two-tab method to be faster, easier, and more attractive.

Scissors are adequate, but the work of cutting paper parts is greatly simplified if one has a robotic paper cutter. The example in Figure 9.5 was made with the aid of an inexpensive computer-controlled paper cutter. These are now marketed for home use and are sufficiently affordable (under $200) that any serious model maker should consider owning one. They allow one to easily make models which would be far too tedious to cut with scissors. I expect such cutters will protect many future paper model aficionados from Carpel Tunnel Scissors Syndrome.

Paper also has the advantage that one can easily print on it. For example, patterns can be found online for globes of the earth projected onto various polyhedra. The most complex GP that I have seen used as a globe of the earth is the truncated icosahedron, which was first used by John Snyder in 1992 as the basis of an equal-area projection. It is mathematically straightforward to project geographic data to the face planes of any polyhedron. More complex GPs would naturally lead to better representations of the sphere.

Wooden models are more difficult to make than it may appear. Similar models can also be made with acrylic or other plastics. The parts must be carefully cut if they are to fit without gaps. The edges must be beveled accurately to the correct dihedral angles. While laser-cutters and other computer controlled technologies can cut out the shapes, current machinery does not easily bevel the edges. So this operation must be done manually and care must be taken not to flip chiral parts or mix up the many different angles in one model.

A second approach to wood models, especially for large ones, is to build up pentagons and hexagons from individual mitered pieces, one for each polygon edge.

Exact models of only the simplest GPs can be made with familiar mathematical kits such as Zometool or Polydron because they have only

certain inventories of lengths and angles, and GPs require many different lengths and angles. A flexible-edge construction set, of the type used by chemistry students for making molecular models, is excellent for making GP models. In these kits, there are rigid trivalent connectors to use at each vertex and slightly flexible tubing to use for edges. The tubing can be cut to make any desired edge lengths. Figure 9.17 shows a 50-inch model of GP(5,1) which I made with the Stony Brook University Chemistry Club. The model is an approximation (the trivalent connectors have all 120 degree angles, the edges are slightly curved, and the polygons in the model are not necessarily planar). In addition, for ease of construction, I chose just two lengths for the edges. It was an educational group project to assemble it and hang it for display in the Chemistry library. Instructions are available online at http://www. georgehart.com/chemistry/C420Fullerene.html/.

Additive fabrication technology removes many geometric constraints. These technologies, sometimes called rapid prototyping, solid freeform fabrication, or 3D printing, can robotically build any describable form. They are ideal for making GP models, with their many slightly different lengths and angles.

There is a long tradition of turning nested ivory spheres on a lathe, where each inner sphere is carved with a tool that is passed through the holes of the outer layers. The tradition started in Europe in the sixteenth century, but in recent centuries has been associated more with Japan and China. In homage to that tradition, I designed a set of nested GP forms which can be produced on additive fabrication machines. Figure 9.18 shows a set of ten different concentric forms: GPs (5,0), (3,3), (4,2), (5,1), (6,0), (4,3), (5,2), (6,1), (4,4), and (5,3), which steadily increase from 252 to 492 faces. In the traditional nested ivory spheres, the pattern of holes in each sphere is identical, because tools must pass through holes in the outer layers to cut into the inner layers. In this modern variation, there is no similar constraint and each layer can be designed independently.

If one applies an affine transformation to a GP, the result is an ellipsoid-like form which still has planar faces. For architectural purposes,

Figure 9.17. 50 inch model of GP(5,1) assembled by the Stony Brook University Chemistry Club.

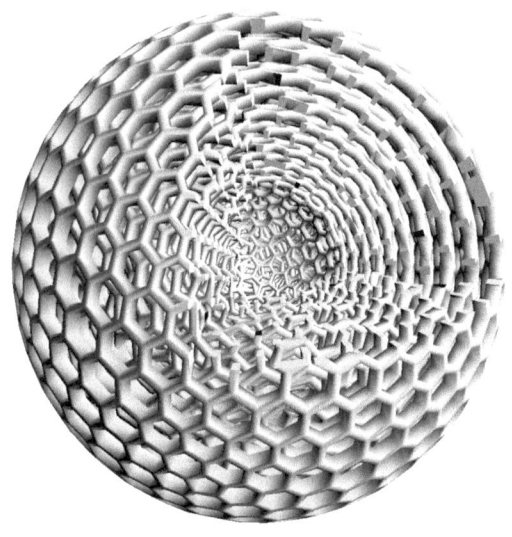

Figure 9.18. Ten different nested GP forms, from GP(5,0) to GP(5,3) shown in cut-away rendering. The outer one is 3 inches in diameter.

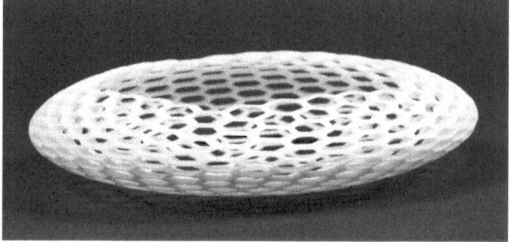

Figure 9.19. Candy dish, 5 inches, nylon, based on oblate GP(5,3).

an oblate form might be used to cover a wide area with a dome that need not be as high as a spherical dome of the same footprint. Figure 9.19 shows an oblate design of this type, but not with architectural intent. It is designed as a decorative bowl. The form began as GP(5,3), one of the two GPs with 492 faces. I compressed it along a

5-fold axis, and removed struts around one pole to create a large pentagonal opening for the top. I considered flattening the bottom slightly so it sits steadily on a table, but then decided to leave the curve so it can rock dynamically. If I get a chance to make a matching dish, I plan to perform the same transformations, but start with the other version, GP(7,0).

Figure 9.20 shows an idea which grew as a development of the puzzle in Figure 9.8. It is a spherical jigsaw puzzle which assembles to make GP(5,3). It is divided into twelve identical parts along edges in such a manner that each piece is centered on a pentagon and has 5-fold symmetry. This is possible for all GP in which the number

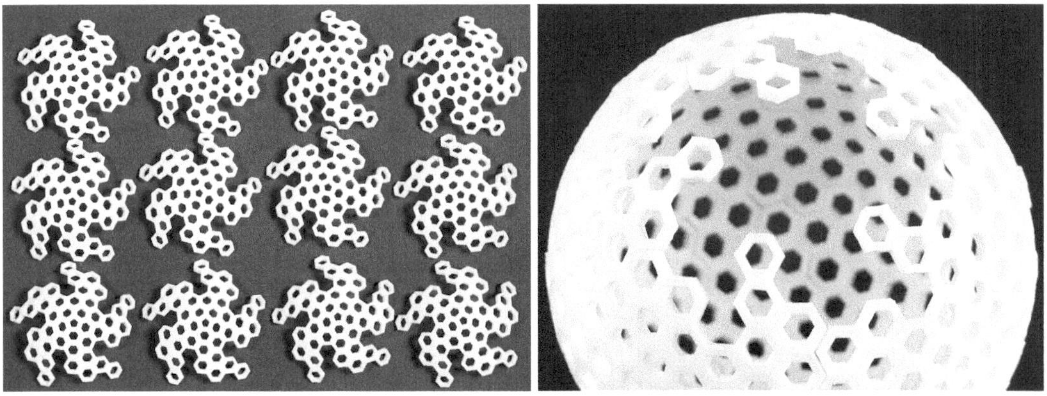

Figure 9.20. Twelve-part assembly Puzzle, 5 inches, nylon, based on GP(5,3).

of faces is a multiple of 12. Because these parts are identical, it is an easy puzzle to solve, but still quite fun to play with. This initial example was intended in part to test the design idea and verify that the parts will hold together by friction alone. Future variations on the concept will incorporate different shapes for the pieces or an egg-shaped solution, to provide more a more challenging puzzle.

The GP(16,0) Expo67 dome mentioned above and all other two-layer domes that I have seen are based on reflexible structures. The same principles apply to chiral forms, but they are more costly to build because a larger inventory of parts lengths is required. As I find the chiral forms to be visually engaging, I designed software to create two-layer domes based on any GP form. Figure 9.21 shows what I believe is the only chiral two-layer dome in existence. The inner layer is GP(3,1) and the outer layer is its dual. Educational, physical models of two-layer spheres can be made with flexible angle kits. While the flexible angles would not make a rigid model of a GP alone, when triangulated with its dual, it becomes rigid.

Finally, the sculpture shown in Figure 9.22 is a kind of chain mail sphere. Starting with GP(7,4), the faces were replaced with small circular rings. Each edge of the dual triangulated structure was replaced with a link that connects adjacent rings. The dimensions of the rings and links were chosen so they do not overlap each other. The resulting chain mail fabric is flexible.

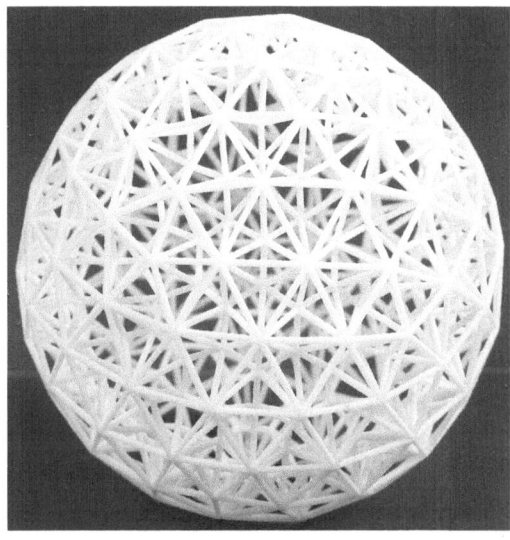

Figure 9.21. Two-layer geodesic sphere based on GP(3,1), 4 inches, nylon. As far as I know, it is the world's only chiral two-layer geodesic structure.

If every link were free to move, the whole form would collapse like an empty sack, so to give the overall sculpture a shape, a subset of the links were modified to lock with their rings, making a rigid skeletal form corresponding to the edges of a dodecahedron. Twelve circular regions of the mesh are free to move within this rigid framework. The total effect is quite remarkable and indicates how many fascinating possibilities might be encountered when one explores such a rich subject as Goldberg polyhedra.

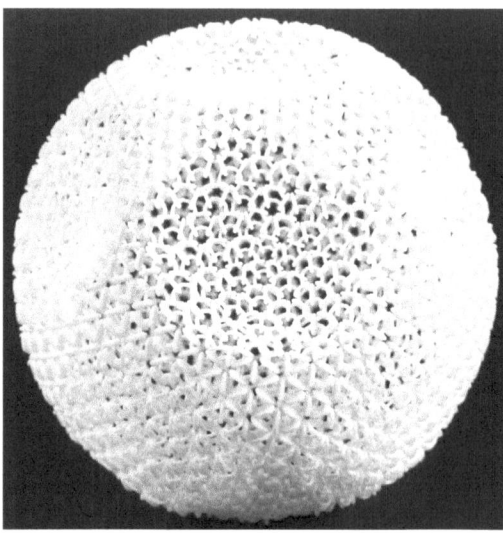

Figure 9.22. Sculpture with chain links based on GP(7,4), 6 inches, nylon.

10

Polyhedra and Crystal Structures

Chung Chieh

I have long been interested in searching for interesting relationships between polyhedra and crystal structures, especially with the application of polyhedra as units for crystal structures. Crystallography uses geometry as a foundation. As a crystal scientist, I am interested in understanding how and why certain crystal structures are built the way they are, particularly from a geometric viewpoint. I am also constantly searching for relationships among the various crystal structures.

Shapes, colors, and geometry have historically been subjects of great interest to philosophers, mathematicians, and scientists. We are here to understand, to construct, to design, to create, to appreciate, and to love the shapes and forms of various kinds. Among various shapes, perhaps polyhedra are especially interesting to us all, and we may also develop an affection for the aesthetic shapes of some very nice crystals such as those shown in Figures 10.1–10.4. My interest is in the arrangements of atoms, molecules, and ions in those crystals, and how the arrangements are related to geometry and polyhedra.

The crystalline state is the most common form of all matter at sufficiently low temperatures. In modern terms, crystals consist of atoms, molecules, or ions arranged in a periodic, repeated manner. For a crystal visible to the naked eye, there may be more than a million repeating units to each of the three directions. Those periodic arrangements may be described by symmetry operations such as one-, two-, three-, four-, and sixfold rotations, mirror and glide planes, screw axes and centers of inversion. (Periodic arrangements cannot have fivefold rotational symmetry, but the discovery of aluminim-mangnese alloys whose diffraction patterns have fivefold symmetry shows that nonperiodic arrangements exist in the solid state.)

Figure 10.1. A wulfenite crystal: an orange flattened octahedron.

C. Chieh
Professor Emeritus, Chemistry Department, University of Waterloo, Waterloo, ON N2L 3G1, Canada
e-mail: cchieh@uwaterloo.ca

M. Senechal (ed.), *Shaping Space*, DOI 10.1007/978-0-387-92714-5_10,
© Marjorie Senechal 2013

Figure 10.4. Chrome alum: perfect octahedron, weighing 867 grams.

Figure 10.2. Crystals of vanadinite: red hexagonal prisms.

Figure 10.3. Hematite-stained quartzoids: polyhedra with threefold symmetry.

Crystal structures are fascinating from both the architectural and geometric viewpoints.

Figures 10.1–10.4 show some representative crystals. Their shapes are certainly related to some familiar polyhedra. The almost-perfect octahedral alum crystal was grown by a high school student, and it weights 867 grams. After the publication of a picture of this crystal by the *Chem 13 News* (a monthly publication of the University of Waterloo Chemistry Department), the editor received pictures of even larger alum crystals from which coffee tables, seats, and other interesting things had been made. (These crystals stand up to normal use when their surfaces are protected by varnish.)

The interesting external shapes of crystals must certainly be related to their internal structures. What are the basic units from which these wonderful and geometrically interesting crystals are built? Perhaps inspired by the beauty of crystals, the philosopher Plato (427–348 B.C.) associated the regular polyhedra with the *primal substances* from which everything is derived. Aristotle agreed that earth, air, fire, and water are the primal substances; however, he disagreed with Plato's associating these substances with four of the regular solids. During the seventeenth century there was lively discussion about the basic units of crystals. Johannes Kepler (1571–1630), Erasmus Bartholin (1625–98), René

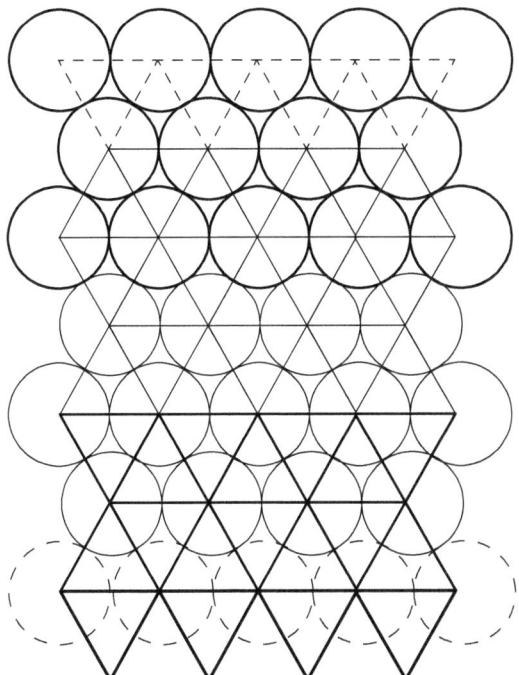

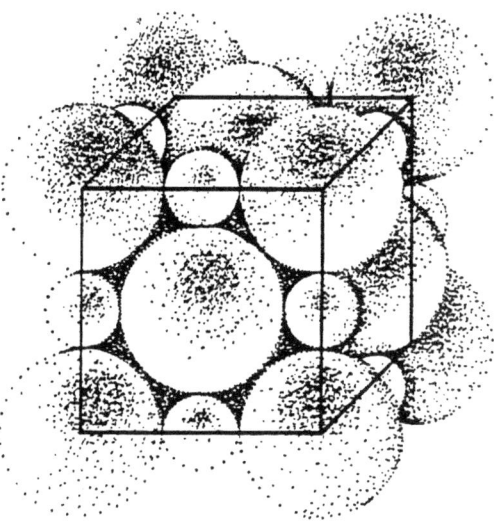

Figure 10.6. The packing of spheres as a model of crystalline sodium chloride, common table salt. The small spheres represent sodium cations, and the large spheres represent chloride anions. The unit cell, a cube, is sketched with straight lines.

Figure 10.5. Transformation of a packing of circles in a plane to a packing of triangles. The equivalent three-dimensional transformation is between spheres and polyhedra.

Descartes (1596–1650), and Robert Hooke (1635–1703) suggested that spheres are the ultimate particles. Today, packings of spheres are used as *models* for the discussion of crystals made up of atoms, but no one knows the real shape of atoms. Certainly the electronic configuration of each constituent atom has the symmetry of the electrostatic environment of the atom. In a crystal that environment is never spherical.

Figure 10.5 shows the packing of circular disks in a two-dimensional space, and the gradual transformation from packing of circles to that of triangles. In three-dimensional space, the transformation can be made from spheres to polyhedra. In Figure 10.6, we see the packing of spheres as a model of the crystal structure of common table salt. This is also a model of the structures of many binary compounds. Yet, we still like to think of the unit as a little cube. We tend to find interesting the relationship between the packing of spheres and the packing of polyhedra. Are crystal structures really packings of spheres, or

are they packings of polyhedra? The choice for crystal science is very much a matter of convenience and a matter of aesthetics. The partition of space into shapes, even within a crystal, is a subject of interdisciplinary interest, involving art, mathematics, and science.

Some highly symmetrical polyhedra such as the Archimedean truncated octahedra have been used by crystallographers and mineralogists to represent complicated crystal structures. Zeolites are natural architectures sometimes employed as chemical ion exchangers and molecular sieves. These natural three-dimensional structures are beautiful in their own right. They are silicates with some silicon atoms replaced by aluminum atoms with a general formula $(Al, Si)_n O_{2n}$. We can easily identify a cagelike unit formed by connecting $(SiO_{4/2})$ groups of atoms. The silicon atoms in the crystal structure are located at the vertices of Archimedean truncated octahedra. These large cages are interconnected in many ways; Figure 10.7 shows one of the open packings or connections of them. The square faces are separated at distances equal to the length of the edges of these polyhedra, thus making the

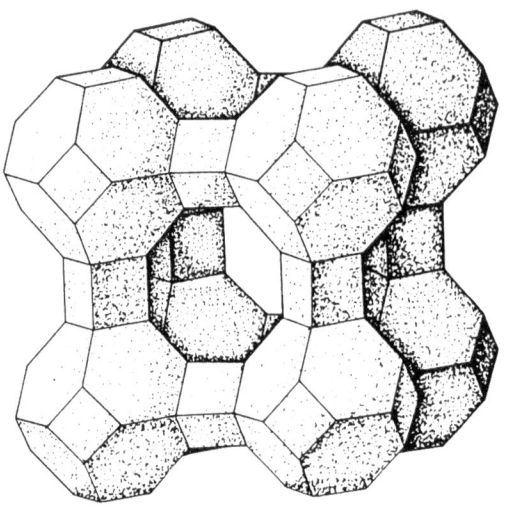

Figure 10.7. A portion of the open-framework packing of Archimedean truncated octahedra, a model that represents the structures of many zeolites.

Figure 10.8. *Fish and Birds* by M.C. Escher.

square faces the faces of the cubes. Of course, the structure may also be considered to be an open packing of cubes. Figure 10.7 shows only part of the framework, and there are millions of these truncated octahedra in any direction in a real crystal. The possible ways that these polyhedra may interconnect is an interesting topological problem.

Mathematicians have contributed greatly to the study of crystallography, and their methods are used for the description of crystal structures. In a periodic crystal, description of the large structure is simplified by considering the crystal to be built up by repetition, in all directions, of the structure enclosed within a parallelepiped (the *unit cell*). Although there are standard conventions for selecting the unit cell, the choice is not unique. As a crystallographer, I am interested in finding out how a particular structure is formed, what are the basic units (not necessarily the unit cells) that build a specific structure, and why a structure type is of common occurrence. I would like to find a general geometric scheme by which crystal structures are formed. Because crystal structures are three-dimensional patterns, they are too complicated to illustrate my approach to the problem of finding the basic units. Thus I shall use some two-dimensional artworks to

demonstrate my search. Let us start by looking at one of Escher's drawings, *Fish and Birds*, reproduced as Figure 10.8.

There are many ways to choose a basic unit that can be used to build these beautiful patterns of fish and birds. Let me show how crystallographers would choose unit cells from a pattern like this. There are various choices as indicated in Figure 10.9. Choices 10.9a–10.9b are arbitrary, and 10.9c is somewhat obscure, yet each is legitimate because each is a parallelogram. None of these choices is unique.

But there is a unique way of defining a different kind of basic unit, first suggested by the German mathematician G. Lejeune Dirichlet (1805–59). In his method, a particular point in the pattern is chosen, for example the eye of a bird, and then it is connected to all other similar points (eyes of birds). We then draw the perpendicular bisectors of these vectors. The smallest area enclosed by these lines is a convex body called a Dirichlet domain (or sometimes a Voronoi cell, after the mathematician George Voronoi who studied them in great detail and in arbitrary dimensions). In two-dimensional space these domains are polygons (Figure 10.9d), whereas in three-dimensional space they are polyhedra. Note that the Dirichlet domain for a two-dimensional

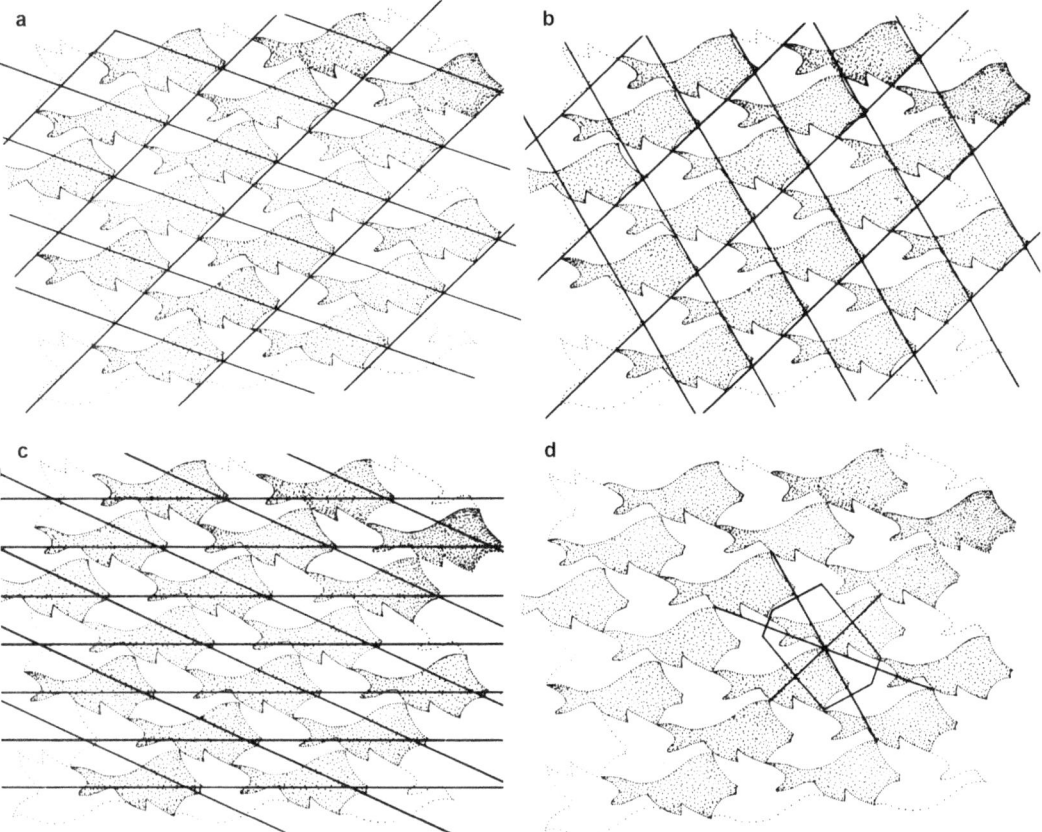

Figure 10.9. A variety of ways (**a–c**) to choose the basic repeating "crystallographic unit cell" in an array of fishes and birds. The Dirichlet domain (**d**) of that array.

pattern usually is not a parallelogram, and that of a three-dimesional crystal structure is usually not a parallelopiped.

The eyes of the birds (or any other set of translationally equivalent points) constitute a two-dimensional *lattice*. Similarly a three-dimensional lattice is a set of points generated by three noncoplanar vectors. Classified by symmetry, there are seven types of coordinate systems. They are often depicted by unit cells (see Figure 10.10). In the nineteenth century, Bravais studied the symmetries of polyhedra and of lattice points, and he came to the remarkable conclusion that there are only fourteen symmetry types of point lattices. Nowadays, we take the fourteen lattices for granted, and in many books, the fourteen Bravais lattices and their relation to the unit cells are displayed together.

(There are fourteen Bravais lattices and only seven symmetry types of unit cells; seven of the unit cells contain more than one lattice point.) Therefore, we ask the question: What are the units if we divide the crystal structures according to the fourteen lattices, instead of by unit cell shape? The Dirichlet method gives unique shapes, but the difficulty is that there are more than fourteen different polyhedra because different axis ratios of the same lattice type give rise to Dirichlet domains with various shapes. I shall return to this problem later.

Let us return to two dimensions. Applying the Dirichlet technique, I shall illustrate the geometric plan for Escher's *Black and White Knights*, shown in Figure 10.11. If we choose a point along a certain line (equivalent to a glide line) on this artwork, we can see that the drawing is

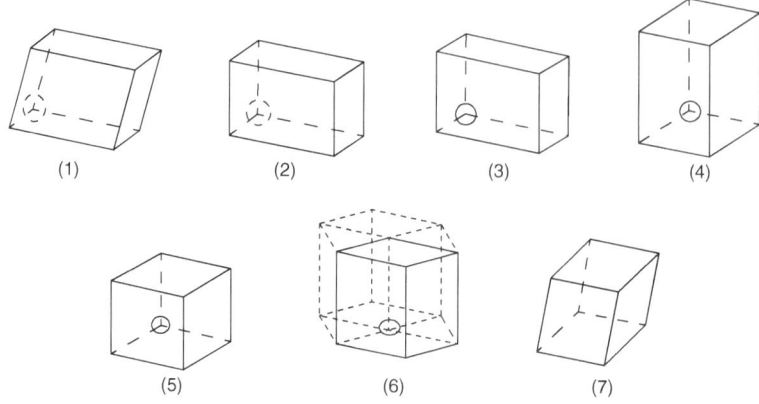

Figure 10.10. The seven polyhedra used as crystallographic unit cells for the seven crystal systems. unique parameters: (**1**) Triclinic—a, b, c, α, β, γ; (**2**) Monoclinic—a, b, c, β or γ depending on choice; (**3**) Orthohombic—a, b, c; (**4**) Tetragonal—a, c (**5**) Cubic—a; (**6**) Hexagonal—a, c ($\gamma = 120°$); (**7**) Rhombohedral—a, α.

Figure 10.11. *Black and White Knights* by M.C. Escher.

made up of two types of units, a black knight and a white knight. There is just one catch, which was pointed out to me by a little boy looking over my shoulder when I placed the Dirichlet domains over the *Black and White Knights*: if the center of the polygon is an arbitrarily chosen

point, then the patterns within the polygon (arrangement of atoms in the polyhedron in case of crystal structures) may no longer be related (by crystallographic and color symmetry).

Let us return to the very beautiful Archimedean truncated octahedron, one of my favorite polyhedra, as the unit for the cubic crystal system. This is a system in which the unit cell is a cube. There are three Bravais lattices in this system. In one case, the "primitive" lattice, the cell does not contain any lattice point in the cube, but the vertices are marked by the lattice points. In the "body-centered" lattice there is a lattice point at each cube center as well as the vertices. In the "face-centered" cubic lattice, there is a lattice point in the center of each cube face, but none in the center of the cube.

For the body-centered cubic lattice, the Dirichlet domain is the Archimedean truncated octahedron. When we used it to represent the structures of zeolites, we did not emphasize the fact that they pack together to fill the entire space, leaving no gaps. But they do. Figure 10.12 illustrates the packing of these semiregular polyhedra. Suppose we make transparent polyhedra of the same size and shape, and put a structural feature in each. Then we may use these polyhedra to build three-dimensional structures, as Escher put fish and birds together to make patterns. For the discussion of crystal structures, the polyhedra are only conceptual units; they represent clusters of

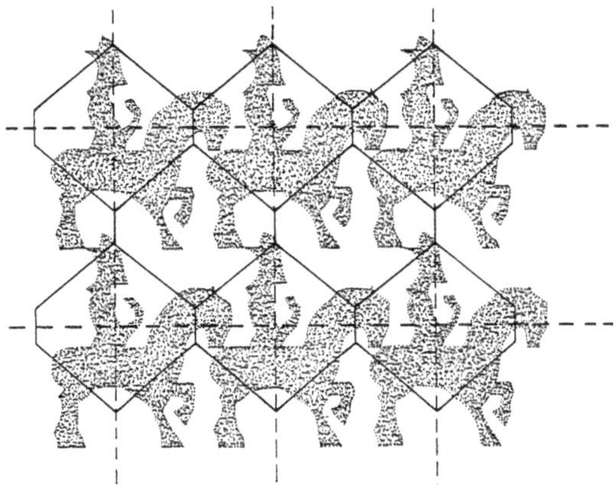

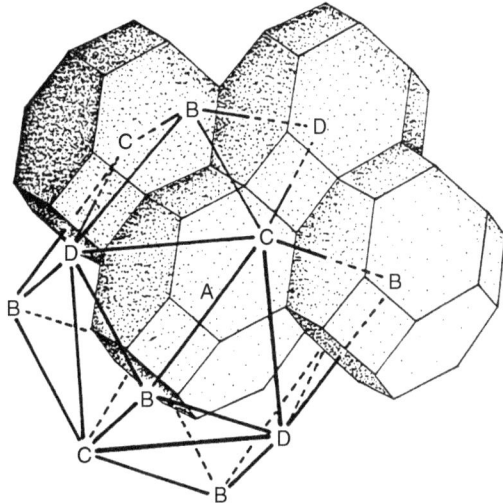

Figure 10.12. The geometric plan for Escher's *Black and White Knights* as depicted by Dirichlet domains.

Figure 10.13. Close packing of four "different" types (different, perhaps, because of color) of Archimedean truncated octahedra in the formation of a face-centered lattice.

packed atoms, ions, or molecules. A packing of one type of Archimedean truncated octahedron gives rise to a cubic body-centered lattice, a packing of two types according to a specific order gives a primitive lattice, and a packing of four types as shown in Figure 10.13 gives a face-centered cubic lattice. If we put some configurations of certain symmetries into these transparent polyhedra, we can build structures having various kinds of symmetry compatible with the symmetry of the cubic lattice.

By assuming that we have one, two, or four types (due to enclosed configuration of atoms) of units all having the shape of Archimedean truncated octahedra, we are able to classify all cubic space groups, and eventually all cubic crystal structures.

We shall now see some examples of atomic arrangements of the cubic crystals. Let us look at the atomic arrangement within one of these polyhedra, within the structure of one geometric unit. This unit comes from the crystal structure of a γ brass, an alloy. Starting from the center of the unit, there are four atoms arranged tetrahedrally as outlined in Figure 10.14a. Given the symmetry, or point group, there should be a limited number of ways to build a geometric unit from the center, keeping in mind that atoms in most metallic crystals maintain definite equilibrium distances among each other. A beautiful way to add atoms

to the unit is to place them at each face of the existing small tetrahedron, resulting in a larger one. This is followed by arranging six atoms in an octahedron outside the two tetrahedra, and finally arranging twelve atoms in a cuboctahedron to complete the unit. The model (using spheres as atoms) in Figure 10.15 shows how these units are fitted together. In one, the two units are slightly separated for clarity. It should be pointed out that these units stack in a three-dimensional fashion, rather than linearly.

All cubic γ brasses belong to one of only three space groups, and they have one, two, and four types of units, respectively. The atomic arrangements in these units are very similar from a purely geometric viewpoint; the differences arise because of the elements which occupy the vertices of the tetrahedra, octahedra, or cuboctahedra.

Metallic crystals are not the only structures which can be described or represented by the idea that the units actually have the shape of an Archimedean truncated octahedron. Organic molecules such as hexamethylenetetramine, $C_6H_{12}N_4$ (see Figure 10.16), are natural units. These units pack in a cubic space group, and the Dirichlet domain for the molecule as a whole is an Archimedean truncated octahedron. This does

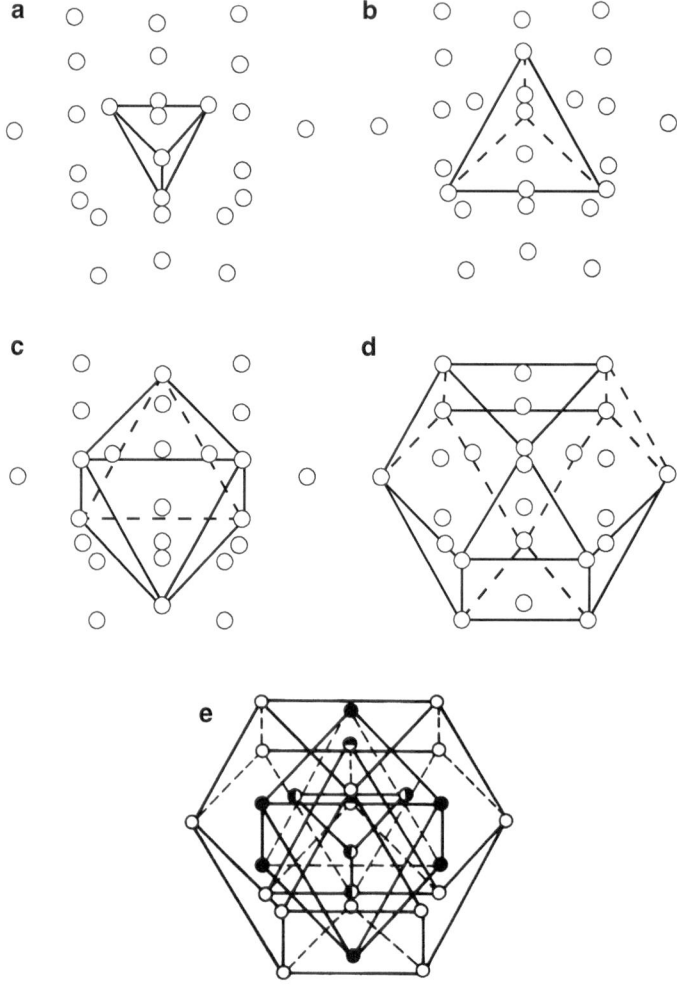

Figure 10.14. A geometric unit of the γ-brass crystal structures, consisting of inner and outer tetrahedra (**a** and **b**), an octahedron (**c**), and a cuboctahedron (**d**). The composite of all these is shown in (**e**).

not mean that the shape of this molecule is that of an Archimedean truncated octahedron; molecules do have bumps and craters on the surface. The birds and fish, or black and white knights in Escher's drawings, are not polygons either, but they do fit together forming a two-dimensional "crystal" in the way polygons do.

Let us turn to other crystal systems, and at this point look at some Dirichlet domains of tetragonal, rhombohedral, and hexagonal lattices. Polyhedra for Dirichlet domains of tetragonal lattices depend on the shape of the unit cell. A tetragonal lattice may be described by the lengths of two vectors a and c. Three of the four possible

shapes of the Dirichlet domains are shown in Figure 10.17a ($c > \sqrt{2a}$), 10.17b ($c = \sqrt{2a}$) and 10.17c ($c < \sqrt{2a}$). Actually, Figure 10.17b shows a cubic face-centered lattice, a special case of both rhombohedral and tetragonal lattices. The fourth shape is the Archimedean truncated octahedron described earlier, where the tetragonal system may be metrically the same as that of a body-centered cubic lattice. Depending on the rhombohedral angle, there are two types of polyhedra. The Dirichlet domain for a rhombohedral lattice in which the angle is less than 60° is shown in Figure 10.17d, whereas Figure 10.17e is the Dirichlet domain for one whose rhombohedral

Figure 10.15. Packing of the geometric units in γ brasses, isolated (*left*) and close-packed (*right*).

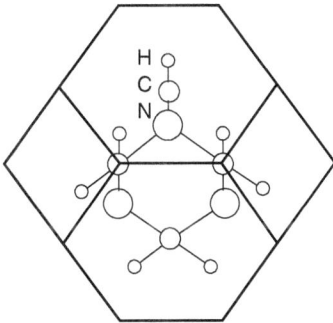

Figure 10.16. The shape of an organic molecule, hexamethylenetetramine. The volume occupied by this molecule in its crystal approximates that of an Archimedean truncated octahedron.

angle is greater than 60°. Of course, the special cases when the angle is 60° or 90° are included in the cubic system. The Dirichlet domain for a hexagonal lattice is the hexagonal prism in Figure 10.17f, but the conventional unit cells are of course parallelopiped.

Something becomes apparent when the Dirichlet domains of various types of lattice are depicted. A three-dimensional point lattice is a set of points generated by translations defined by three noncoplanar vectors. It has six parameters, three each of angles and magnitudes

of the vectors. The variations of these parameters generate an infinite number of lattices, but they fall into the 14 Bravais symmetry types. Thus, each of these six parameters may vary independently, but the six-dimensional space may be divided into 14 regions, within each of which the symmetry of the Dirichlet domains is the same. However, there is still a variation of the shapes of the polyhedra representing the Dirichlet domains in each region.

The partition of a crystal structure into units according to Dirichlet domains is an interesting strategy for their study. However, there are many ways to choose the points from which the Dirichlet domains are derived. For example, the lattice points (which are not unique) could be chosen as the centers of atoms, or the gravitational centers of molecules. A reasonable method should keep all the units for a structure the same shape and size; the packing patterns of these units should apply to perhaps many structures for easy memory. Furthermore, the arrangement of atoms or molecules in this unit takes advantage of the symmetry properties, and all units for a structure have the same symmetry. Keeping these criteria in mind, we may derive the Dirichlet domains from points of the highest symmetry in a structure and call them *geometric units*. I have been concerned about the ways these geometric units

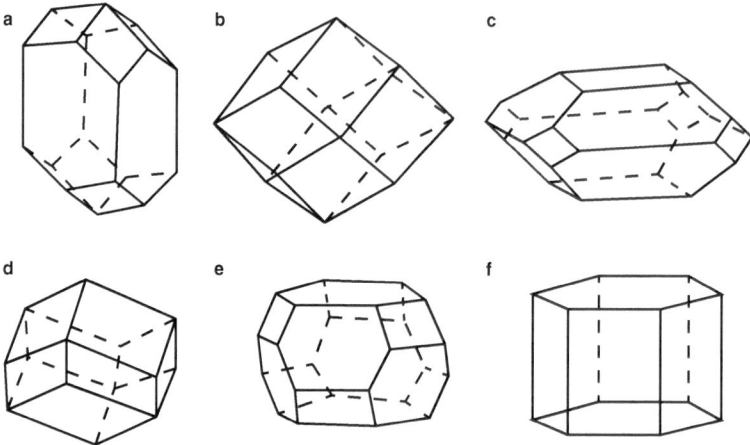

Figure 10.17. Dirichlet domains of tetragonal (**a–c**), rhombohedral (**d** and **e**), and hexagonal (**f**) lattices.

pack, and the possibility of classifying crystal structures using geometric units and their packing patterns. This concern led me to study the space groups. I tried to classify them according to the packing of geometric units, since they represent hypothetical crystal structures. I termed the study of this scheme "geometric properties or geometric plans of space groups." Space groups theoretically classify all crystal structures according to their symmetry, and the study of space groups for their geometric plans is therefore a study of crystal structures for the same.

Using Dirichlet polyhedra derived from sites of highest symmetry of the tetragonal crystal system, we may proceed in a similar way to those of the cubic space groups to work out the geometric plan of the tetragonal space groups. The polyhedra used as units may vary in shape due to the axial ratio of the tetragonal system. The arrangements of units are shown using a plane perpendicular to the $x + y$, and x directions respectively (represented by (110) and (100)) as given in Figure 10.18. There are nine types of arrangement for 68 space groups, and the nine packing patterns are also given in Figure 10.18.

As an illustration, I will choose a series of organometallic compounds, tetraphenyl derivatives of the group IV elements C, Si, Ge, and Pb. The molecules of these compounds belong to the same point group, and they occupy sites of the same symmetry in the space group. The molecules are natural geometric units. The packing of these molecules in solid state is shown in Figure 10.20.

After working on the analysis of spatial arrangements in the tetragonal system, we have been able to apply the knowledge gained for the solution of an interesting tetragonal crystal structure, that of anhydrous zinc bromide. The geometric units can be seen as cubes or large tetrahedra, $Zn_4Br_6Br_{4/2}$ (see Figure 10.19a). The real structure belongs to a space group that is too complicated for theoretical calculation in the analysis of itsinfrared spectrum. At this point, I realized that these units may be packed together in more than one way, as exemplified by illustration in Figure 10.19b, and c, which shows two types of stacking related to two very common crystal structures, ice and diamond. In ice, the tetrahedron is made up of an oxygen atom in the center, and four shared hydrogen atoms at the vertices $OH_{4/2}$, whereas in diamond, each tetrahedron represents a carbon atom, which is connected to four others in a tetrahedral fashion. The ice structure belongs to a hexagonal lattice, but the diamond structure belongs to a cubic one. Let us return to the $ZnBr_2$ structure. By a simple change in the orientation of these large tetrahedral units,

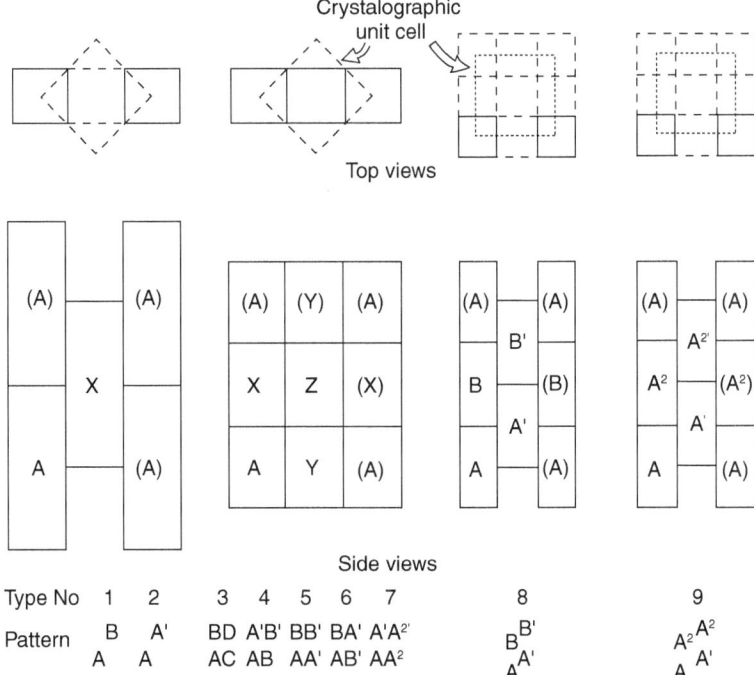

Figure 10.18. Nine packing types of geometric units in the tetragonal crystal system.

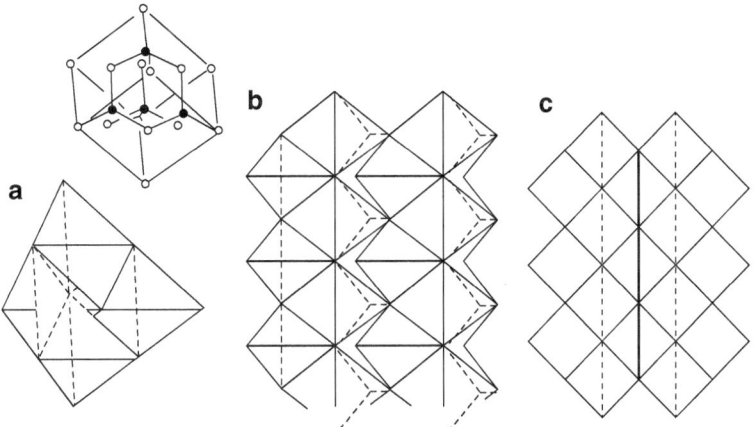

Figure 10.19. The crystal structure of $ZnBr_2$. The geometric unit of this compound consists of a large unit with a formula $Zn_4Br_6Br_{4/2}$ that can be viewed as a large tetrahedron made up of four small ones (**a**). There are many ways to connect tetrahedra, and crystal structures may be represented by these connections: (**b**) ice, (snow), (**c**) diamond.

$Zn_4Br_6Br_{4/2}$, we manage to reduce the complexity of the calculation by using another space group, to which the structure approximates.

Further variations of packing of tetrahedra are indicated of Figure 10.21. The interconnected tetrahedra at three of their vertices form a

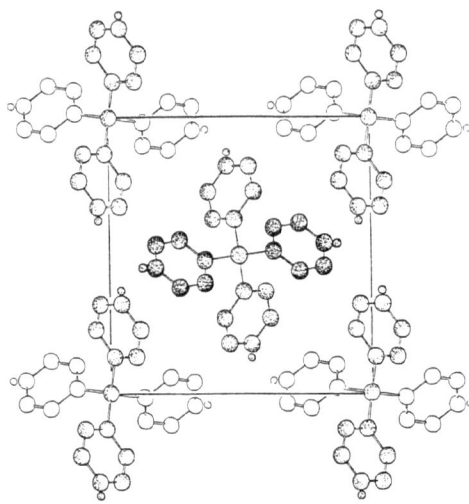

Figure 10.20. Molecular packing of tetraphenyl derivatives of the group IV elements (C, Si, Ge, and Pb). The molecule at the center of the diagram is moved up by a half period in the direction perpendicular to the paper, and because of the orientation difference of this molecule with respect to one at the origin, the structure belongs to a primitive lattice.

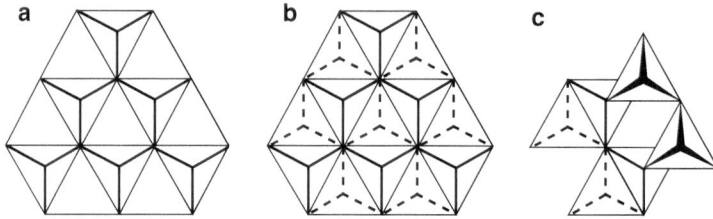

Figure 10.21. Layer packing of interconnected tetrahedra: (**a**) a layer of interconnected tetrahedra, (**b**) a double layer, and (**c**) a triple layer.

layer 10.21a; the fourth vertex and every three-connected point are for interlayer connections. If another layer of the same type is used but turned upside-down, then we have a double layer 10.21b. The ice structure is simply a back-to-back stacking of these double layers on top of each other. However, a third layer may be added to the double layer in a fashion shown in Figure 10.21c. By varying the positions of interlayer connections, we can stack a sequence of any length. This aspect of geometry is richly demonstrated in the natural crystal structures.

The foregoing discussion about crystal structures indicates that I agree with Plato's argument that *all matter is the result of combinations and* *permutations of a few basic (polyhedral) units.* Nowadays, we know a lot more about crystal structures than Plato did. My association of these structures with polyhedra is partly for the ease of recognition and partly for providing a minimum inventory to get the maximum diversity in application. The intricate shapes of crystals stimulate us to study geometry, but geometry is the most important tool for the understanding and systematic classification of crystal structures.

My heartbeat increases whenever I see art by M.C. Escher. I have tried to find out how some of his exciting works were created. Escher has fantasized geometry and symmetry into visually stimulating forms. It can also be said that crystal structures are the artwork of God or nature. I am as curious about the formation of crystal structures, and about the geometric design of those structures, as I am about Escher's art.

In conclusion, I am excited to see so many people enthusiastically making contributions in terms of models, in terms of educational materials, and in terms of teaching me how to understand and appreciate geometric units. Your effort has made it a little easier for me to understand the crystal structures or natural three-dimensional patterns in terms of their geometric plan, something that I wanted to comprehend.

11

Polyhedral Molecular Geometries

Magdolna Hargittai and Istvan Hargittai

H.S.M. Coxeter has said that "the chief reason for studying regular polyhedra is still the same as in the times of the Pythagoreans, namely, that their symmetrical shapes appeal to one's artistic sense." The success of modern molecular chemistry affirms the validity of this statement; there is no doubt that aesthetic appeal has contributed to the rapid development of what could be termed polyhedral chemistry. The chemist Earl Muetterties movingly described his attraction to boron hydride chemistry, comparing it to Escher's devotion to periodic drawings:

> When I retrace my early attraction to boron hydride chemistry, Escher's poetic introspections strike a familiar note. As a student intrigued by early descriptions of the extraordinary hydrides, I had not the prescience to see the future synthesis developments nor did I have then a scientific appreciation of symmetry, symmetry operations, and group theory. Nevertheless, some inner force also seemed to drive me but in the direction of boron hydride chemistry. In my initial synthesis efforts, I was not the master of these molecules; they seemed to have destinies unperturbed by my then amateurish tactics. Later as the developments in polyhedral borane chemistry were evident on the horizon, I found my general outlook changed in a characteristic fashion. For example, my doodling, an inevitable activity of mine during meetings, changed from characters of nondescript form to polyhedra, fused polyhedra, and graphs.

I (and others, my own discoveries were not unique nor were they the first) was profoundly impressed by the ubiquitous character of the three-center relationship in bonding (e.g., the boranes) and non-bonding situations. I found a singular uniformity in geometric relationships throughout organic, inorganic, and organometallic chemistry: The favored geometry in coordination compounds, boron hydrides, and metal clusters is the polyhedron that has all faces equilateral or near equilateral triangles.

Molecular geometry describes the relative positions of atomic nuclei. Although positions may be given by position vectors or coordinates of all nuclei in the molecule, chemists usually give the positions by bond lengths, bond angles, and angles of internal rotation. This second way greatly facilitates the understanding and comparison of various structures. The most qualitative but nonetheless a very important feature of molecular geometry is the shape of the molecule. Polyhedra are especially useful in expressing molecular shapes for molecules with a certain amount of symmetry.

The molecules As_4 and CH_4 both have tetrahedral shapes (Figure 11.1) and T_d symmetry, but there is an important difference in their structures. In As_4 all nuclei are at vertices of a regular tetrahedron and each edge of this tetrahedron is a chemical bond. Methane has a central carbon atom, with four chemical bonds directed from it to vertices of a tetrahedron where the protons are located; no edge is a chemical bond. The As_4 and CH_4 molecules are clear-cut examples of the two distinctly different arrangements. Such distinctions are not always so unambiguous.

M. Hargittai • I. Hargittai
Department of Inorganic and Analytical Chemistry,
Budapest University of Technology and Economics,
POBox 91, 1521 Budapest, Hungary
e-mail: hargittaim@mail.bme.hu;
istvan.hargittai@gmail.com

M. Senechal (ed.), *Shaping Space*, DOI 10.1007/978-0-387-92714-5_11,
© Marjorie Senechal 2013

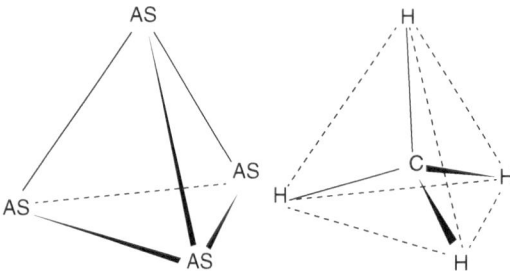

Figure 11.1. The molecular shapes of As$_4$ and CH$_4$.

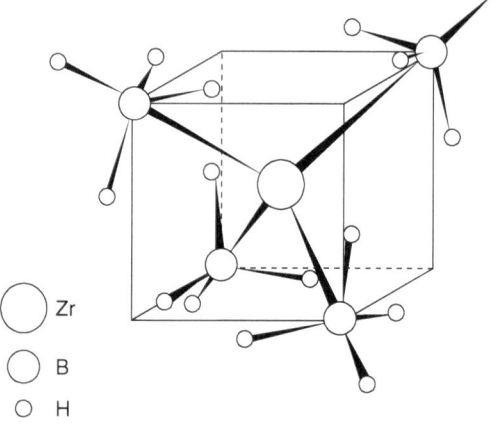

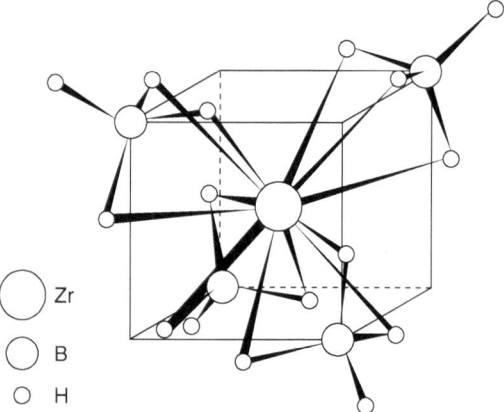

Figure 11.2. The molecular configuration of zirconium borohydride, Zr(BH$_4$)$_4$, in two interpretations but described by the same polyhedral shapes.

An interesting example is zirconium borohydride, Zr(BH$_4$)$_4$. Two independent studies describe its structure by the same polyhedral configuration, but give different interpretations (Figure 11.2) of the bonding between the central zirconium atom and the four boron atoms at the vertices of a regular tetrahedron. In one interpretation, there are four Zr–B bonds; according to the other, each boron atom is linked to zirconium by three hydrogen bridges, and there is no direct Zr–B bond.

Real molecules ceaselessly perform intramolecular vibrations. In even small-amplitude vibrations, nuclear displacements amount to several percent of the internuclear separations; large-amplitude vibrations may permute atomic nuclei in a molecule. In describing a molecule by a highly symmetric polyhedron, we refer to the hypothetical motionless molecule. The importance and consequences of intramolecular motion in the polyhedral description of molecules are discussed in the final section.

Boron Hydride Cages

All faces of boron hydride polyhedra are equilateral or nearly equilateral triangles. Boron hydrides with a complete polyhedral shape are called *closo* boranes (Greek *closo:* 'closed'). One of the most symmetrical and most stable of the polyhedral boranes is the B$_{12}$H$_{12}^{2-}$ ion; its regular icosahedral configuration is shown in Figure 11.3. Table 11.1 presents structural systematics of B$_n$H$_n^{2-}$ *closo* boranes and the

related C$_2$B$_{n-2}$H$_n$ carboranes in which some boron sites are taken by carbon atoms. The so-called quasi-*closo* boranes are derived from the *closo* boranes by replacing a framework atom with a pair of electrons.

Figure 11.4 shows the systematics of borane polyhedral fragments obtained by removing one or more polyhedral sites from *closo* boranes. Since all faces of the polyhedral skeletons are triangular, they are deltahedra. The derived deltahedral fragments are the tetrahedron, trigonal bipyramid, octahedron, pentagonal bipyramid, bisdisphenoid, symmetrically tricapped trigonal prism, bicapped square

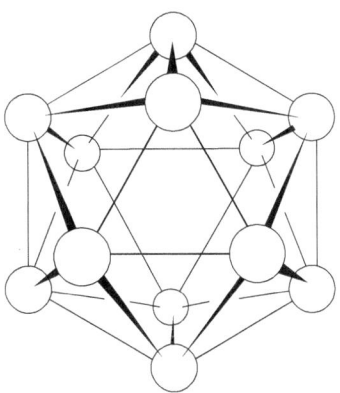

Figure 11.3. The regular icosahedral configuration of the $B_{12}H_{12}^{2-}$ ion. Only the boron skeleton is shown.

Table 11.1. Structural systematics of $B_nH_n^{2-}$ *closo* Boranes and $C_2B_{n-2}H_n$ *closo* Carboranes after Muetterties

Polyhedron and point group	Boranes	Dicarboranes
Tetrahedron, T_d	$(B_4Cl_4)^*$	–
Trigonal bipyramid, D_{3h}	–	$C_2B_3H_5$
Octahedron, O_h	$B_6H_6^{2-}$	$C_2B_4H_6$
Pentagonal bipyramid, D_{5h}	$B_7H_7^{2-}$	$C_2B_5H_7$
Dodecahedron (triangulated), D_{2d}	$B_8H_8^{2-}$	$C_2B_6H_8$
Tricapped trigonal prism, D_{3h}	$B_9H_9^{2-}$	$C_2B_7H_9$
Bicapped square antiprism, D_{4d}	$B_{10}H_{10}^{2-}$	$C_2B_8H_{10}$
Octadecahedron, C_{2v}	$B_{11}H_{11}^{2-}$	$C_2B_9H_{11}$
Icosahedron, I_h	$B_{12}H_{12}^{2-}$	$C_2B_{10}H_{12}$

*No boron hydride

antiprism, octadecahedron, and icosahedron. Only the octadecahedron is not a convex polyhedron.

A *nido* (nestlike) boron hydride is derived from a *closo* borane by removal of one skeleton atom. An *arachno* (weblike) boron hydride is derived from a *closo* borane by removal of two adjacent skeletal atoms. In either case, if the starting *closo* borane is not a regular polyhedron, then the atom removed is the one at a vertex with the highest connectivity. Complete *nido* and *arachno* structures are shown together with starting boranes in Figure 11.5.

Polycyclic Hydrocarbons

Fundamental polyhedral shapes are realized among polycyclic hydrocarbons where the edges are C—C bonds and there is no central atom. Such bond arrangements may be far from the energetically most advantageous, and particular arrangements may be too unstable to exist. Yet the fundamental character of these shapes, their high symmetry, and their aesthetic appeal make them an attractive and challenging playground for organic chemists. These substances also have practical importance as building blocks for such natural products as steroids, alkaloids, vitamins, carbohydrates, and antibiotics.

Tetrahedrane (Figure 11.6a) is the simplest regular polycyclic hydrocarbon. The synthesis of this highly strained molecule may not be possible. Its derivative, tetra-*tert*-butyltetrahedrane (Figure 11.6b), is amazingly stable, perhaps because the substituents help "clasp" the molecule together. Cubane (Figure 11.6c) has been known for some time. Dodecahedrane (Figure 11.6d), prepared more recently, was predicted to have "almost ideal geometry . . . practically a miniature ball bearing!" Its carbanion was predicted to be stabilized by a "rolling charge" effect, delocalizing the extra electron over twenty equivalent carbon atoms.

In the $(CH)_n$ convex polyhedral hydrocarbons, each carbon atom is bonded to three other carbon atoms; the fourth bond is directed externally to a hydrogen atom. Around the all-carbon polyhedron is thus a similar polyhedron whose vertices are protons. The edges of the inner polyhedron are C—C bonds. Because four bonds would meet a carbon atom at the vertices of an octahedron, and five in an icosahedron, the enveloping-polyhedra structure is not possible for these Platonic solids. For similar reasons, only seven of the 14 Archimedean polyhedra can be considered in the $(CH)_n$ polyhedral series.

Cubane may also be described as tetraprismane, composed of eight identical methine units arranged at the corners of a regular tetragonal prism with O_h symmetry, bound into two parallel four-membered rings cojoined by four four-

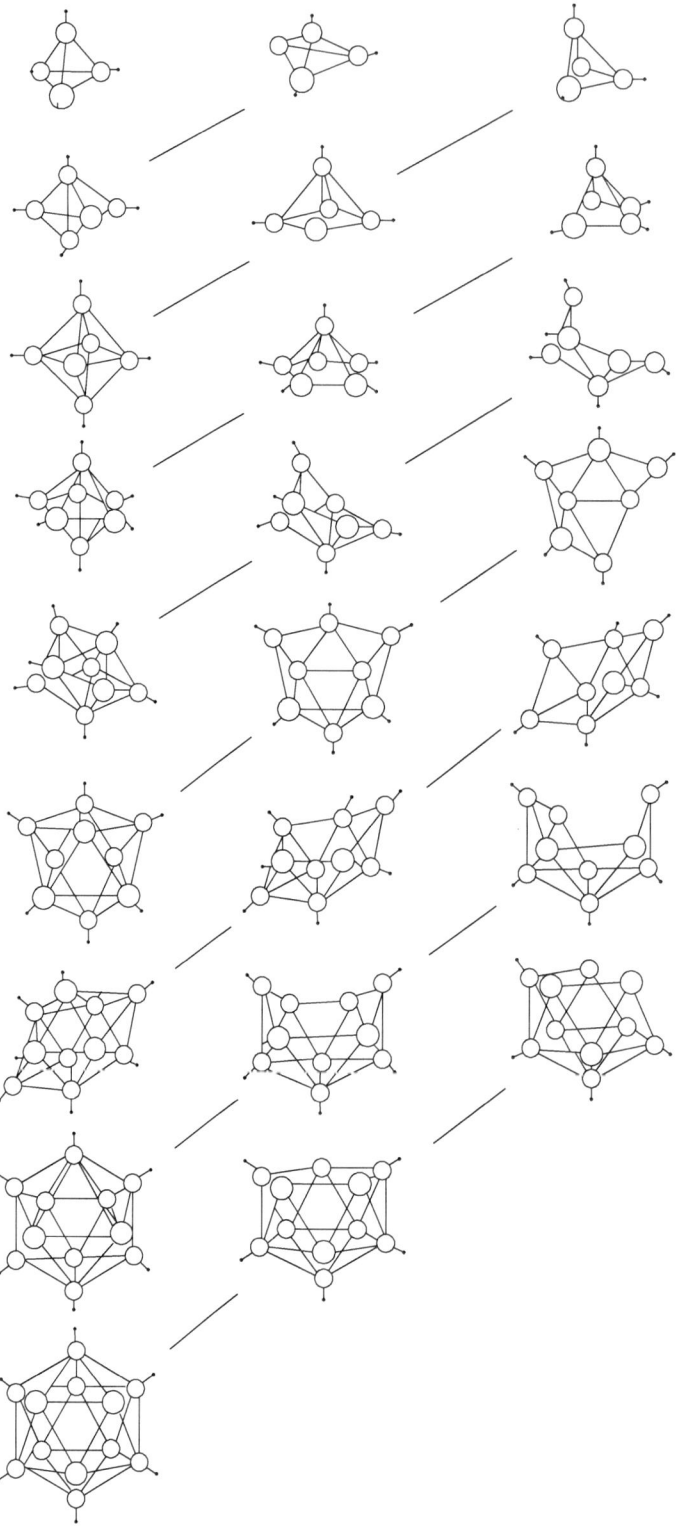

Figure 11.4. Closo, nido, and arachno boranes. The genetic relationships are indicated by diagonal lines.

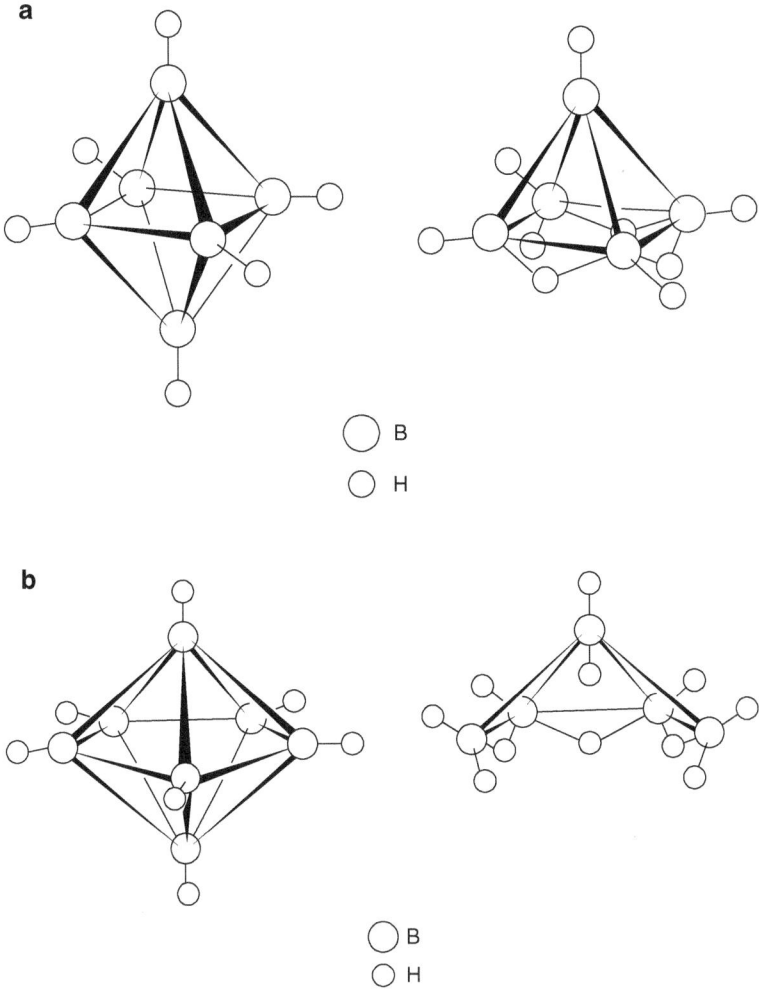

Figure 11.5. Examples of *closo/nido* and *closo/arachno* structural relationships. (**a**) *Closo*-$B_6H_6^{2-}$ and *nido*-B_5H_9. (**b**) *Closo*-$B_7H_7^{2-}$ and *arachno*-B_5H_{11}.

membered rings. Triprismane, $(CH)_6$, has D_{3h} symmetry and pentaprismane, $(CH)_{10}$, has D_{5h} symmetry. The quest for pentaprismane is a long story. Hexaprismane, $(CH)_{12}$, the face-to-face dimer of benzene, has yet to be prepared. The molecular models are shown in Figure 11.7.

Structural varieties become virtually endless if one reaches beyond the most symmetrical convex polyhedra. There are 5,291 isomeric tetracyclic structures of $C_{12}H_{18}$ hydrocarbons; only a few are stable. One is iceane (Figure 11.8), which may be visualized as two chair cyclohexanes connected by three axial bonds; it can also be described as three fused boat cyclohexanes. The

name *iceane* was proposed by Fieser almost a decade before its preparation. Considering water molecules in an ice crystal, he noticed three *vertical* hexagons with boat conformations. The emerging *horizontal* $(H_2O)_6$ units possess three equatorial hydrogen atoms and three equatorial hydrogen bonds available for horizontal building. He noted that this structure

suggests the possible existence of a hydrocarbon of analogous conformation of the formula $C_{12}H_{18}$, which might be named "iceane." The model indicates a stable strain-free structure analogous to adamantane and twistane. "Iceane" thus presents a challenging target for synthesis.

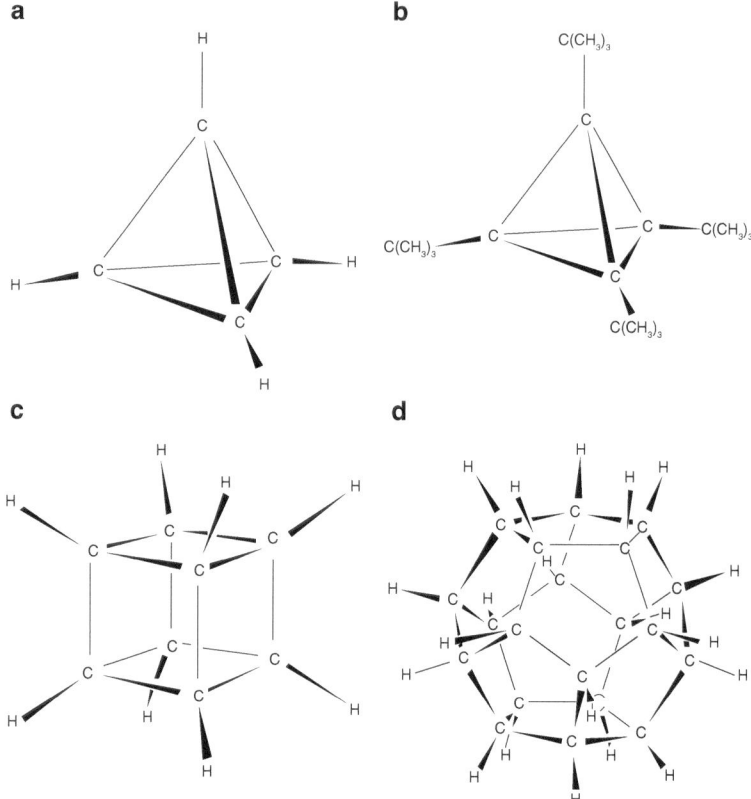

Figure 11.6. (**a**) Tetrahedrane, $(CH)_4$. It has very high strain energy and has not (yet?) been prepared. (**b**) Tetra-*tert*-butyltetrahedrane, $\{C[C(CH_3)_3]\}_4$. (**c**) Cubane, $(CH)_8$. (**d**) Dodecahedrane, $(CH)_{20}$.

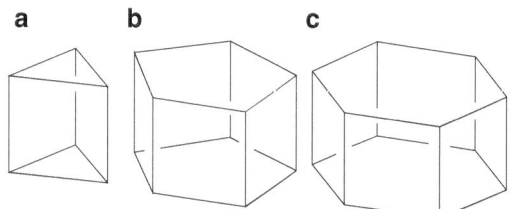

Figure 11.7. (**a**) Triprismane, $(CH)_6$. (**b**) Pentaprismane, $(CH)_{10}$. (**c**) Hexaprismane, $(CH)_{12}$, not yet prepared.

The challenge was met.

The adamantane molecule, $C_{10}H_{16}$, and the diamond crystal are closely related. Diamond has even been called the "infinite adamantylogue of adamantane." The high symmetry of adamantane is emphasized when its structure is described by four imaginary cubes packed one inside the other; two are shown in Figure 11.9.

Structures with a Central Atom

Tetrahedral AX_4 molecules belong to the point group T_d. Successive substitution of the X ligands by B ligands leads to other tetrahedral configurations of the following symmetries:

$$AX_4 \quad AX_3B \quad AX_2B_2 \quad AXB_3 \quad AB_4$$
$$T_d \quad C_{3v} \quad C_{2v} \quad C_{3v} \quad T_d$$

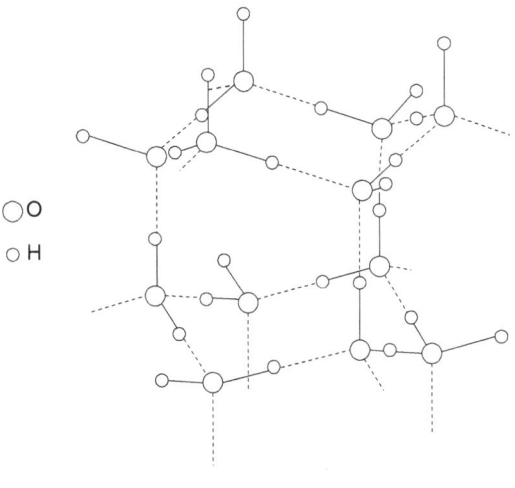

○ O
○ H

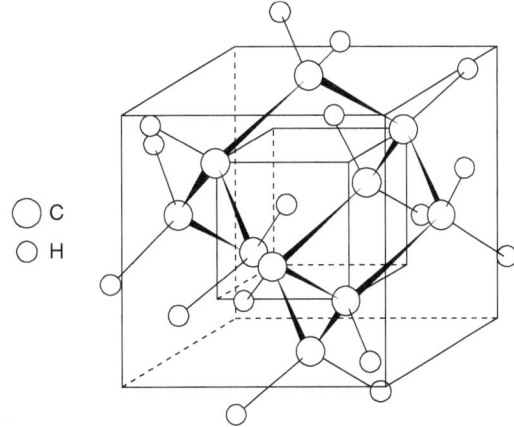

○ C
○ H

Figure 11.9. Adamantane, $C_{10}H_{16}$ or $(CH)_4(CH_2)_6$.

$$AX_4 \quad AX_3B \quad AX_2BC \quad AXBCD$$
$$T_d \quad C_{3v} \quad C_s \quad C_1$$

Important structures may be derived by joining two tetrahedra or two octahedra at a common vertex, edge, or face (Figure 11.10). Ethane ($H_3C - CH_3$), ethene ($H_2C = CH_2$) and acetylene ($HC \equiv CH$) may be derived formally in such a way. Joined tetrahedra are even more obvious in some metal halide structures with halogen bridges.

Complex formation has similar consequences in molecular shape and symmetry. The $H_3N \cdot AlCl_3$ donor–acceptor complex has a triangular antiprismatic shape with C_{3v} symmetry (Figure 11.11). Complex formation can be viewed as completion of the tetrahedral

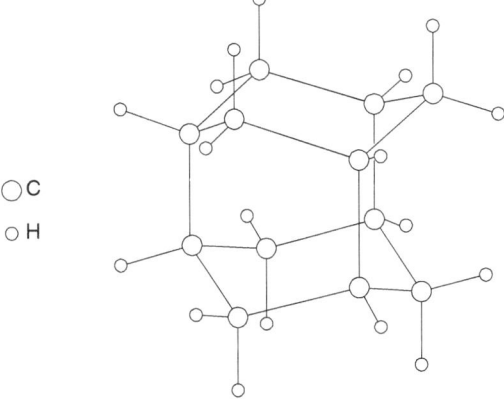

○ C
○ H

Figure 11.8. Ice crystal structure (*top*) and the iceane hyrdocarbon, $C_{12}H_{18}$ (*bottom*).

If each substitution introduces a new kind of ligand, then the resulting tetrahedral configurations will have the following symmetries:

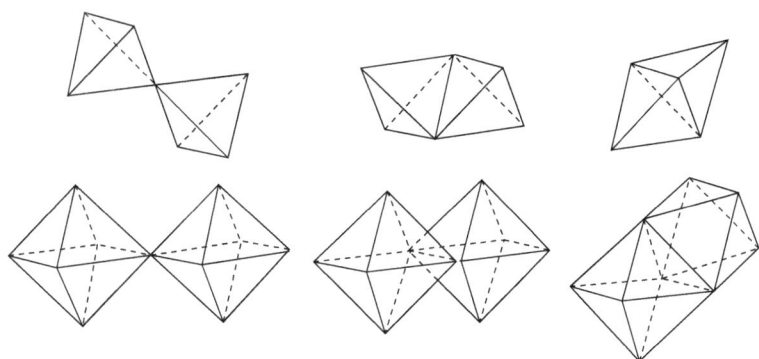

Figure 11.10. Joining two tetrahedra (and two octahedra) at a common vertex, edge, or face.

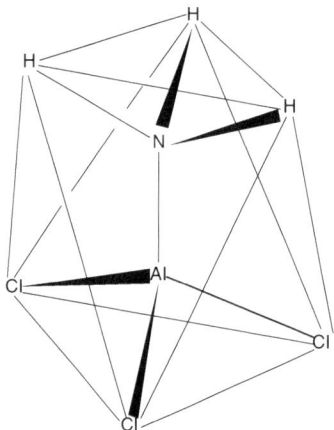

Figure 11.11. The triangular antiprismatic shape of the $H_3N \cdot AlCl_3$ donor–acceptor complex.

bond configuration around the central atoms of the donor (NH_3, C_{3v}) and the acceptor ($AlCl_3, D_{3h}$).

The structure of the mixed metal–halogen complex potassium tetrafluoroaluminate, $KAlF_4$, can be viewed as formed from KF and AlF_3, with completion of the aluminum tetrahedron. The tetrahedral tetrafluoroaluminate structural unit is relatively rigid, whereas the position of the potassium atom around the AlF_4 tetrahedron is rather loose. The most plausible models are shown in Figure 11.12; the model with two halogen bridges best approximates the experimental data. The $KAlF_4$ molecule is representative of a class of compounds of growing practical importance: the mixed halides have much higher volatility than the individual metal halides.

The prismatic cyclopentadienyl and benzene complexes of transition metals are reminiscent of the prismanes. Figure 11.13a shows ferrocene, $(C_5H_5)_2Fe$, for which both the barrier to rotation and free-energy difference between the prismatic (eclipsed) and antiprismatic (staggered) conformations are very small. Figure 11.13b presents a prismatic model with D_{6h} symmetry for dibenzene chromium, $(C_6H_6)_2Cr$. Molecules with multiple bonds between metal atoms often have structures with beautiful and highly symmetric polyhedral shapes. One example is the

square prismatic $[Re_2Cl_8]^{2-}$ ion which played an important role in the discovery of metal-metal multiple bonds (Figure 11.14). Another is the paddlelike structure of dimolybdenum tetra-acetate, $Mo_2(O_2CCH_3)_4$ (Figure 11.15).

There are hydrocarbons called *paddlanes* for their similarity to the shape of riverboat paddles. The most symmetrical, highly strained [2.2.2.2.]-paddlane (Figure 11.16a) has not yet been prepared. The most unusual parent hydrocarbon known is the related [1.1.1]-propellane (Figure 11.16b) in which interactions between bridgehead carbons have been interpreted by three-center, two-electron orbitals. The hydrocarbon skeleton seems to be electron deficient, while extra electron density is on the outside of the skeleton.

Regularities in Nonbonded Distances

There is no chemical bond between bridgehead carbons of [1.1.1]-propellane, even though the atoms are in a pseudobonding situation with proper bonding geometry. A reverse situation is seen in the ONF_3 molecule (Figure 11.17), an essentially regular tetrahedron formed by three fluorines and one oxygen, each bonded to the central nitrogen atom. The nonbonded $F \cdots F$ and $F \cdots O$ distances are equal within experimental error.

Certain intramolecular 1,3 separations (the 1,3 label referring to two atoms each bonded to a third) are constant throughout a series of related molecules. The 1,3 distance may remain constant even though bond distances and bond angles in the rest of the molecule change considerably. A controversy between two structure determinations of tetrafluoro-1,3-dithietane

$$F_2C \underset{S}{\overset{S}{\diamondsuit}} CF_2$$

was settled by considering the $F \cdots F$ nonbonded distances. The mean of the $F \cdots F$ 1,3 distances

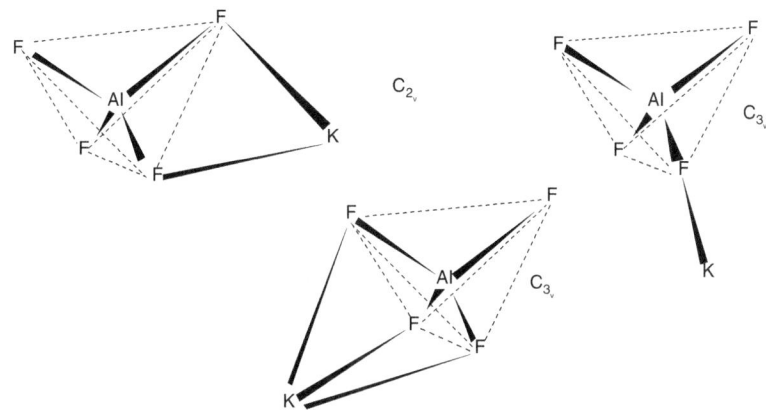

Figure 11.12. Alternative models of the KAlF$_4$ molecule.

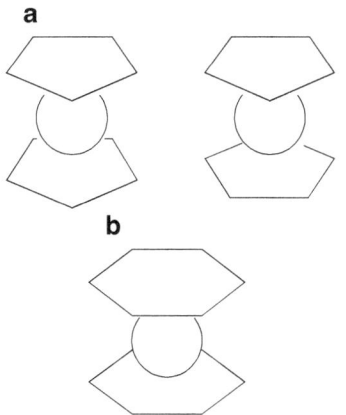

Figure 11.13. (**a**) Prismatic (D_{5h}) and antiprismatic (D_{5d}) models of ferrocene. (**b**) Prismatic model (D_{6h}) of dibenzene chromium.

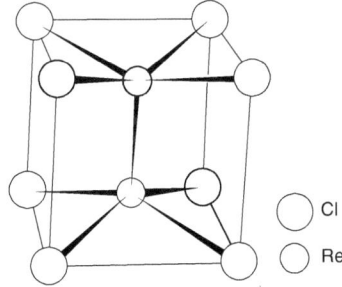

Figure 11.14. The square prismatic model of the $[\text{Re}_2\text{Cl}_8]^{2-}$ ion.

in 40 molecules containing a CF$_3$ group was found to be 2.162 Å, with a standard deviation of 0.008 Å!

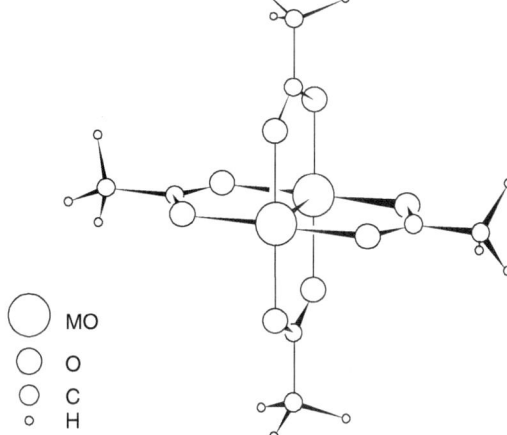

Figure 11.15. The paddlelike structure of the anhydrous dimolybdenum tetra-acetate, $\text{Mo}_2(\text{O}_2\text{CCH}_3)_4$.

The $\text{O} \cdots \text{O}$ nonbonded distances in XSO$_2$Y sulfones are remarkably constant at 2.48 Å in a series of compounds in which the S=O bond lengths vary by 0.05 Å and the O=S=O bond angles by 5°. Geometric variations in the sulfone series can be visualized (Figure 11.18a) as if the oxygen ligands were firmly attached to two vertices of the ligand tetrahedron around the sulfur atom, and this central atom were moving along the bisector of the O=S=O angle, depending on the X and Y ligands.

The oxygen atoms in a sulfuric acid molecule, (HO)SO$_2$(OH), form a nearly regular tetrahedron around the sulfur (Figure 11.18b). The largest difa

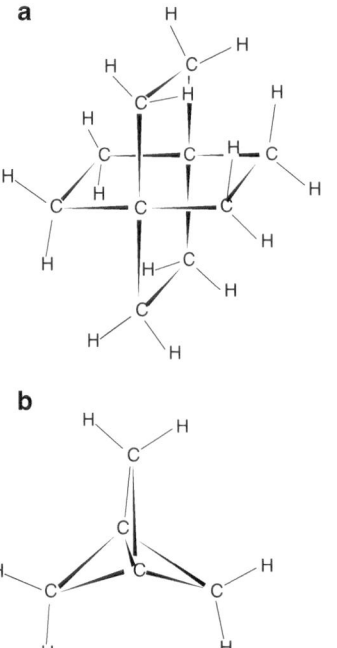

b

Figure 11.16. (**a**) [2.2.2.2]-paddlane, $C_{10}H_{16}$, not yet prepared. (**b**) [1.1.1]-propellane, C_5H_6.

a

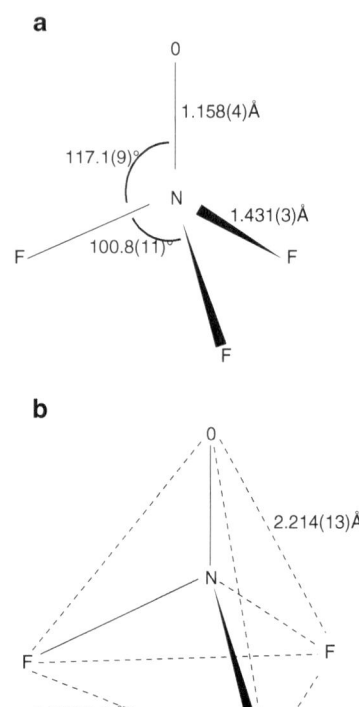

b

Figure 11.17. The molecular geometry of ONF_3. (**a**) Bond lengths and bond angles. (**b**) Nonbonded distances.

ference between the $O \cdots O$ distances is 0.07 Å, even though the SO bond distances differ by 0.15 Å, and the OSO bond angles by 20°. Structures of alkali sulfate molecules ere written in old textbooks as

$$Na - O \overset{\displaystyle O}{\underset{\displaystyle O}{\diagdown \, S \, \diagup}} \\ Na - O \diagup \quad \diagdown O$$

In fact, the SO_4 groups in such molecules are nearly regular tetrahedral, and the metal atoms are located on axes perpendicular to the edges of the tetrahedral; this structure is bicyclic (Figure 11.18c). Sulfate and tetrafluoroaluminate structures are markedly similar; each has a well-defined tetrahedral nucleus around which atoms occupy relatively loose positions.

The VSEPR Model

Why is a methane molecule tetrahedral, whereas xenon tetrafluoride is planar? Why is ammonia pyramidal rather than planar? Why is water bent, rather than linear?

A simple and successful model, designed to answer just such questions about molecules with a central atom, is based on the following postulate: *The geometry of the molecule is determined by the repulsions among the electron pairs in the valence shell of its central atom.* We shall illustrate the utility of this valence shell electron pair repulsion (VSEPR) model, showing the importance of the polyhedral description of molecular structure.

If the electron distribution around a central atom has spherical symmetry, then all the electron pairs in its valence shell will be equidistant from the nucleus. Distances among the electron pairs will be maximized in the following arrangements:

The arrangements are shown in Figure 11.19 where electron pairs are represented by points on a sphere. For four or more electron pairs, the arrangements are polyhedral. Of the three

a

b

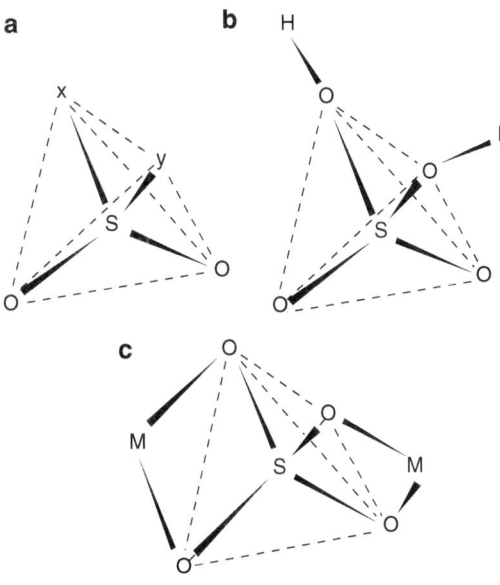

c

Figure 11.18. The molecular geometry of (**a**) sulfones, XSO_2Y, (**b**) sulfuric acid, H_2SO_4, and (**c**) metal sulfates, M_2SO_4.

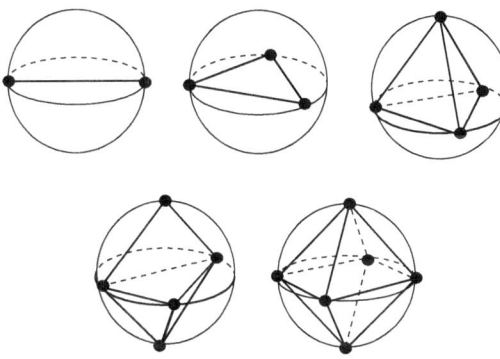

Figure 11.19. Points-on-a-sphere configurations.

Figure 11.20. Shapes of groups of balloons.

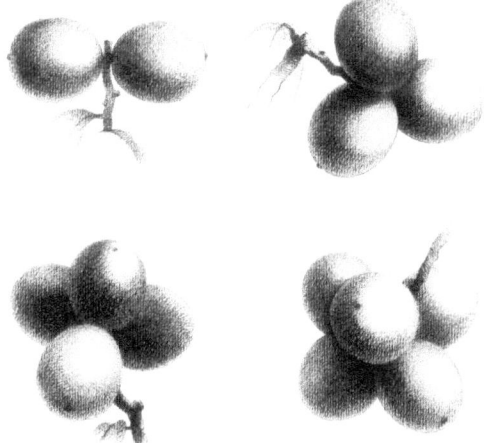

Figure 11.21. Walnut clusters drawn by the artist Ferenc Lantos.

polyhedra in Figure 11.19, two are regular. The trigonal bipyramid is not a strongly unique solution; the square pyramidal configuration for five electron pairs is only slightly less advantageous. The space requirement and mutual repulsion of electron pairs are nicely simulated by balloons; a natural simulation is provided by nut clusters on walnut trees (see Figures 11.20 and 11.21).

To predict the bond configurations around a central atom, the number of valence shell electron pairs must be known. The formula of a binary compound AX_n may be given as AX_nE_m, E denoting a lone pair of electrons. For methane (CH_4 or CH_4E_0), ammonia (NH_3 or NH_3E_1), water (OH_2 or OH_2E_2) and hydrogen fluoride (FH or FH_1E_3), $n + m = 4$; accordingly, each has a tetrahedral electron-pair configuration. The corresponding nuclear configurations are tetrahedral, pyramidal, bent, and linear. The ideal tetrahedral bond angles of $109°28'$ occur only when all electron pairs are equivalent. A double bond or a lone electron pair requires more space than a single bond, repelling neighboring electron pairs more strongly. A bond to a strongly electronegative ligand such as fluorine has less electron density and repels electron pairs more weakly than a bond to a less electronegative ligand such as hydrogen.

Do differences in electron-pair repulsions influence the symmetry of a molecule? The AX_4, AX_3E, and AX_2E_2 molecules have T_d, C_{3v} and C_{2v} symmetries, regardless of the ligands. For trigonal bipyramidal systems where $n + m = 5$, however, the nature of the ligands may be decisive in determining the symmetry. The axial and equatorial positions in the D_{3h} trigonal bipyra-

AX$_5$

F
|
F — P$_{F}^{F}$
|
F
D_{3h}

AX$_4$E

F
|
⊂S$_{F}^{F}$
|
F
C_{2v}

AX$_3$E$_2$

F
|
⊂Cl$_{F}$
|
F
C_{2v}

AX$_2$E$_3$

F
|
⊂Xe
|
F
$D_{\infty h}$

F
|
O=S$_{F}^{F}$
|
F
C_{2v}

F
|
⊂Cl$_{F}^{O}$
|
F
C_s

F
|
O=Xe$_{O}^{O}$
|
F
D_{3h}

F
|
⊂Xe$_{O}^{O}$
|
F
C_{2v}

Figure 11.22. Molecules with trigonal bipyramidal and related configurations.

midal configuration are not equivalent. While the PF$_5$ molecule as an AX$_5$E$_0$ system has D_{3h} symmetry, it is not trivial to predict the symmetry of the SF$_4$ molecule as an AX$_4$E$_1$ system. The question is: In which of the two possible positions will the lone electron pair occur? The lone pair has the larger space requirement, and the equatorial position is more spacious than the axial; thus the lone-pair position is equatorial, and the SF$_4$ structure has C_{2v} symmetry. For the same reason, lone pairs are equatorial in ClF$_3$(AX$_3$E$_2$) and XeF$_2$(AX$_2$E$_3$). Double bonds require more space than single bonds, and behave in the VSEPR model similarly to lone pairs (Figure 11.22).

Consider octahedral arrangements in which a central atom has six electron pairs in its valence shell. The symmetry is unambiguously O_h for AX$_6$; an example is SF$_6$. The IF$_5$ molecule (AX$_5$E$_1$) is a tetragonal pyramid; the electron pair may be at any of the six equivalent sites. When there are two lone pairs, they occupy positions maximally distant; thus XeF$_4$(AX$_4$E$_2$) is square planar, D_{4h} (Figure 11.23). Difficulties encountered with five-electron-pair valence shells are intensified for the case of seven electron pairs. Seven vertices cannot describe a regular

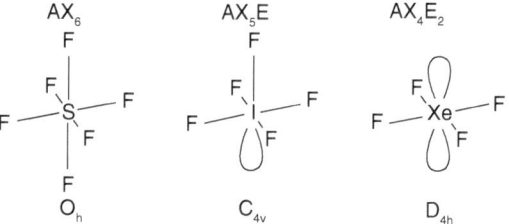

Figure 11.23. Molecules with octahedral and related configurations.

polyhedron; the number of nonisomorphic polyhedra with seven vertices is large, but no one is relatively very stable. One of the early successes of the VSEPR model was that it correctly predicted the nonoctahedral structure of XeF$_6$, as it is indeed a seven-electron-pair case (AX$_6$E$_1$).

Complete geometrical characterization of the valence shell configuration for a molecule with more than one lone pair requires more than specification of the bond angles. Sometimes, though by far not always, the angles made by lone pairs may be attainable from experimental bond angles. For example, the E—P—F angle of PF$_3$ can be calculated from the F—P—F angle by

Table 11.2. Arrangement of electron pairs

Number of Electron Pairs	Arrangement
2	Linear
3	Equilateral triangle
4	Regular tetrahedron
5	Trigonal bipyramid
6	Octahedron

Table 11.3. Calculated angles in a series of tetrahedral molecules

	AF_4E_0	AF_3E_1	AF_2E_2	AFE_3	AF_0E_4
	SiF_4	PF_3	SF_2	ClF	Ar
FAF	109.5°*	96.9°	98.1°	–	–
FAE	–	120.2°	104.3°	101.6°	–
EAE	–	–	135.8°	116.1°	109.5°

*By virtue of T_d symmetry

virtue of the C_{3v} symmetry. On the other hand, the E—S—E and E—S—F angles of the C_{2v} symmetry SF_2 molecule cannot be calculated from the F—S—F bond angle.

Even when angles between lone pairs can be calculated from experimental data or deduced from quantum mechanical calculations, they are often ignored. Proper application of the VSEPR model should direct at least as much attention to angles of lone pairs and their variations as to bond angles. As an example of the consistency of variations in all angles of a series of tetrahedral molecules, Table 11.3 presents a set of results from quantum chemical calculations.

Consequences of Intramolecular Motion

Imagine watching the dynamic dance shown in Matisse's famous painting *Dance*. As choreographed, one dancer jumps out of the plane of the other four. As soon as this dancer returns into the plane of the others, it is the role of the next to jump, and so on. The exchange of roles from one dancer to another throughout the five-membered troupe is so quick that a photograph with slow shutter speed would give a blurred picture; only a short exposure can identify a well-defined configuration of dancers. Matisse's *Dance* simulates the *pseudorotation* of the cyclopentane molecule, $(CH_2)_5$, which has a special degree of freedom in which the out-of-plane carbon atom exchanges roles with one of the four in-plane atoms. The process is equivalent to a permutation of two carbon atoms (and their hydrogen ligands), and is also equivalent to a rotation by $2\pi/5$ about the axis perpendicular to the plane of the four coplanar carbons.

It is an extreme approach to disregard intramolecular motion. The motionless state, although hypothetical, is well defined: it is the equilibrium structure, the structure with minimum potential energy, a structure that emerges from quantum chemical calculations. Yet real molecules are never motionless, and experimentalists study real molecules. As with Matisse's *Dance*, the relationship between the lifetime of a configuration and the time scale of the investigating technique has crucial importance.

Large-amplitude, low-frequency intramolecular vibrations may lower the molecular symmetry of the average structure versus the higher symmetry equilibrium structure. Some examples from metal halide molecules are shown in Figure 11.24. Today, it is already possible to compute the equilibrium structures with high precision. For these molecules the lower-symmetry experimentally determined structures occur because of averaging over all molecular vibrations. No permutations of the nuclei are involved in these intramolecular vibrations on the time scale considered.

Rapid interconversion of nuclei takes place in a molecule of bullvalene, $(CH)_{10}$, under very mild conditions (Figure 11.25). Bonds are made and broken, but nuclei shift only slightly. Four different kinds of carbon positions interconvert simultaneously. Hypostrophene (Greek *hypostrophe:* 'turning about', 'recurrence') is a $(CH)_{10}$ hydrocarbon whose trivial name was chosen to reflect its behavior. The molecule ceaselessly undergoes intramolecular rearrangements indicated in Figure 11.26. The atoms have a complete time-average equivalence, yet hypostrophene could not be turned into pentaprismane.

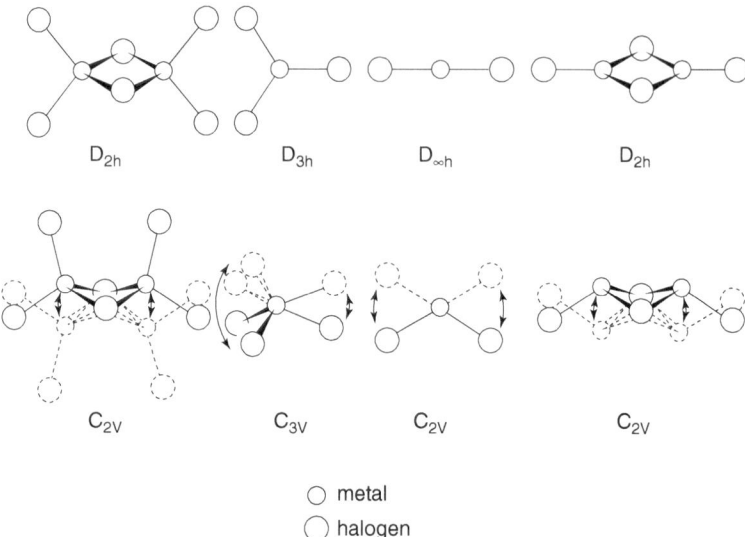

Figure 11.24. Equilibrium versus average structures of metal halide molecules with low-frequency, large-amplitude deformation vibrations.

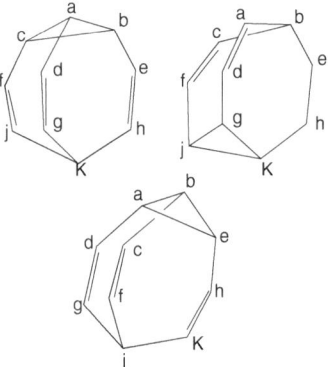

Figure 11.25. Interconversion of nuclear positions in bullvalene.

Permutational isomerism among inorganic trigonal bipyramidal structures was discovered by Berry. Although the D_{3h} trigonal bipyramid and the C_{4v} tetragonal pyramid have very different symmetries, they are easily interconverted by bending vibrations (Figure 11.27). Permutations in an AX_5 molecule (e.g., PF_5) are easy to visualize as the two axial ligands replacing two equatorial ones, while the third equatorial ligand becomes axial in the transitional tetragonal pyramidal structure. Rearrangements quickly follow one another, with no position being unique. The C_{4v} form originates from a D_{3h} structure and yields another D_{3h} form.

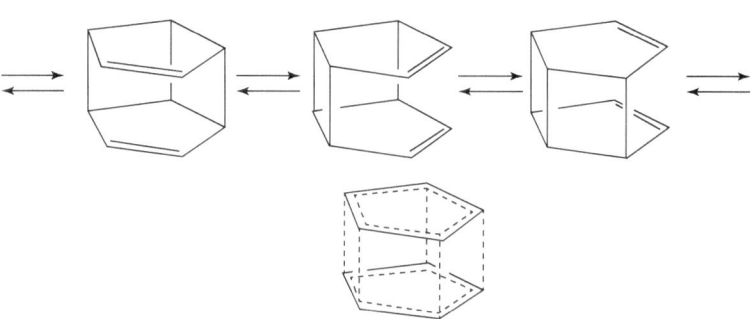

Figure 11.26. Interconversion of nuclear positions in hypostrophene.

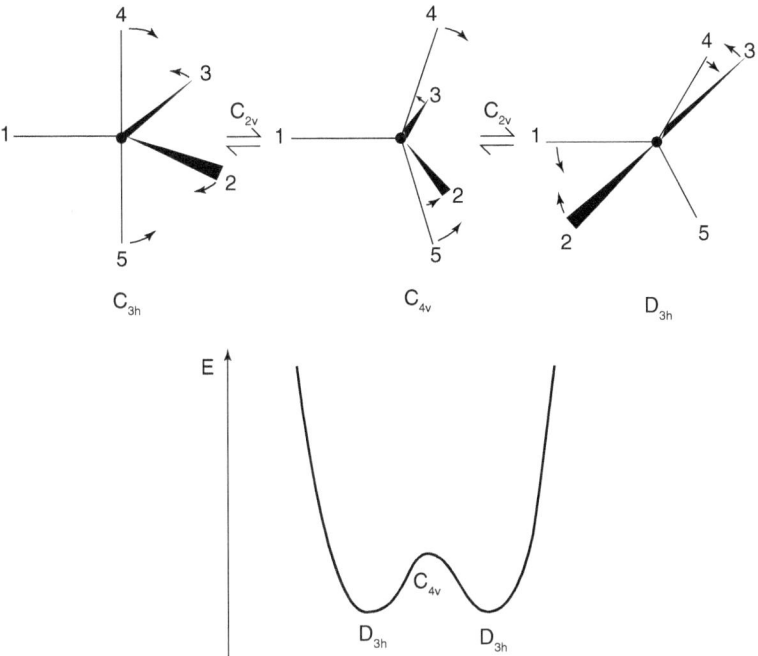

Figure 11.27. Berry-pseudorotation of PF$_5$-type molecules.

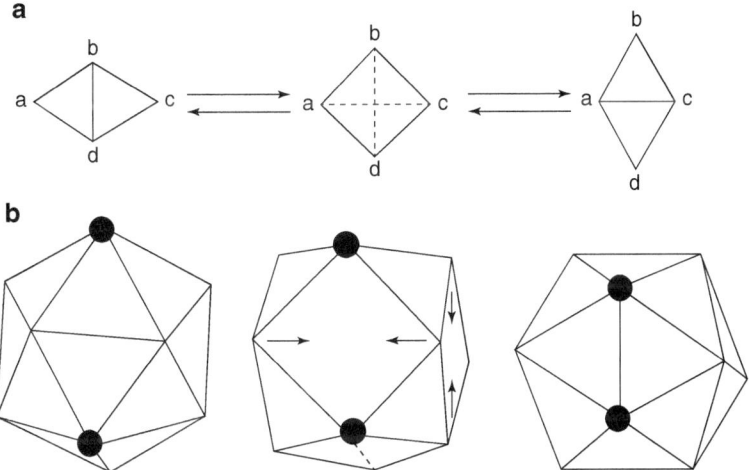

Figure 11.28. (a) The Lipscomb model of the rearrangements in polyhedral boranes, and (b) an example of icosahedron/cuboctahedron/icosahedron rearrangement.

A similar pathway was established for the $(CH_3)_2NPF_4$ molecule in which the dimethyl group is permanently locked into an equatorial position whereas the fluorines exchange in pairs all the time. The PF$_5$ rearrangement also well describes the permutation of nuclei in five-atom polyhedral boranes. In one mechanism for re-arrangements of polyhedral boranes, two common triangular faces are stretched to a square face. This intermediate may revert to the original polyhedron with no net change, or may turn into a structure isomeric with the original (Figure 11.28). This mechanism is illustrated by interconversion of the *ortho* and *meta* isomers of

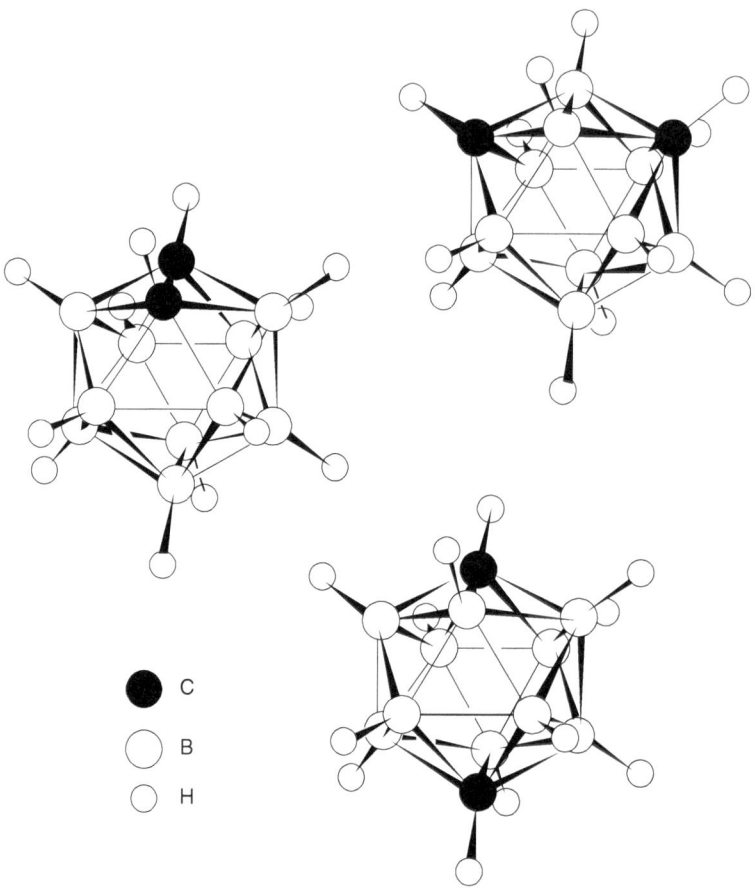

C

B

H

Figure 11.29. Ortho-, meta-, and para-dicarba-*closo*-dodecaboranes. Whereas the ortho isomer easily transforms into the meta, the para isomer is obtained only under more drastic conditions and only in small amounts.

dicarba-*closo*-dodecaborane (Figure 11.29); the *para* isomer is obtained under more drastic conditions and only in small amounts. A similar model has been proposed for carbonyl-scrambling in $Co_4(CO)_{12}$, $Rh_4(CO)_{12}$, and $Ir_4(CO)_{12}$.

Rapid interconversion among different modes of carbonyl coordination is possible, even in the solid state, in transition-metal carbonyl molecules of the form $M_m(CO)_n$. The usually small m-atom metal cluster polyhedron is enveloped by another polyhedron whose vertices are occupied by carbonyl oxygens. An attractive example is $[Co_6(CO)_{14}]^{4-}$ in which the octahedral metal cluster has six terminal and eight triply bridging carbonyl groups (Figure 11.30a). This structure may also be represented by an omnicapped cube enveloping an octahedron (Figure 11.30b). These

models are a reminder of another model in which polyhedra envelop other polyhedra; that model is Kepler's planetary system (Figure 3.6).

Post Script 20 Years Later

Shortly after the Shaping Space Conference at Smith College in 1984, the beautiful molecule of truncated icosahedral shape, buckminsterfullerene, C_{60}, was discovered. First it was merely observed in mass spectra in 1984; the structure was assigned to it in 1985, and in 1990 the substance itself was produced for the first time—enabling chemists to carry out all kinds of experiments on it. All fullerenes are Goldberg

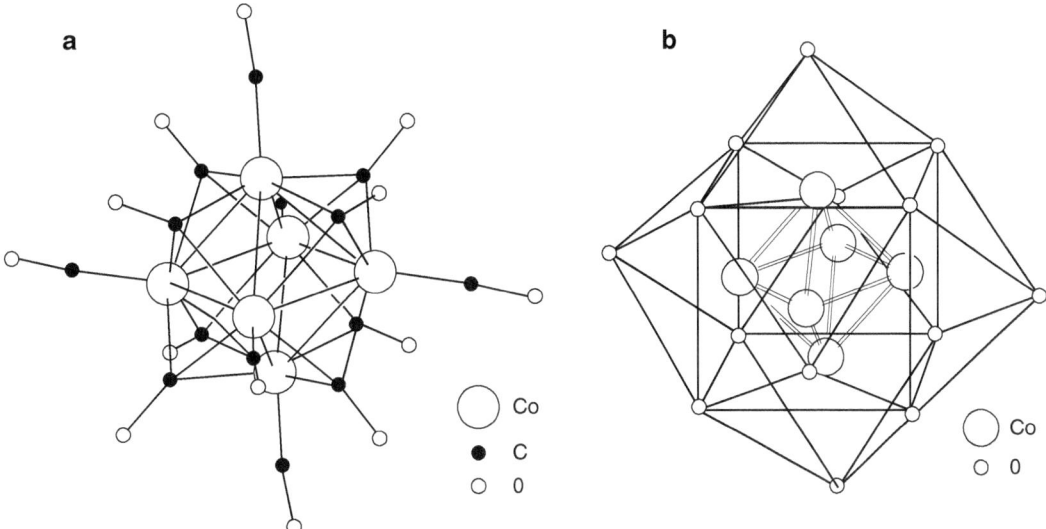

Figure 11.30. The structure of $[Co_6(CO)_{14}]^{4-}$ in two representations. (**a**) The octahedron of the cobalt cluster possesses six terminal and eight triply bridging carbonyl groups. (**b**) An omnicapped cube of the carbonyl oxygens envelopes the cobalt octahedron.

polyhedra; see Chapter 9 for illustrations and details. Lending Buckminster Fuller's name to this new family of molecules invoked a connection between molecular structure and design science. In fact, the assignment of the truncated icosahedral structure to C_{60} was facilitated by a visit of some of the discoverers to the U.S. Pavilion at the Montreal Expo in 1967—that geodesic dome was built following R. Buckminster Fuller's original ideas. Figure 11.31 shows a close up of the Montreal Geodesic Dome, a decoration at the Topkapi Sarayi in Istanbul, and the sphere under the paw of a dragon sculpture in Beijing—all are characterized by containing pentagons among the more numerous hexagons. The discovery of buckminsterfullerene gave birth to fullerene science, which in turn enriched the emerging nano-science and nano-technology.

Figure 11.31. (**a**) Close up of the Montreal Geodesic Dome—the *arrow* points to one of the pentagons among the hexagons; (**b**) decoration above the entrance in the Topkapi Sarayi, Istanbul; (**c**) decorated sphere under the paw of the gold-plated dragon sculpture in front of the Gate of Heavenly Purity in the Forbidden City, Beijing (Photographs by the authors).

12

Form, Function, and Functioning

George Fleck

Polyhedra are objects worthy of study and admiration in their own right. They have been inspirations for mathematicians, artists, and architects, and have also served as models for abstract notions about the biological and physical world. The sophistication of such modeling has evolved over the centuries, influencing both physical and mathematical theories. In studying polyhedra we see over and over again ways in which theory is inspired by nature, and ways in which science is inspired by theory.

The Polyhedron Kingdom lies within the realm of mathematics, and polyhedron theory deals with precise ways of talking about polyhedra, ways which seem comfortable to mathematicians (see Part III). Some of polyhedron theory treats properties of space. Much of polyhedron theory has developed within the minds of mathematicians as old problems have suggested new ones. This theory sometimes appears connected only tenuously with the natural world.

Science and engineering are also fields of investigation in which abstract theory is formulated within people's minds, but the connections with the real world seem—to many non-mathematicians at any rate—both more necessary and more various than in mathematics.

We shall look at some contemporary investigations in botany, microbiology, robotics, and chemistry in which polyhedra play central roles, investigations which illustrate ways that the geometry formulated by mathematicians is related to the geometry used by scientists and engineers. This symbiotic relationship is dynamic and mutually enriching.

Does Form Explain Function? Science Looks to Geometry for Models

Geometry has been considered a fundamental source of insight into the nature of the universe since the time of Pythagoras. We shall note how geometric ideas, and polyhedra especially, have been employed by some of the most creative and influential contributors to the development of natural philosophy (as natural science used to be called) and the contemporary sciences, providing models for atoms, viruses, robots ... even the solar system.

As many contributors to this book have noted, Plato modeled a richly featured, diverse, and constantly changing world in terms of a small number of geometric solids whose features derived only from numbers, lines, and triangles. He made use of the polyhedron theory of his day, and it is likely that his use inspired the variations later introduced into the theory by such persons as Archimedes (287–212 B.C.E.). And,

G. Fleck
Department of Chemistry, Smith College,
Northampton, MA 01063, USA
e-mail: gfleck@smith.edu

M. Senechal (ed.), *Shaping Space*, DOI 10.1007/978-0-387-92714-5_12,

though manifestly incorrect, Kepler's model of the solar system is detailed, quantitative, and provocative. Most importantly, the model is visually stimulating, giving a pictorial vocabulary for discussing ideas that might otherwise be too abstract for easy communication. The model captured the imaginations of many persons beyond those with special expertise in quantitative astronomy.

It is appealing to think of the material world as composed of very small building units. With little information about the nature of those building blocks, early theorists speculated quite freely about them. We noted that Plato thought of them as the regular solids. Since earliest days, some theorists of the nature of matter have explicitly described the shapes of such building units as spheres, and others have described them as polyhedra. Sometimes particular units were chosen for convenience, without regard for a perfect correspondence between theory and reality, though surely some investigators intended their models to be faithful to the natural world.

Another issue which has been of concern for two millennia in various guises is whether matter must fill all space, or whether there can be interstitial voids between the building units. Plato seems vague about whether he thought there can be empty space, but Aristotle (384–322 B.C.E.) rejected empty space as inconsistent with his theory of motion. However, Epicurus (342–270 B.C.E.) and Lucretius (99–50 B.C.E.) argued that motion is impossible unless there is empty space between particles of matter.

If there is no void, spheres alone cannot be the building units of condensed matter; two spheres can come no closer together than to touch at one point, and consequently some space in any packing of spheres is empty. Some combinations of polyhedra fill space completely; certain of those combinations have been known since antiquity. But a satisfactory theory of matter must be able to account for change. Thus if the units do fully fill space without void, they must somehow be able to deform and transform to permit both motion and chemical change. We shall see that detailed modeling of such transformations is an important part of current scientific research.

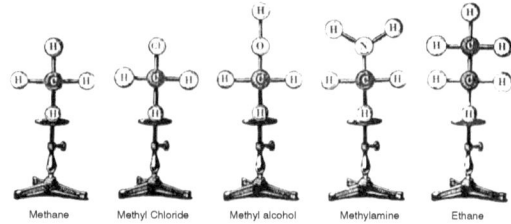

Figure 12.1. Models of molecules, made from croquet balls, used to illustrate a lecture in 1865.

Spheres and Whirlpools as Models for Atoms and Molecules

Modern chemistry dates from the late eighteenth century. One of the first modern chemists was John Dalton (1766–1844) who, as early as 1810, constructed physical models which he could hold in his hands, modeling atoms and compounds of atoms with spheres joined with connecting rods. Dalton used these models as teaching tools, but we do not know how faithful to reality he believed these spheres to be. There is no evidence that he believed there to be any relationship between the shapes of his models and the shapes of what we now call molecules.

Ball-and-stick models became popular with many chemists during the last half of the nineteenth century. Chemists were rapidly acquiring structural information about molecules from the laboratory, and as the century closed their models were increasingly intended to portray the three-dimensional geometry of molecules. August Wilhelm Hofmann (1818–1892) used a collection of elaborate croquet-ball models to illustrate his 1865 lectures at the Royal Institution in London (see Figure 12.1). In his models, the centers of the croquet balls were coplanar; Hofmann seems to have used his models to represent only connectivity of atoms, not their three-dimensional geometry. Benjamin Collins Brodie (1817–1880), the controversial Oxford chemist, strongly urged his colleagues during the 1860s and 1870s to avoid use of ball-and-stick models, warning that the models depicted much more detail about molecules than was warranted by experimental data. His warnings drew a mixed response. The majority of physical chemists in

the late nineteenth century explicitly rejected geometric ideas about molecules, but most organic
chemists of the same period found ball-and-stick
models to be increasingly useful. The balls depicted atoms, the sticks were bonds, and the
shapes of the models were usually thought to
represent the shapes of real molecules. By the
turn of the century it was an almost universal article of faith in organic chemistry that all possible
molecules can be modeled faithfully with balls
and sticks.

Many late nineteenth century scientists
modeled the molecules of gases by billiard-
ball spheres, but most of these theoreticians
considered the billiard balls to be models
of only certain properties of the gases. For
example, the balls were good models of idealized
temperature-pressure-volume behavior, but not
of the chemical reactivity of the components of
the gases.

The arguments of Epicurus and Lucretius
about the void became in the nineteenth
century arguments about the existence (and the
properties) of various types of aether, and this
controversy extended into the twentieth century.
The billiard-ball model of a gas seems to require
the notion of empty space between particles of
a gas. Yet from the beginning of the nineteenth
century proponents of the wave theory of light
argued persuasively for a plenum (the subtle fluid
which they called the *aether*) to transmit those
waves, and by 1880 the luminiferous aether had
become dogma.

The nature of the space within ordinary matter was widely and enthusiastically debated. As
the nineteenth century closed, the luminiferous
aether was joined by a whole collection of electromagnetic and dielectric aethers invented to explain phenomena such as radio waves, magnetic
waves, and even gravitational waves. Specula-
tions about the properties of aether produced a
theory of vortex atoms which attempted to combine features of the continuous and discontinu-
ous theories of matter and space. William Rankine (1820–1872) proposed a theory of molecular
vortices in 1849, and Hermann von Helmholtz
(1821–1894) derived mathematical expressions
which show that in a frictionless, isotropic fluid

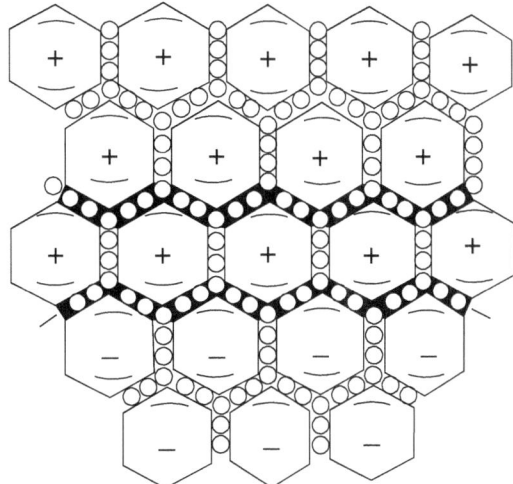

Figure 12.2. Maxwell's mechanical model for the
aether, using small idle wheels to permit all vortices to
revolve in the same direction. The idle wheels represent
electrical particles.

of uniform density, vortices once formed would
retain their identity forever. Both Lord Kelvin
(1824–1907) and Peter Tait (1831–1901) developed these ideas about whirlpools in the aether
further. In his investigations of vortices, Tait combined mathematical theory with physical models.
It has been said that a lecture demonstration
of smoke rings by Tait (to illustrate Helmholtz
vortex motion) early in 1867 gave Kelvin the idea
of a vortex atom. Tait described an apparatus to
produce smoke rings, telling about various ways
those rings could model properties of atoms.
Kelvin's theory, in turn, led Tait to extend his
investigations on the analytic geometry of knots,
Tait believing that a mathematical theory of in-
tertwining and knotting of vortices was necessary
for understanding vortex atoms.

James Clerk Maxwell (1831–1879) proposed
a gear-and-idle-wheel mechanical model of vortices in aether (Figure 12.2), remarking that his
model "serves to bring out the actual mechanical
connections between the known electromagnetic
phenomena; so that I venture to say that any one
who understands the provisional and temporary
character of this hypothesis, will find himself
rather helped than hindered by it in his search
after the true interpretation of the phenomena."

The quantitative models developed for the aether yielded predictions for its properties, and some scientists set out to test the models by experimental measurements. A series of experiments designed to measure predicted drift of the aether, conducted between 1880 and 1930 by Albert Michelson, Edward Morley, and Dayton Miller, were widely interpreted as demonstrating that there is no aether. Without a detectable aether to swirl around, the vortex-atom theory died, but the questions that it attempted to answer remain with us.

Polyhedra as Models for Atoms, Molecules, and Viruses

One of the most productive ideas of modern chemistry has been the model of an atom as a polyhedron. This model has been central to structural chemistry since the last decades of the nineteenth century when it was introduced into the mainstream of European chemistry independently by Joseph Le Bel (1847–1930) and Jacobus Henricus van 't Hoff (1852–1911). Plato had considered the ultimate units of matter to be polyhedra, but Le Bel and van 't Hoff extended this idea by showing how organic substances could be modeled by joining polyhedra in systematic ways to form molecules of great variety and complexity. The notion of three-dimensional molecular geometry was popularized by van 't Hoff, who considered carbon atoms to be situated at the centers of tetrahedra. He encouraged chemists to construct cardboard tetrahedra to examine the various geometrically possible arrangements of atoms. The balls of the Daltonian models become the vertices of tetrahedra. But the polyhedral model did more than simulate the environment around single carbon atoms. A carbon-carbon single bond was modeled by two tetrahedra interpenetrating at vertices (Figure 12.3), a carbon-carbon double bond was modeled by two tetrahedra interpenetrating along edges (Figure 12.4), and a carbon-carbon triple bond was modeled by two tetrahedra sharing a common face (Figure 12.5). These ideas were extended to include a wide range of molecules

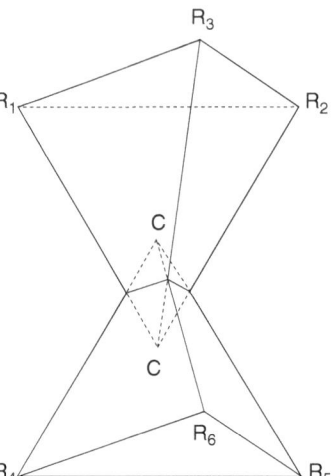

Figure 12.3. Two interpenetrating tetrahedra presented by J.H. van 't Hoff as a model for a molecule with a carbon-carbon single bond and six different groups, $R_1 \ldots R_6$, bonded to the carbons. Each carbon atom is at the center of one tetrahedron and at the vertex of the other. Van 't Hoff suggested that such models could be constructed from hard rubber tubes as edges, hollow hard rubber balls as vertex connectors, and sealing wax to bond the parts together

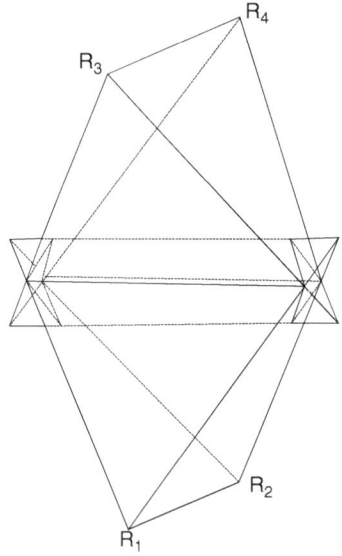

Figure 12.4. A tetrahedral model of a compound with a carbon-carbon double bond.

(see Chapter 10), and "the tetrahedral atom" became a central unifying concept of organic chemistry. Joined tetrahedra are still used by chemists

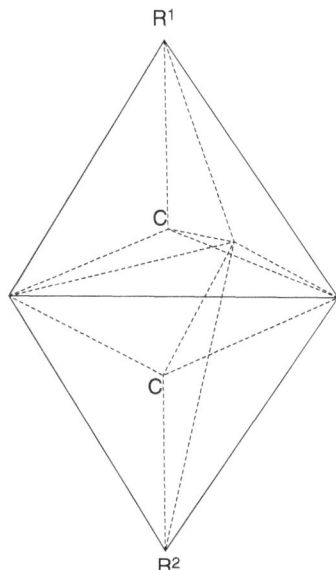

Figure 12.5. A tetrahedral model of the compound R_1—$C \equiv C$—R_2 with three equivalent bonds joining the carbon atoms.

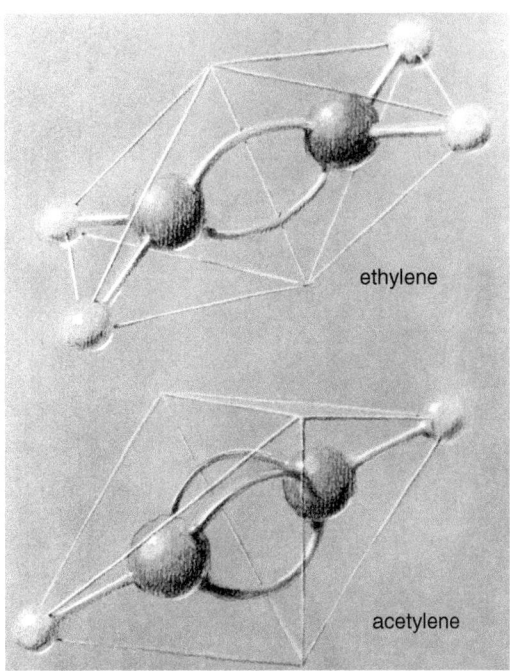

ethylene

acetylene

Figure 12.6. Artistic conception of the forms of ethylene and acetylene molecules.

for visualizing molecular form; see the representations by Hargittai and Hargittai in Chapter 11 and by Pauling and Hayward (Figure 12.6). Indeed, a major contemporary journal of organic chemistry is titled *Tetrahedron* (see Figure 12.7).

Polyhedral models are widely used also in inorganic chemistry. The spatial theory of molecular structure, based on polyhedra, was readily adaptable to the compounds of many elements. Nitrogen, depending on its oxidation state, could be modeled with tetrahedra or cubes. Alfred Werner (1866–1919) used octahedra to model metal complexes. In 1902 Gilbert Newton Lewis (1875–1946) found it useful in teaching general chemistry to model atoms with cubes, and later developed the cubic model in a more formal manner. This polyhedral model of the atom placed an electron at each vertex of the cube and provided a context for discussing the role of electrons in chemical bonding. The Lewis model became known as the *octet* theory for chemical bonding, and the eight dots at the vertices became known as Lewis dots.

More recently, in the 1930s and later, Linus Pauling developed, utilized, and popularized a theory of coordinated polyhedra to predict structures for crystals. To visualize the consequences of this theory, he often built elaborate models. An example is Pauling's model for the structure of the mineral sodalite shown in Figure 12.8.

Polyhedra play a significant role in contemporary research. Very recently, a simple and successful polyhedral model for molecules—the valence shell electron pair repulsion (VSEPR) model—has been developed. As well as a guide for chemical researchers, the VSEPR model has become an important pedagogical tool for teaching about molecular structure. The geometric problem posed by the VSEPR model is well-known to geometers as "The Problem of Tammes." VSEPR theory is described by Hargittai and Hargittai. An indication of the perceived importance of polyhedral chemical models is that a major journal of inorganic chemistry is titled *Polyhedron* (see Figure 12.7). The structures of certain molecules reinforce that perception; especially interesting is the structure (see Figure 12.9) of the 60-carbon cluster molecule named Buckminsterfullerene!

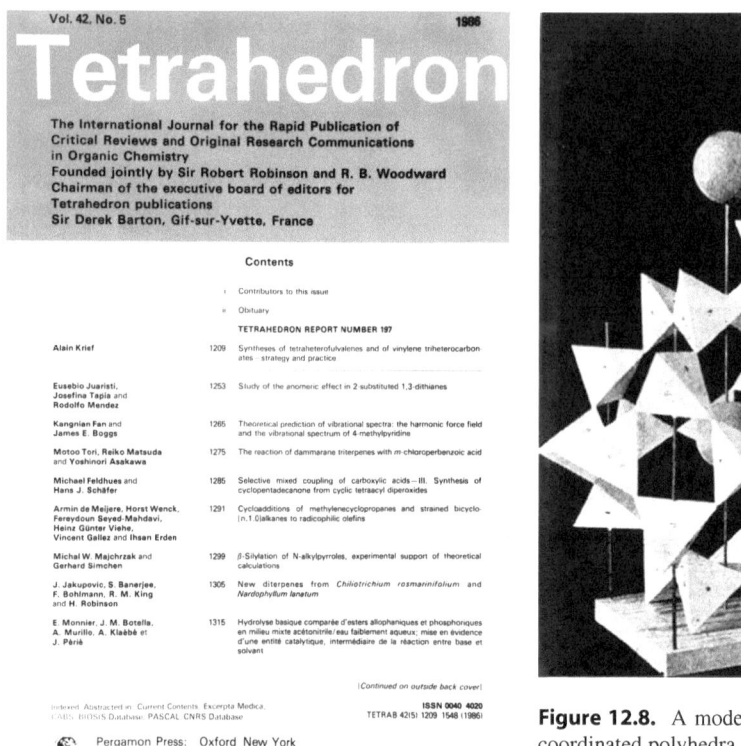

Vol. 42, No. 5   1986

Tetrahedron

The International Journal for the Rapid Publication of
Critical Reviews and Original Research Communications
in Organic Chemistry
Founded jointly by Sir Robert Robinson and R. B. Woodward
Chairman of the executive board of editors for
Tetrahedron publications
Sir Derek Barton, Gif-sur-Yvette, France

Contents

Figure 12.7. International journals of chemistry have the titles *Tetrahedron* and *Polyhedron*.

Figure 12.8. A model of the mineral sodalite, utilizing coordinated polyhedra.

Polyhedral models, which have proved so valuable in discussing atoms and molecules, have also been useful in investigating structures larger than single molecules. In 1956 Francis Crick and James Watson suggested that the design of "spherical" viruses, built from large numbers of identical protein subunits, would be based on the symmetry of the Platonic polyhedra. That same year Donald Caspar obtained experimental evidence in Cambridge for an icosahedral structure of one of the small isometric plant viruses. Icosahedral symmetry requires sixty identical parts, and at that time the symmetry of icosahedral viruses was thought to be a consequence of specific bonding among identical units. Such a structure is illustrated in Figure 12.10.

Caspar and his collegue Aaron Klug argued that the essential idea in the design of viral structures is that they "build themselves," that

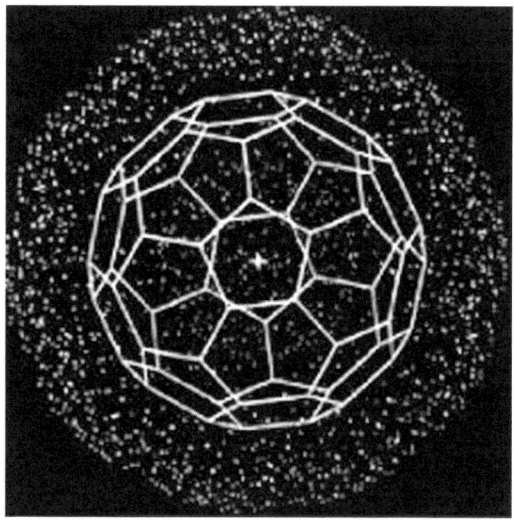

Figure 12.9. A computer-generated depiction of the truncated icosahedral structure suggested for the $C_{60}L$ a molecule.

the design is embodied in the specific bonding properties of the parts. Given the bonding rules, the units combine to form the structure automatically. The problem faced by the investigators

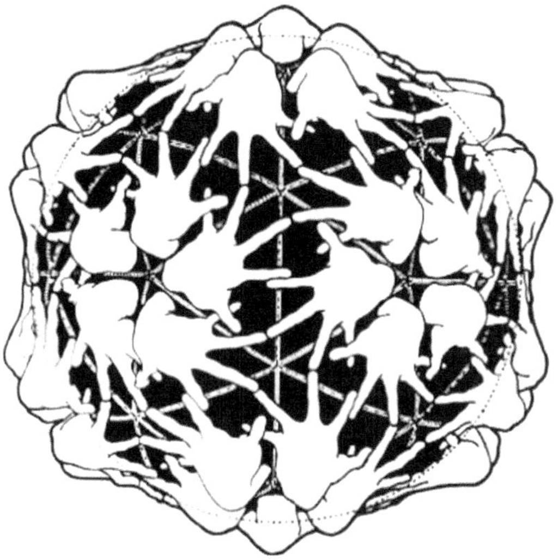

Figure 12.10. A drawing by D.L.D. Caspar illustrating strict equivalence in a shell with icosahedral symmetry constructed from sixty identical left-handed units. The three classes of connections in this surface lattice are represented by the specific bonding relations: thumb-to-pinkie = pentamer bond; ring finger-to-middle finger = trimer bond; and index finger-to-index finger = dimer bond. Any two of these classes of bonds would hold the structure together. The triangles drawn under the hands define equivalent subdivisions defined by the three- and fivefold axes at their intersections.

Figure 12.11. A geodesic dome built on a plan of a $T = 12$ icosahedral surface lattice.

was that most of the icosahedral viruses are not built of sixty identical subunits; indeed, by 1962 electron microscopy had revealed regular surface arrays of morphologic units that were neither multiples nor submultiples of 60. Their question was then twofold: Why icosahedral symmetry?

What are the design possibilities for such icosahedrally symmetric structures?

The clue to the answer formulated by Caspar and Klug in 1962 was based on an icosageodesic dome (see Figure 12.11) which, were it a complete sphere, would be divided into 720 truncated triangular facets grouped to form 12 pentamers and 110 hexamers. If a hand (such as one of those in Figure 12.10) were placed in each of these triangles, one could imagine (with a certain amount of flexibility either in bonding or in the structure units themselves) that identical units could be connected according to the requirements of bond specificity, but allowing for some departure from strict equivalency. The term that Caspar and Klug proposed for this was "quasi-equivalence." The units would be deformed in slightly different ways in symmetrically distinct but quasi-equivalent positions.

The design for such structures can be described by the ways that the plane hexagonal net can be folded into polyhedra. These designs can be enumerated completely and nonredundantly by the triangulation numbers

$$T = (h^2 + hk + k^2),$$

which designate the number of symmetrically distinct but quasi-equivalent situations for the $60T$ units in a design. The indices h and k can be any positive integers; one may be zero. These designs for indices 0, 1, and 2 (that is for $T = 1, T = 3$, and $T = 7$) have been recognized for a number of icosahedral viruses. Some are built on the $T = 1$ plan. The $T = 3$ plan is also very common.

These surface lattices can be represented as icosadeltahedra, polyhedra consisting of $20T$ equilateral triangular facets. Capsid models can be built of $60T$ identical subunits grouped to form 12 pentamers and $10(T - 1)$ hexamers with quasi-equivalent bonding in the T symmetrically distinct environments. It is informative to build both rigid and flexible models of these structures.

A number of small tumor viruses are built on the $T = 7$ plan. But, to everyone's great surprise, a radical departure from the idea of quasi-equivalence was revealed with the discovery that the $T = 7$ icosahedral polyoma virus capsid is built of 72 pentamers instead of the predicted 12 pentamers and 60 hexamers. Bonding specificity apparently is not conserved in this structure. Caspar described a polyhedral model of the polyoma capsid constructed from 72 pentagons connected using three more equivalent types of contacts which correspond to switching of bonding specificity. He said that this idea was so incompatible with the expectations of his theory that the referees who reviewed the paper said the model was not suitable for publication. "The theory was so good. Why throw away a good theory with experiments that haven't been thoroughly tested?"

However, further experiments have confirmed the structure, a design which appears inscrutable in geometric terms but which obviously has biological logic. Caspar commented: "The theories we have formulted in the past have given a very good explanation of why icosahedral viruses are icosahedral. Now we don't know!"

Modeling Condensed Matter

Polyhedra have been used not only in modeling molecules but also in modeling the space-filling qualities of whatever it is that forms the condensed states of matter. In crystals, polyhedra have been considered to be the units that repeat in three dimensions to fill space. As we have seen, a continuum model (a model in which there is no empty space) for crystals of a pure substance (an element or a compound) cannot be achieved with spheres. But there are only a few polyhedra that fill space by periodic repetitions of themselves. There are even greater difficulties in modeling arbitrary mixtures of different substances with polyhedra, since most collections of *different* polyhedra do not fill space.

Theories about structure of the solid state have long been involved with the question of how polyhedra can be packed to fill three-dimensional space. Bricklayers have known about the packing of parallelopipeds since antiquity, but they have generally not been concerned about theo-

ries of the structure of matter. Aristotle (refuting Plato) asserted that, of the regular solids, both the cube and the tetrahedron fill space; he was wrong, however. Kepler considered the shapes of space-filling polyhedra which would be obtained if closely packed spheres were uniformly compressed; we shall see that this strategy for investigating the shaping has been fruitful in recent years. The quite complicated general question of how space can be filled by repetition of identical polyhedra has not yet been solved.

One experimental way to prepare space-filling collections of different polyhedra is to use the method of Kepler and compress a collection of plasticene balls. Another way is to start with a packing of spherical objects, and increase the size of the spheres without increasing the volume of the collection; Stephen Hales (1677–1761) used this strategy in studying the shapes of peas which were swollen in a closed container. It appears that the unfinished task of describing general collections belongs largely to the mathematicians, but their intuitions have benefited from these physical experiments.

Clearly the solid state cannot be described by a single geometric model. Some solids are almost perfectly ordered in a simple manner at the atomic scale, and others are almost completely random. Most solids have structures between these two extremes. Only the perfect crystalline structures—which do not really exist—are well understood. The structure of the solid state has been a subject of inquiry in which the interaction between geometry and the physical sciences has been particularly fruitful. Lord Kelvin took a very empirical approach in investigating "the division of space with minimum partitional area." Extending experimental methods used by Plateau, he observed and manipulated intersecting soap films in imaginative ways, and his paper is an instructive example of how a physical model can guide a geometric investigtion, and how mathematics can permit generalizations from a few simple physical observations. Lord Kelvin concluded that space could reasonably be divided into modified truncated octahedra with 14 faces (eight hexagons and six squares); he called these polyhedra *tetrakaidecahedra*, and argued that they would pack to fill space with minimum partition area. Kelvin's shapes are not classical polyhedra; the edges are not straight lines and the faces are not plane surfaces.

Static space-filling units probably cannot model simultaneously both the space-filling qualities (such as the quasi-crystalline regions sometimes found near apparently chaotic regions in liquids, or the orientation of polar solvent molecules near solute ions) and the dynamic aspects (such as fluid flow, or diffusion of solvent and solutes) of liquid solutions. It would be interesting to try to model such systems with dynamically transforming polyhedra. Meanwhile, sphere-packing models continue to be useful.

Packing of Spheres of Various Sorts

Twentieth century models of atoms, with a very small nucleus surrounded by wavelike electrons, have spherical symmetry (probably instantaneous spherical symmetry, but certainly at least time-averaged spherical symmetry) in isolation. The sphere is an excellent model for isolated atoms and has been developed by many investigators as a model for atoms in molecules, and for molecules in liquids and solids.

When pressed, few chemists insist that a collection of spheres with interstices is a realistic model for combinations of atoms, whether crystalline arrays or discrete molecules. They say that their ball-and-stick models are merely conveniences, that their ball-and-stick drawings are artistic conventions, and that their tables of covalent and ionic radii are just conventional ways of presenting data concisely. Yet it would be naive to believe that the ubiquitous presence of spherical models for two centuries has had no influence on the concepts held by chemists.

Much effort has been spent examining the consequences of spherical models for atoms, calculating atomic-radii for these spheres, and discussing how spheres of various sizes can pack together. William Barlow (1845–1934) observed that there are two "closest-packed" ways of arranging identical spheres in space, one with cubic symmetry and the other with hexagonal symme-

try. He was convinced that a thorough understanding of crystalline solids would necessarily involve both a geometrical theory of space groups (to which he contributed) and a complementary mechanical theory of crystal structures (toward which he worked for several decades). Lord Kelvin examined the problem of close-packing spheres with oriented binding sites. Pauling developed in detail the model of atoms-as-spheres, using what he called "covalent radii" to correlate distances within molecules and crystals among quite diverse compounds of particular elements.

Spherical models for atoms have been useful, even though atoms within an environment of "touching" nearest-neighbor atoms are not spherical, and even though the electrons in such environments fill space without interstices. All this has stimulated mathematical studies of sphere packings in spaces of three and higher dimensions. These studies, in turn, have suprising and important applications in coding theory.

Polyhedra as Models for Plant Structures

Ralph Erickson has used polyhera as models for plant cells. These models suggested novel experiments, involving such unlikely materials as lead spheres and soap bubbles. The experiments in turn have stimulated mathematical speculation.

Parenchymal tissues such as are found in a plant stem have cells that pack closely together. The cells are not classical polyhedra; there probably are no plane faces or any straight edges in these cells. Erickson described experiments in which James Marvin obtained single-cell geometric data about pith cells from the Joe-pye weed. Marvin then constructed paper polyhedral scale models of those cells. Previous investigators had taken Kelvin's tetrakaidecahedra as a model for such space-filling cells, but the data of Matzke and co-workers showed that the real botanical world was more complex. A fundamental question is the extent to which the shapes of such cells are determined by geometry alone, and the extent to which the shapes are the results of other factors.

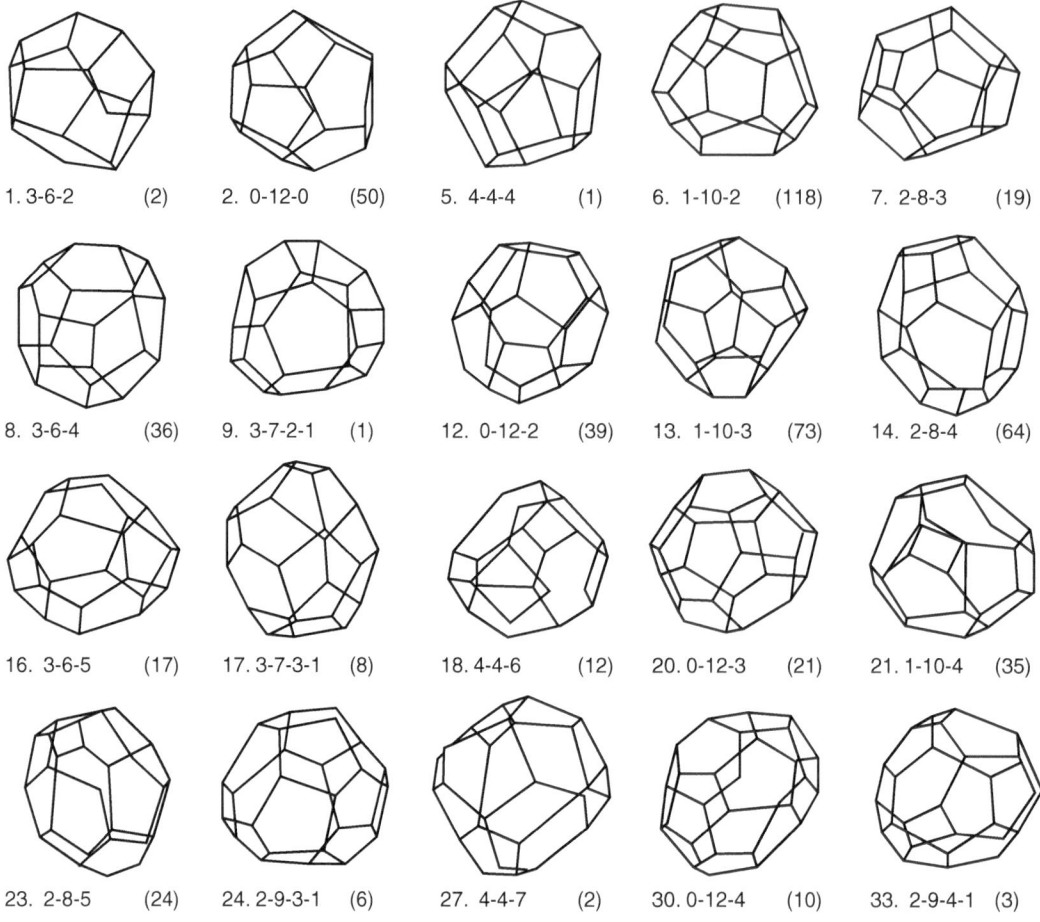

1. 3-6-2   (2)   2. 0-12-0   (50)   5. 4-4-4   (1)   6. 1-10-2   (118)   7. 2-8-3   (19)

8. 3-6-4   (36)   9. 3-7-2-1   (1)   12. 0-12-2   (39)   13. 1-10-3   (73)   14. 2-8-4   (64)

16. 3-6-5   (17)   17. 3-7-3-1   (8)   18. 4-4-6   (12)   20. 0-12-3   (21)   21. 1-10-4   (35)

23. 2-8-5   (24)   24. 2-9-3-1   (6)   27. 4-4-7   (2)   30. 0-12-4   (10)   33. 2-9-4-1   (3)

Figure 12.12. Camera lucida drawings of representative soap bubbles from the center of a foam. Each drawing is specified by listing the numbers of rectangular, pentagonal, hexagonal (and heptagonal) faces. In parentheses is the frequency of occurence of the polyhedron class in a group of 600 bubbles studied.

Erickson described studies in which Matzke and Marvin used lead shot and soap bubbles to model plant cells. In a series of Kepler-type experiments, Matzke and Marvin compressed collections of spherical lead shot with enough pressure to force the initially spherical pieces of lead into shapes that together fill spaces without interstices. As Kepler had shown centuries earlier, spheres transform into polyhedra when forced to be space filling. If uniform spheres were packed as a face-centered cubic array and compressed, rhombic 12-hedra resulted. If the lead shot were poured into the container "randomly," a distribution of irregular polyhedra (averaging about 14 faces) resulted. Matzke and

Marvin undertook their studies in attempts to model cell shapes in plant tissue in terms of the polyhedra observed in the shot-deformation studies.

Matzke also used another model, reminiscent of the 1880s' work of Lord Kelvin. Several thousand soap bubbles were assembled in a transparent container, and the interior bubbles were examined microscopically (Figure 12.12). These soap films partition space. We can distinguish between the *filling of space* by a collection of geometric shapes and the *partition of space* by surfaces that can be considered to be the faces of polyhedra. Matzke made detailed comparisons of these polyhedra, the

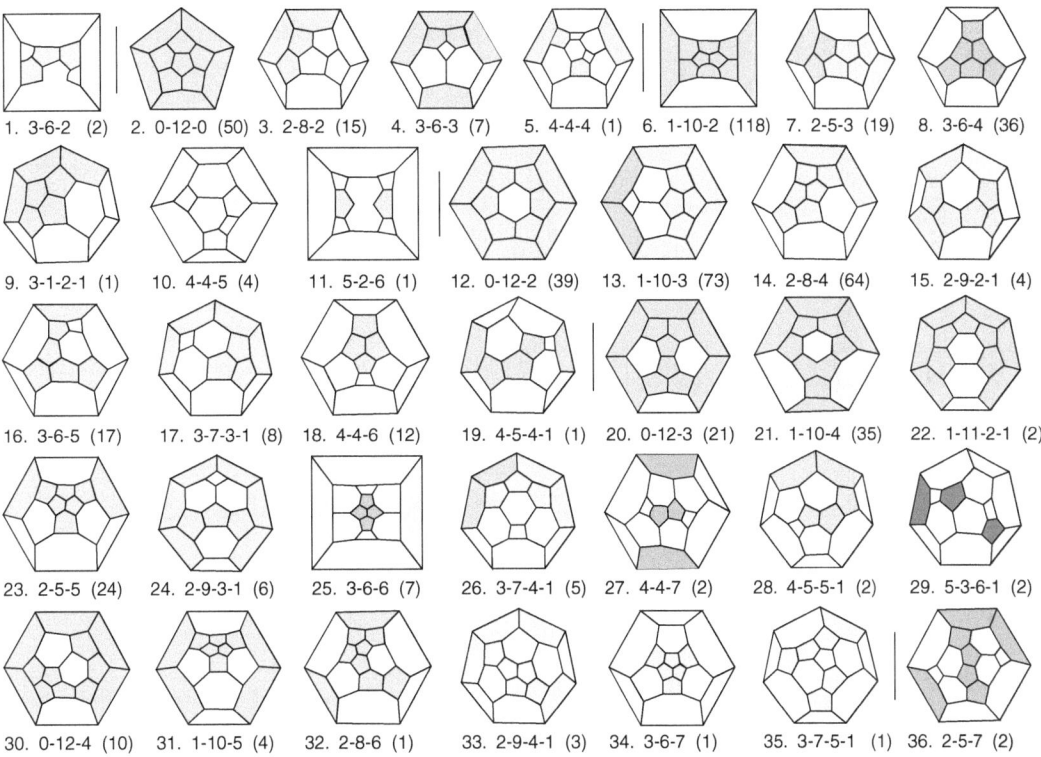

1. 3-6-2 (2)   2. 0-12-0 (50) 3. 2-8-2 (15) 4. 3-6-3 (7)   5. 4-4-4 (1)   6. 1-10-2 (118) 7. 2-5-3 (19) 8. 3-6-4 (36)

9. 3-1-2-1 (1) 10. 4-4-5 (4) 11. 5-2-6 (1) 12. 0-12-2 (39) 13. 1-10-3 (73) 14. 2-8-4 (64) 15. 2-9-2-1 (4)

16. 3-6-5 (17) 17. 3-7-3-1 (8) 18. 4-4-6 (12) 19. 4-5-4-1 (1) 20. 0-12-3 (21) 21. 1-10-4 (35) 22. 1-11-2-1 (2)

23. 2-5-5 (24) 24. 2-9-3-1 (6) 25. 3-6-6 (7) 26. 3-7-4-1 (5) 27. 4-4-7 (2) 28. 4-5-5-1 (2) 29. 5-3-6-1 (2)

30. 0-12-4 (10) 31. 1-10-5 (4) 32. 2-8-6 (1) 33. 2-9-4-1 (3) 34. 3-6-7 (1) 35. 3-7-5-1 (1) 36. 2-5-7 (2)

Figure 12.13. Schlegel diagrams of central bubbles (some from Figure 12.12) tabulated by Matzke. The numbering corresponds to the numbering in Figure 12.12. Pentagons are shaded. Viewpoints are chosen to demonstrate symmetries when possible.

polyhedra from the lead-shot experiements, and the polyhedra observed in plant cells.

Erickson noted that accurate visualization of models is difficult, even when the models (such as polyhedra) are apparently tangible. A difficulty in using two-dimensional representations of stick models or solid models of polyhedra on the printed page is that all faces cannot be shown simultaneously. To understand a three-dimensional model, a person must pick up the physical object and turn it around. Erickson constructed paper models of Matzke's soap bubbles to aid visualization and he drew Schlegel diagrams (Figure 12.13) to aid classification. Schlegel diagrams distort shape, but they allow simultaneous viewing of all faces and of their connectivity. Many of these Schlegel diagrams have obvious symmetry. Consider number $0 - 12 - 2$. It is highly symmetrical, with two

hexagonal faces opposite each other, more or less as in an antiprism. But it is not an antiprism. The sides are pentagons. Symmetries are very prominent in these polyhedra; only 44 of the 600 bubbles are in the symmetry group that contains just the identity element.

Visualization becomes less of a problem when the investigators use several different schemes which appear to model the same geometry. Erickson constructed paper models so that he could look at them from all directions. He also used stick structures (representing the same polyhedra) built from semiflexible plastic tubing and four-arm connectors, the arms at the 109.5° tetrahedral angle.

All edge intersections of soap films are necessarily tetrahedral intersections. So by building models with these rigid connectors and semiflexible tubes, the model approximates the minimum-surface models. Lord Kelvin went to some pains to point out that in the packing of soap bubbles the edges of the cells which he visualized were plane and curved, alternate edges curving in alternating ways. This very delicate structure of the Kelvin 14-hedron is only approximated by these skeletal models.

Kelvin's tetrakaidecahedron alone is not a sufficient model for either plant cells or soap bubbles. The tetrakaidecahedron has no pentagonal faces, whereas the majority of natural cells do have pentagonal faces. Matzke's data on 600 bubbles, probably the largest sample of such cells that exists in literature, reveal 36 polyhedral forms; some contain one heptagonal face. Erickson built them all both as stick models and also as cardboard models.

Figure 12.14 displays the fantastic differences in the frequency of the polyhedra. Some of them occurred only once in Matzke's 600 cells. Amazingly, a 13-sided polyhedron $(1 - 10 - 2)$ occurred 118 times.

Does Form Explain Dynamic Functioning? Science Looks to Geometry for Mechanistic Models

Growth of a rigid plant stem, self-assembly of a virus, functioning of a robot arm: such dynamic processes are being simulated with dynamic polyhedral models which focus on transformations. Ralph Erickson, Donald Caspar, and Godfried Toussaint spoke at the Shaping Space Conference on aspects of their research in this frontier area in which the symbiotic relationships among theory, mechanical models, and the natural world are strikingly evident. We shall examine some of their work.

Plant Growth and Polyhedral Transformation

A rigid arrangement of parenchymal plant cells is a dynamic system with very interesting spatial

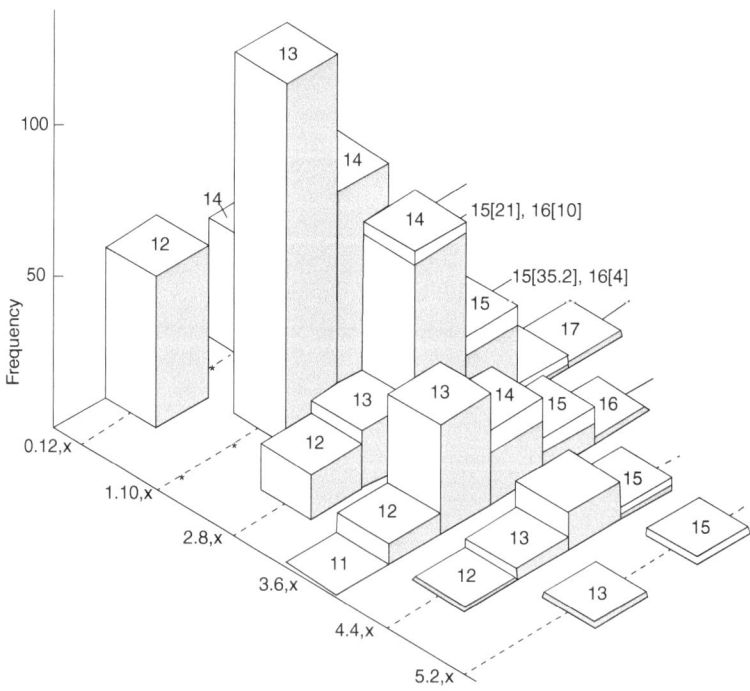

Figure 12.14. Frequencies of polyhedral forms among 600 central bubbles. Each bar is labeled with the number of faces of the polyhedron. *Unshaded* portions of bars indicate frequencies of polyhedra having one heptagonal face. Asterisks indicate forms [(0-12-1), (1-10-0); (1-10-1)] which cannot be built. Five bars are hidden: (0-12-3); (0-12-4); (1-10-4); (1-11-2-1); and (1-10-5).

properties. The plant cells divide as the tissue grows. If those cells are space-filling polyhedra, the polyhedra must be capable of local transformations which result in changing the number of packed polyhedra without weakening the plant structure. We would expect that division of a single cell would occur with minimal disruption of intersections and edges in adjacent cells.

An advantage of the skeletal, stick models discussed by Erickson for modeling transformation of plant cells is that these models can be manipulated and transformed rather easily. A face can be added by breaking a couple of connections and inserting three edges and two connectors. Another operation which is very helpful in exploring the possible transformational forms consists of disconnecting and then reconnecting an edge and its four neighboring edges. The result is promotion of square faces to pentagons and demotion of two hexagons to pentagons. Other 14-hedra can be found by carrying out similar manipulations on the edges. R.E. Williams has proposed a model for a cell packing based on repetition of this neighbor-switching operation, creating cells with two square faces, eight pentagons, and four hexagons.

Erickson noted that a puckered surface can be traced through the packing of the truncated octahedra. It consists of hexagons and squares, going up and down like valleys and ridges. The squares, most interestingly, are oriented at 90° to each other along a path. Neighbor-switching operations within this packing can be used to discuss dislocations in cellular structures as well as creation and annihilation of cells. He proposed that it should be possible to orient two puckered surfaces properly, connect them appropriately, and create a uniform packing of polyhedra which will fill space. With a different orientation and edge-connections, another space-filling packing of 14-hedra would be created. This can be done extensively, if not exhaustively. The convincing way to do that is with stick models; there are too many possibilities and transformations for cardboard models to be feasible for this task. Erickson said that his models covered half of the kitchen floor!

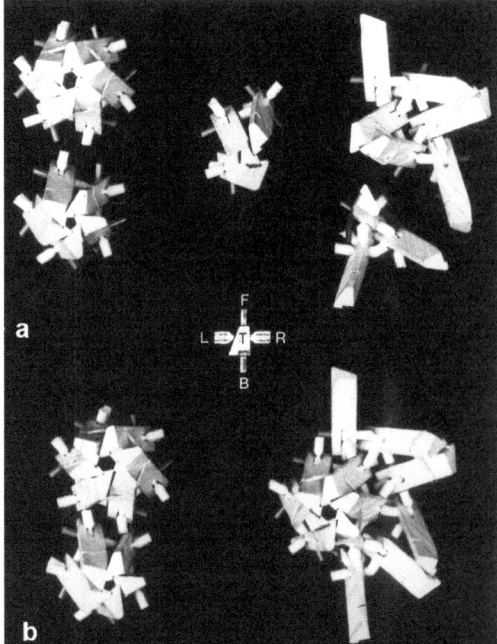

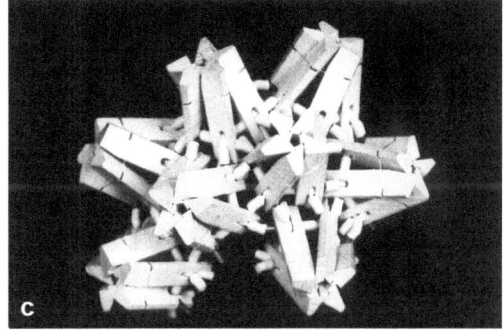

Figure 12.15. Constituent parts of Caspar's self-assembly model. (**a**) On the *left*: a pentamer and hexamer; *center*, part of a hexamer or pentamer; *right*, a trimer and two trimers bonded together. (**b**) The same units with more bonds formed. (**c**) Misassembly of eight pentamers attempting to form a $T = 1$ shell.

Polyhedral Models for Self-Assembly of Viruses

Caspar used a variety of model-building strategies to explore both the geometry and the energetics of protein assemblies. Caspar and Klug first illustrated their idea of self-assembly with a dynamic model with wooden-peg subunits designed to assemble in the $T = 3$ icosahedral surface lattice. The structural unit, and various stages in assembly, are shown in Figure 12.15.

This model shows that it is possible to design a single unit with bonds that can be switched by interaction with identical copies of itself to bond in three different environments of the $T = 3$ lattice. Such a model is too simple to explain all features of the control mechanisms that were postulated for icosahedral virus self-assembly, but it does display some of the essential interaction properties of a structural unit designed to form an icosahedral capsid.

The model shown in Figures 12.15 and 12.16 illustrates some of Caspar's further ideas about designing mechanical models to represent the dynamics of macromolecules. Such models should illustrate how the energy of interaction is distributed throughout the structure. One strategy has been to underdesign, so that the structure is not unintentionally overdetermined. The construction of a mechanical model is often a trial-and-error process. This is, in fact, Nature's way for adaptation and evolutionary development. Analogies with human technology and behavior have indeed provided essential keys for understanding the operation of biological systems, and, conversely, analogizing Nature's methods is a natural way to make analogs of Nature's machines. In nature where there is regularity, with structures built of identical parts, there are likely to be regular plans, says Caspar. Geometric considerations are always important in these plans, and sometimes they predominate. However, satisfactory a priori predictions about what in fact happens in nature cannot be made. The only way to find out is to look.

Robotics and Motions of Polyhedra

Dynamic computational geometry is an area in computer science that has evolved from work in graphics and visual design, inspired by robotics and problems of movement. A fundamental problem in robotics theory involves the ways in which a set of objects can be moved without collisions.

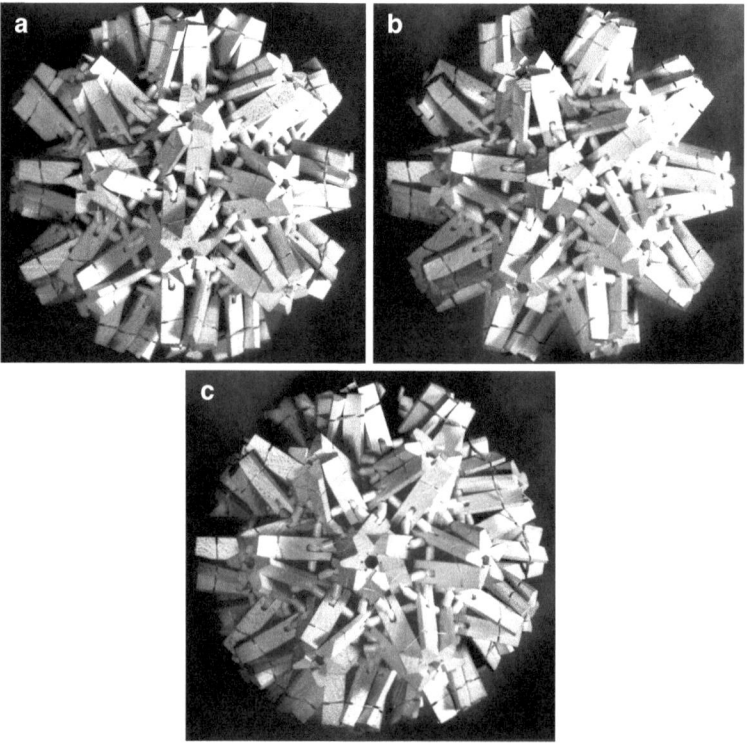

Figure 12.16. Completed $T = 3$ self-assembly model viewed down: (**a**) twofold axis. (**b**) Fivefold axis. (**c**) Threefold axis.

This section is a snapshot, taken at the Shaping Space Conference in 1984, of work by Godfried Toussaint and others on how sets of polygons in the plane can be translated without collisions and generalizing such studies to three-dimensional movement of polyhedra.

Toussaint began with a problem: Can all members of a set of nonintersecting rectangles in the plane with their edges parallel to the x and y axes (such rectangles are called *isothetic*) be translated in the same direction by some common vector to a final destination, subject to the constraints that the rectangles are moved only one at a time, and that no collisions occur? (A collision occurs whenever the interiors of two rectangles intersect.)

The answer is yes. There always exists (for any direction) at least one order in which the polygons can be moved to permit such a translation. This property holds for every finite set of convex polygons, and the ordering can be determined by computation. We say that convex polygons in the plane exhibit the translation-ordering property.

Efficient computation of translation ordering is important for a robot who is given the task of separating such polygons. A simple, intuitive algorithm which sometimes works is the *line-sweep heuristic* illustrated in Figure 12.17. The vector l is the desired direction of translation. Lines perpendicular to l are constructed to intersect the rectangles at *support vertices* a, b, c, and d. According to the line-sweep heuristic, the translation ordering is the order of the projections of the support vertices on l. The algorithm fails in this particular case since the method requires D to move second, even though it is blocked by C.

But is there a class of objects for which the line-sweep heuristic always works? It works if the objects are all circles of the same size, and with a slight modification (use of centers instead of support points) for any set of circles of arbitrary sizes (see Figure 12.18). In three dimensions, this method is called the *plane-sweep heuristic*; it works for sets of spheres and for some sets of isothetic polyhedra.

This problem can be generalized for other types of polygons, and for motions other than simple translations. When the convexity

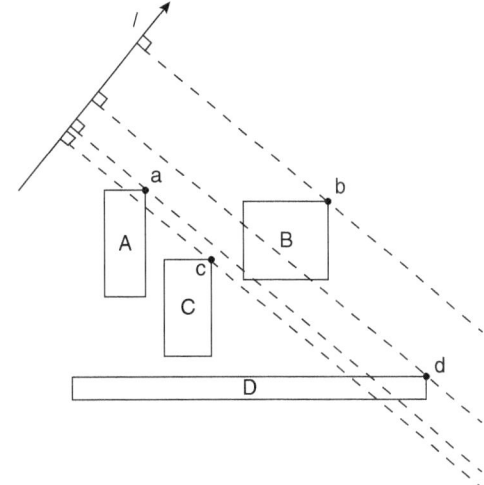

Figure 12.17. An illustration of a failure of the *line-sweep heuristic* for a set of isothetic rectangles. Note that rectangle D cannot be translated before rectangle C in direction l, even though d occurs before c on line l.

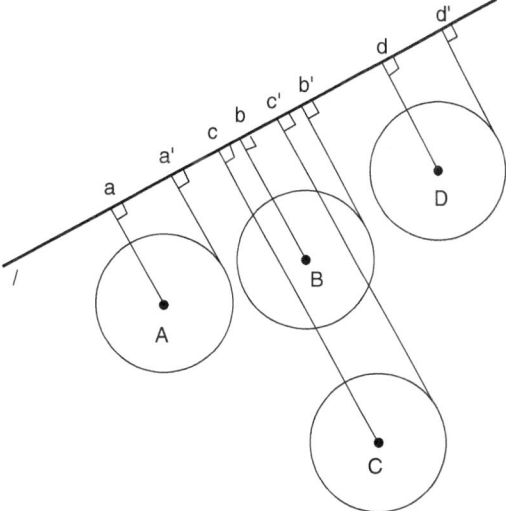

Figure 12.18. When the objects are circles of the same size, their centers yield the same ordering as their support points. The line-sweep heuristic gives a valid ordering for this set.

constraint is relaxed, there results a class of problems concerning interlocking polygons. It becomes interesting to ask whether a collection of polygons is "movably separable" in a specified sense.

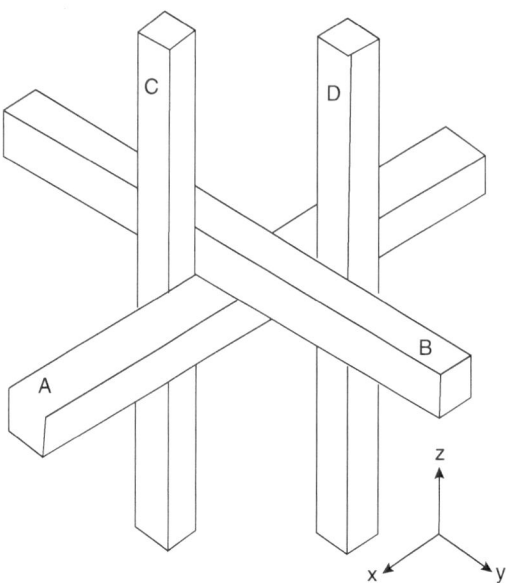

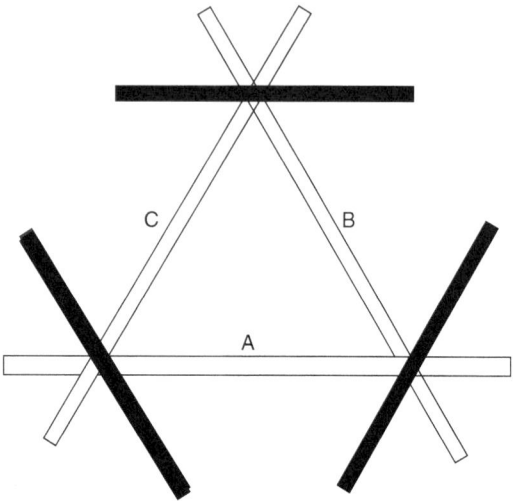

Figure 12.21. Three more triplets added to the basic one.

Figure 12.19. A set of isothetic convex polyhedra that does not allow a translation ordering in the direction $x + y$. This example was discussed by Guibas and Yao.

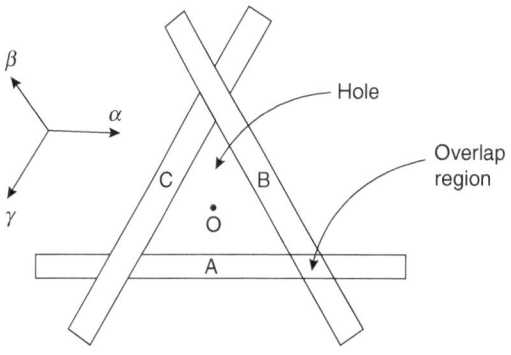

Figure 12.20. The basic interleaving triplet.

In three-dimensional space, four isothetic rectangular polyhedra can be arranged so that no translation ordering exists for some directions. Such an arrangement is shown in Figure 12.19. Some sets of convex polyhedra interlock in all directions, with no translation ordering in any direction. Such a configuration can be built from 12 long, flat, very-thin sticks. The first step is to construct a set of three interlaced sticks A, B, and C (a *triplet*) on the xy plane as shown in Figure 12.20, with their lengths parallel to $\alpha, \beta,$ and γ directions. Define a "hole" at the

center of the configuration, and three "overlap" regions where the sticks touch each other. Three more triplets are constructed (see Figure 12.21) on planes perpendicular to the xy plane, planes individually parallel to the $\alpha, \beta,$ and γ directions. The key feature of the final configuration is that each new triplet added embraces an overlap region of two sticks in the original triplet; each overlap region of the original triplet lies in the central hole of an embracing triplet.

Inspection of Figure 12.21 reveals that there is no direction in which more than two of the dozen sticks can be translated. Even though each polyhedral stick can be individually translated away from the configuration without disturbing the others, there exists no translation ordering in any direction.

Are there sets of polyhedra in which no member can be moved out without disturbing the others? If the constraint that the polyhedra be rectangular solids is relaxed, a configuration of twelve convex polyhedra can be constructed in which no polyhedra can be translated in any direction without disturbing the others. K. A. Post found an example of six convex polyhedra that interlock in such a way that no one can be moved without disturbing the others.

Toussaint discussed problems which arise in generalizing results from polygons in the plane

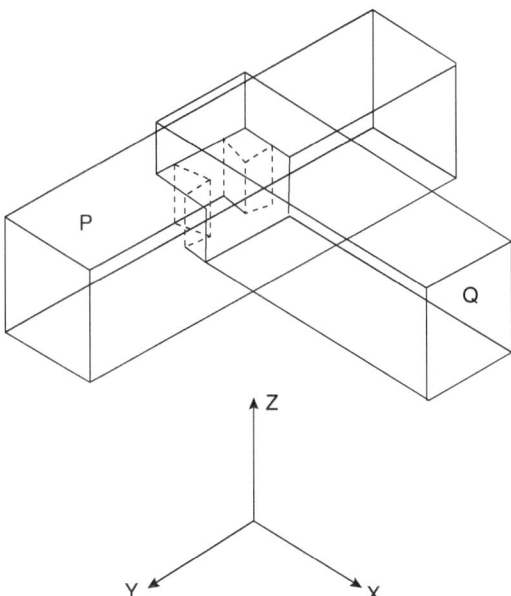

Figure 12.22. Two polyhedra (components of an arikake joint) strongly monotonic with respect to PL(z). The only way to separate this pair is to translate either P or Q in the z direction.

der all motions. Further, a polyhedron is *strongly monotonic* with respect to a direction PL(l) if the polygons which result from the intersection of planes parallel to l are all monotonic in a direction orthogonal to l. Figure 12.22 illustrates two polyhedra, P and Q, which are strongly monotonic with respect to PL(z). It is known in Japanese carpentry as the ari-kake joint, a dovetail which can be separated only by a translation in the z direction. Such polyhedra are always separable if they share a common direction of monotonicity. It is an open question whether they are necessarily separable when they do not share a common direction.

Polyhedron Theory Accommodates Changing Expectations

What can we expect when time and change are included in polyhedron theory? What are the new questions, and what might be the form of the answers?

We have seen that static polyhedral units probably do not suffice as bases for chemical modeling of the *dynamic* aspects of solution structure. More needs to be known about the transformational properties of three-dimensional arrays of nonidentical polyhedra. Some important questions are: What local transformations can be achieved without disrupting long-range structure? In an array, can a rotation of one polyhedron be achieved by interactions only with its nearest neighbors? Under what conditions can a polyhedron migrate through a space-filling array of transformable polyhedra?

We have looked at skeletal models for transformations of the polyhedra which serve as models for plant cells in growing tissue. Their detachable, rigid tetrahedral connectors and flexible tubing model both the individual plant cells and the transformations of individual cells, as well as the packing of cells into large arrays and transformations within arrays. These arrays have long-range patterns which can be seen, for instance, as paths along puckered surfaces. Their short-range transformational possibilities are seen in the neighbor-switching

to polyhedra in 3-space. Star-shaped polygons generalize in a straightforward way to star-shaped polyhedra. Any two star-shaped polygons can be separated with a single translation, and this property also holds for star-shaped polyhedra in 3-space. The situation is more complicated with *monotonic* figures. (A polygon is monotone if there exist two extreme vertices in some direction connected by two polygonal chains in which the vertices of the chain occur in the same order as their projections onto a line in that direction.) Any two monotone polygons can be separated with a single translation in at least one direction.

The property of monotonicity does not generalize straightforwardly and uniquely to three dimensions. Toussaint defined a polyhedron as *weakly monotonic* in a direction if the intersection (with the polyhedron) of each plane perpendicular to that direction is a simple polygon, a line segment, or a point. These weakly monotonic polyhedra can be classified in terms of the properties of the polygons of intersections. It turns out that two polyhedra weakly monotonic with respect to a common direction can interlock un-operations which produce dislocation, creation, or annihilation of cells. Such tangible models are suggestive, but general conclusions seem elusive. In this fundamental research further collaboration between botanists and geometers will surely be fruitful.

Nineteenth-century scientists effectively used billiard-ball models to discuss gases, without necessarily abandoning belief in an aether that pervaded the space in which those billiard balls moved. Since no model of reality is complete, realists should notice that complementary models, although inconsistent in detail, may yield complementary information about the natural world. Thus a chemist, or a botanist, or a biophysicist may see in Toussaint's dynamic computational geometry applications beyond robotics to more general problems about how objects move and how they become bound to one another. Biochemists are concerned about how complex, stable structures are assembled by bringing together (in the proper orientations) individual molecules which one might have expected to be moving randomly in solution. Geometers with various perspectives may see beyond these qualitative connections.

One central fact of biology is that living systems build structures in the midst of apparent molecular chaos. Biochemical explanations of such structure creation are, at best, incomplete and inadequate. Figure 12.23, in which a representation of Plato is held before us by a mechanical model of a chemical system, may well be an appropriate image for an appropriate strategy for future symbiotic progress in natural science, engineering and mathematics. It is clear that many fields of inquiry are enriched by both the results and the questions from other disciplines. It is also clear that the possibilities for learning from the work of others are vast.

Figure 12.23. A mechanical model illustrating rigidity in a $T = 3$ icosahedral lattice, built of 90 identical pieces of $\frac{1}{32}$-inch-thick sheet aluminum bent to represent dimers connected by pentamer/hexamer bonds. The diameter along the threefold axis is 11.5 inches, and the model weighs 2 pounds. Its shape approximates a truncated Platonic icosahedron, and it supports without distortion a 32-pound plaster bust of Plato.

Addendum

This chapter, written more than two decades ago, reflects my bias toward mechanisms and physical models, a bias that is shared by most of the conference participants. This is, in part, a generational bias. Today, many younger investigators are actively and enthusiastically engaged in manipulating virtual models, using powerful computer-based methods that many persons believe creatively supplement, and some persons believe may supplant, reliance on analysis of physical objects. It is too early to place these approaches in perspective, but a tentative attempt seems in order.

Among my own first toys were a Meccano set (created by Frank Hornby in 1901), an Erector set (created by A.C. Gilbert in 1911) and a Tinkertoy set (created in 1914 by Charles H. Pajeau and Robert Pettit). Similar toys were part of the culture in which most of the contributors to Shaping Space learned about construction. Meccano and Erector toys are literally hands-on, nuts-and-bolts mechanical building tools. Not every idea of a young inventor is realizable, using only the Meccano and Erector parts; the concept of unbuildable develops naturally. I remember my fascination watching a machinist fabricate mechanisms that I had thought to be unbuildable. I continue to be fascinated by explorations of the unbuildable realm by graphic artists.

Tinkertoys employ cylindrical wooden spool connectors with holes drilled around the perimeter at 45° angles. Tinkertoy parts can be used to construct a Ferris wheel, but not a molecular model of methane. I was already teaching chemistry when sets (with appropriate connectors) became commercially available, so that students could construct molecular models of simple compounds. Those models, and all subsequent student model sets, are faithful to some (but only some) aspects of the geometry of the molecules of nature. Chemists have increasingly used virtual constructions to explore molecular structures as models of nature, with the calculated results displayed on computer screens and printed as two-dimensional representations. Virtual structures of giant molecules can be constructed in this way. The striking apparent reality of these representations is seductive.

Computer-generated images, from Photoshop constructions to the graphics of computer games, have demonstrated how easy it is to produce convincing images that bear little relationship to "the real world." Total reliance on virtual images of molecules is dangerous for any chemist. Comparison of virtual images with physical models for identical small molecules is a necessary intellectual calibration step in chemical education. My classroom experience, for example, convinced me that few students can internalize the concept of mirror-image isomers without physically handling a pair of nonsuperposable molecular models.

Every model incorporates assumptions, and these assumptions impose limitations. The constituent parts of Donald Caspar's self-assembly model are readily examined, and implications of his model building can be interpreted in the light of restrictions imposed by those parts. It is a formidable challenge for interpreters of virtual models to verify the full range of assumptions and limitations inherent in each of those models.

Ralph Erickson has described many sorts of physical objects that model the partitioning of space. Lord Kelvin in 1887 proposed body-centered cubic tetrakaidecahedra with six square faces and eight slightly curved hexagonal faces as optimal space-fillers with minimal surface area. H.M. Princen and P. Levinson examined (by numerical calculations) the extent to which the facial curvature reduces the surface area: about 0.2 percent. D. Weaire and R. Phelan proposed an alternative unit cell composed of six 14-sided polyhedra and two 12-sided polyhedra, with the curvature of faces optimized by computer calculations. Weaire and Phelan assert that their counterexample has "a remarkably large margin of superiority." The Kelvin Problem is an intriguing case study both in methodology and in presentation of conclusions about "symbiotic relationships among theory, mechanical models and the natural world."

Part III

Polyhedra in the Geometrical Imagination

13

The Polyhedron Kingdom Tomorrow

Marjorie Senechal

Now you've reached the kingdom's wild frontier: polyhedron theory at the research level. The natives in these parts speak a slightly different dialect from the artists and scientists you've met already, so listen carefully. (If you only grasp a phrase here and there, or even nothing at all, read on anyway: the sights—the chapters that follow—are not to be missed.)

What Is a Polyhedron (Yet Again)?

You have met enough polyhedra by now to be able to know that there is not one answer to this question, there are many.

The problem in choosing a definition was explained many years ago by Robert Frost in his poem,"Mending Wall": "Before I'd build a wall, I'd want to know what I was walling in, and what I was walling out." The delightful book *Proofs and Refutations*, written in the form of a classroom discussion, is required reading for all polyhedron enthusiasts. In it the author, Imre Lakatos, shows how a careful examination of the implications of a definition forces us to construct our walls very carefully. Consider, for example, the following excerpt:

GAMMA: A polyhedron is a solid whose surface consists of polygonal faces . . .

DELTA: Your definition is incorrect. A polyhedron must be a *surface*: it has faces, edges, vertices, it can be deformed, stretched out on a blackboard, and has nothing to do with the concept of "solid." *A polyhedron is a surface consisting of a system of polygons.*

TEACHER: For the moment let us accept Delta's definition. Can you refute our conjecture (Euler's formula, $F - E + V = 2$, which the class is trying to prove) now if by polyhedron we mean a surface?

ALPHA: Certainly. Take two tetrahedra which have an edge in common. Or, take two tetrahedra which have a vertex in common. Both these twins are connected, both constitute one single surface. And, you may check that for both $V - E + F = 3$.

DELTA: I admire your perverted imagination, but of course I did not mean that *any* system of polygons is a polyhedron. By polyhedron I meant a *system of polygons arranged in such a way that* (1) *exactly two polygons meet at every edge* and (2) *it is possible to get from the inside of any polygon to the inside of any other polygon by a route which never crosses any edge at a vertex.* Your first twins will be excluded by the first criterion in my definition, your second twins by the second criterion.

Delta believed that Alpha's twin tetrahedra are not polyhedra, but "monsters" which can and must be barred by a proper definition. Monster-barring, argued Lakatos, is often the reason that complicated, abstract definitions like Delta's second one appear in mathematics. (Unfortunately

M. Senechal
Department of Mathematics and Statistics,
Smith College, Northampton, MA 01063, USA,
e-mail: senechal@smith.edu; http://math.smith.edu/~senechal; http://www.marjoriesenechal.com

M. Senechal (ed.), *Shaping Space*, DOI 10.1007/978-0-387-92714-5_13,
© Marjorie Senechal 2013

they are usually presented to the student in a take-it-or-leave-it way, with no explanation of how or why anyone would ever come up with them.)

So, if you wish to define "polyhedron," you should think about the kinds of objects that you are willing to accept as polyhedra. For example, you might (or might not) want to include

- Star polyhedra
- Toroidal polyhedra
- Infinite polyhedra
- Polyhedra whose faces are skew polygons
- Either pair of Delta's tetrahedral twins
- A finite capped cylinder (it has three faces, two edges, and no vertices!)

Of course, you might want to include some of these in some cases, and exclude them in others, depending on what properties you are studying.

Branko Grünbaum's definition of "polyhedron" is widely accepted. First, he defines a polygon:

Definition 1 *A finite polygon is a figure formed by a finite sequence of vertices in E^3, $V_1, V_2, \ldots, V_n$, together with the edges $[V_i, V_{i+1}]$, $i = 1, 2, \ldots, n - 1$, and $[V_n, V_1]$. (An infinite polygon is defined in a suitably analogous way.)*

Next, he defines "polyhedron" this way:

Definition 2 *A polyhedron P is any family of polygons (called faces of P) that has the following properties: (i) Each edge of one of the faces is an edge of just one other face. (ii) The family of polygons is connected; that is, for any two edges E and E′ of P there exists a chain $E = E_0, P_1, E_1, P_2, E_2, \ldots, P_n, E_n = E′$, of edges and faces of P, in which each P_i is incident with E_{i-1} and with E_i. (iii) Each compact set meets only finitely many faces.*

(The term *compact* is defined in the phrasebook below.)

How does Grünbaum's definition compare with Delta's second one?

Notice that Definition 1 does not require the vertices of a polygon to be coplanar. Thus the polyhedra permitted by Definition 2 may have

skew nonplanar "faces"—and there may be infinitely many of them! Definition 2 is broad enough to include most of the polyhedra you have met in this book, and then some. On the other hand, it may be too broad for some purposes: for example, we may want to restrict ourselves to convex polyhedra. It might not be broad enough for other purposes, however. Do we really want to exclude the twin tetrahedra which share only a vertex, for example? As you saw in Chapter 11, they are used to interpret molecular structures. Other important applications call for polyhedra with movable parts. Will this definition be able to accommodate the demands of scientists for a broader theory? (See Chapter 14.)

A Polyhedral Phrasebook

Specialized discussions of problems in polyhedron theory use technical mathematical terminology; a nonspecialist needs a phrasebook. Here, in intuitive language, is a description some of the less familiar mathematical terms that you will encounter. (You should skip the next two paragraphs if you already understand the italicized words.)

We already defined the word *polygon* (Definition 1). Since this word has metric connotations (e.g., angles and edge-lengths can be defined) which are not needed in the purely combinatorial theory of polyhedra, some authors prefer to speak of a *2-cell*, a floppy polygon that can be deformed into a circle. You can think of a *2-manifold* as a *two-dimensional surface in three-dimensional space*, such as a plane, an infinitely long cylinder, a sphere, or a torus, possibly deformed. Similarly, the *2-sphere* is the ordinary sphere in three dimensions (surface only, of course). The prefix "2-" emphasizes the fact that the surface is two-dimensional, as opposed to three-, four-, or n-dimensional. A *2-cell complex* is a 2-manifold every point of which belongs to the interior or boundary of a 2-cell.

The terms "closed," "bounded," and "boundary" can lead to confusion unless you remember that each of them has a precise mathematical meaning. A subset of space is *bounded* if it is of finite extent, that is, if it can be entirely enclosed in some sphere of finite radius (even if the radius has to be very large). On the other hand, to say that a set is bounded does not mean that it contains or even has a boundary! The word *boundary* refers to an intrinsic property of the set, whereas "bounded" refers to the space in which the set lies. A set of points is *closed* if it contains all its boundary points. For example, the set of points in the plane *interior* to a circle is not closed, although it is bounded, because it contains points arbitrarily close to the circle itself, but the points of the circle do not belong to it. A set is *compact* if it is closed and bounded; for example a circle together with its interior points is a compact set. But the (infinite) plane is not compact because it is not bounded. Finally, you can think of an *orientable* manifold as a two-sided surface; the plane, the sphere, and the torus are orientable, but the Möbius band is not. As mentioned earlier, the *genus* of a manifold is its number of holes: "hole" is used here in the sense that a torus has a hole; it is not the kind of hole which punctures the manifold.

You can test your understanding of this terminology by trying to decipher and interpret the definitions of the word "polyhedron" that are given below (all taken from papers in this section). We leave it to you to determine which shapes are walled in by each of them, and which are walled out. Study the definitions carefully before deciding! It is easy to be misled. For example, Delta did not seem to realize that her definition did not exclude polyhedra like our toroidal hat, which she despised (she called them "non-Eulerian pests")!

Definition 3 *A (convex) polyhedron is a bounded subset of E^3 which can be expressed as the intersection of a finite number of closed half-spaces.*

Definition 4 *A polyhedron is a cell complex whose point set is a closed orientable 2-manifold, each of whose 2-cells is an affine polygon that is not coplanar with any adjacent 2-cell.*

Definition 5 *A polyhedron (in three-dimensional space) is a compact 2-manifold that has no boundary and can be expressed as a finite union of plane polygonal regions.*

Why a Theory of Polyhedra?

What's all this then? Why don't we simply take polyhedra as we find them, admire them for their beauty and the wonderful things that they represent? There are many reasons, besides intellectual curiosity and aesthetic pleasure, why mathematicians are engaged in polyhedral research.

Many practical problems involve polyhedra built to specification, either by nature or by man. To understand these structures, and to be able to create new ones, we must know what the specifications are, why they are necessary, and what sorts of objects are characterized by them. For example, if we want to design bridges and buildings that stay up, we must study the form and dynamics of trusses and braces, and this leads to questions about polyhedral stability (see Chapters 14, 18, and 21). Other problems motivated by the needs of science and technology are discussed in this part of the book.

Moreover, the need for a theory of polyhedra arises almost spontaneously when we try to make clear to colleagues and students what we are talking about. For example, if someone asks what we mean by the word "polyhedron" we might begin, like Gamma and Delta, by specifying certain characteristics we think all polyhedra have in common; these are usually characteristics of the polyhedra that we already know. Having listed them, it is then natural to wonder, like Alpha, whether all objects which have these characteristics are necessarily things that we want to call polyhedra. Or, we might ask whether there are new, as yet undiscovered, structures which also have these properties.

This line of questioning often leads us to generalize familiar concepts. For example, Definition 1 implies that there are many kinds of regular polygons in addition to the convex and star plane polygons: it implicitly permits regular prismatic and antiprismatic polygons (a prismatic, or antiprismatic, polygon is a finite nonplanar zigzag polygon whose vertices are the vertices of a prism or antiprism, respectively). It also permits regular infinite polygons whose edges lie on straight lines, or zigzag, or are helical. By a careful analysis of the possible regular polyhedra with such polygons as faces, Grünbaum found that, in addition to the regular convex and star polyhedra, there are six additional families: the infinite regular plane tessellations, the Petrie–Coxeter polyhedra, a class of nine regular polyhedra with finite skew polygons as faces, infinite regular polyhedra with finite skew polygons as faces, regular polyhedra whose faces are infinite zigzag polygons, and regular polyhedra whose faces are infinite helical polygons! In this way our understanding of regular three-dimensional polyhedra has been greatly enriched.

Polyhedra can be generalized in other ways. One of these is to investigate their analogues—"polytopes"—in higher dimensions. The *regular convex* polytopes in n dimensions were discovered by Ludwig Schläfli in the early 1850s. In spaces of five or more dimensions, there are only three of them: the higher dimensional analogues of the tetrahedron, the cube, and the octahedron. But in four-dimensional space, there are six regular convex polytopes: the three just mentioned, one with 120 dodecahedral cells (cells are three-dimensional "faces"), one with 600 tetrahedral cells, and one with 24 octahedral cells. These six polytopes are not easy for us to visualize, but there are various geometric and algebraic techniques which, together with computer graphics, can help a great deal. One of the most interesting of these is described in Chapter 20.

Or we may generalize the definition of regularity to apply to a broader class of polyhedra. Chapter 17 concerns such a generalization of the regular solids, called Platonohedra. Here the polyhedra are "equivelar"; that is, their faces are regular k-gons and q meet at each vertex, but the polyhedra can be toroidal or indeed of any genus (the genus of a polyhedron is the number of its holes. Convex polyhedra have genus zero, like the sphere; toroidal polyhedra have genus one, and so forth.). The "symmetries" of these objects include some transformations which preserve their combinatorial structures but, strictly speaking, not their metrical properties.

Sometimes problems are generalized because the original problem is too hard. The ancient question: "Which polyhedra fill space?" is one of these. The difficulties are so severe that it makes good sense to begin with the more tractable one: "Which combinatorial types of polyhedra fill space?" You will read more about this problem in (Chapters 16 and 22).

Another impetus for the development of a theory of polyhedra is the need to clarify fundamental concepts. As we have seen, the history of polyhedra is long and it has many roots. On close examination, we sometimes find that well-entrenched definitions and classifications are not as clear as we once thought. Or, as new classes of polyhedra are discovered, we may find old characterizations inadequate. This confusion leads to new questions about what it is we are talking about, and these questions generate research. For example (see Chapter 15), there is no problem reconciling the several widely held (but distinct) concepts of duality as long as we are talking about convex polyhedra, but with more general types of polyhedra it is no longer even clear what "dual" is supposed to mean.

Finally, it often happens that in the course of investigating one problem, suprising and illuminating links are found with others. The equivalence of the seemingly unrelated concepts of convex polyhedra, Dirichlet tessellations, and "spider webs" discussed in Chapter 18 is an intriguing example.

Polyhedral Themes

In addition to the variety of motivations for studying polyhedra theory, several themes run throughout the following papers and problems.

Symmetry. The aesthetic link between symmetry, beauty, and perfection was undoubtedly the reason why the regular polyhedra were first noticed and singled out for attention millennia ago. A great deal has already been said about symmetry in this book, so we will not review the basic concepts here. The following remarks, however, may be helpful to keep in mind while reading the following chapters.

Symmetry theory is not a museum piece, but a valuable tool in the study of polyhedra. We have just seen that symmetry often suggests interesting generalizations. It can also be a guide in searching for new kinds of polyhedra. The semiregular polyhedra discovered by Archimedes were (at that time) a new class of highly symmetrical polyhedra, and Archimedes probably used symmetry considerations to ensure that this would be so. In fact, the eleven that can be obtained from the Platonic "solids" by truncation are obtained by truncating symmetrically: first a vertex (or edge) of a solid is truncated in such a way that its contribution to the symmetry of the polyhedron is not destroyed, and then all the other vertices are truncated in exactly the same way. This ensures that the truncated polyhedron has all the symmetries of the original regular one.

A requirement of symmetry can also help us to restrict a problem to a reasonable size. Obviously all sorts of idiosyncratic constructions are admitted under Definition 2; we cannot hope to survey them all. By focusing on those that have some symmetry properties, however, we obtain a manageable class of objects. And because symmetry is a hierarchical concept, we can broaden our study later by omitting its restrictions one by one. Symmetry can help to organize and present complex information. Several contributors (see Chapters 3, 5, and 19) point out the effectiveness of symmetry arguments in characterizing the coordinates of certain polyhedra; in his short chapter Barry Monson points out an interesting connection between this problem and the theory of numbers. (Indeed, the theory of polyhedra has connections with, and implications for, almost every branch of mathematics.)

Symmetry theory can also be used to study properties of polyhedra that are inadequately characterized by their geometry. For example, we have seen that a carbon atom is often represented as a regular tetrahedron because it is 4-valent. But when an atom joins with other atoms in a molecule or crystal, its four bonds may no longer be equivalent. To incorporate this information into the geometry of the tetrahedron, we can color its vertices or faces in such a way that equivalence is properly indicated. This leads us to the concept of "color symmetry," which has been extensively studied both for polyhedra and tessellations of the plane. The colored tilings of interlocking creatures designed by M. C. Escher are typical examples of patterns with color symmetry, but less orderly colorings are important too. Again we must decide what we want to wall in and what we want to wall out. There are many interesting colored polyhedra which do not satisfy even the least restrictive definitions that have been proposed. Clearly the theory is still evolving.

Networks. By now you have memorized Euler's deceptively simple formula for the vertices, edges, and faces of a convex polyhedron, $V - E + F = 2$. This formula turns out to have a wealth of implications, even though it says nothing at all about angles, edge lengths, or other metric properties of polyhedra, being concerned only with the networks formed by edges and vertices. For example, it implies that there are at most five regular polyhedra (in addition to the so-called digonal polyhedra and dihedra). It is very suprising that this ancient and famous result does not in fact depend on either the symmetry or metric properties of the polyhedra, but only on their combinatorial properties.

The numbers of faces, vertices, and (consequently) edges of a polyhedron constitute its *f−vector* (V, E, F). Some of the questions one might ask about f-vectors are: If V, E, F are integers which satisfy the Euler relation, are there any corresponding polyhedral networks of edges and vertices? For example, there is no polyhedron with f-vector $(0, 4, 6)$, but there are two very different polyhedra with f-vector $(8, 12, 6)$. (One is the cube; what is the other?) Thus the relation $V - E + F = 2$ is a necessary condition for the existence of a polyhedron with f-vector (V, E, F) but it is not always sufficient. (The additional conditions that must be satisfied by V, E, and F have been found for the three-dimensional case, but the analogous problem in higher dimensions remains unsolved.)

Another important question is: If a polyhedron with f-vector (V, E, F) does exist, what kinds of faces does it have? (How many triangles, how many quadrilaterals, and so forth?) In other words, what is its *face sequence* $[f_k]$, $k = 3, 4, 5, 6, \ldots$, where f_k is the number of $k-$gonal faces of the polyhedron? A famous equality, which is a direct consequence of Euler's formula, states that if the polyhedron is trivalent (that is, if three edges meet at each vertex), then

$$3f_3 + 2f_4 + f_5 = 12 + \sum_{k>6}(k-6)f_k. \quad (13.1)$$

This is a condition that the face sequence of a polyhedron must satisfy, but it does not guarantee that a polyhedron with such a face sequence exists.

Euler's formula has been generalized to polyhedra of higher dimension, and of other genera. For a polyhedron of genus g, the formula becomes

$$V - E + F = 2 - 2g. \quad (13.2)$$

Thus for the torus the right-hand side is zero. A great deal of effort has been, and is being, devoted to finding analogues, for polyhedra of genus greater than one, of *Eberhard's theorem*, which is closely related to Equation 13.1: For every finite sequence of nonnegative integers $[f_k, k \geq 3, k \neq 6]$ satisfying Equation 13.1 there are values of f_6 such that a polyhedron with face sequence $[f_k]$ exists.

Next we might ask: "How many distinct combinatorial types of polyhedra belong to each sequence $[f_k]$?" This is an unsolved problem. The numbers of combinatorial types belonging to each f-vector are known only for polyhedra with eleven or fewer faces. There is no apparent pattern to these numbers, but there are surprisingly low bounds for the number of combinatorially distinct polytopes (of a certain type) with n vertices in $d-$dimensional space.

Geometric Realization. Even when a polyhedron network exists, it may happen that polyhedra with straight edges and planar faces cannot be constructed according to these plans. If the edges of the digonal networks, for example, are straightened out, they all collapse to a single line. This raises the question of determining the conditions

under which a polyhedron with certain properties can be *realized geometrically.*

In studying the problem of realization, a fundamental theorem is that of Steinitz, which characterizes the types of planar graphs that correspond to convex polyhedra in three-dimensional space. (A planar graph is a network, or 1-*skeleton* of edges and vertices. The 1 indicates the one-dimensionality of the edges, which can be drawn in the plane without any unintended crossing of edges. Schlegel diagrams are planar graphs; star polygons are not.) Among the unsolved problems in polyhedron theory are several concerning realizations of combinatorial polyhedra of higher genera. The long-range goal of such research is of course to find appropriate analogues of Steinitz's theorem.

Rigidity. Realization questions lead to the study of rigidity and stability. Because of the power and comparative simplicity of the combinatorial approach, the metrical theory was relatively neglected for many years. But high-speed computation has changed the picture. We now can ask, and sometimes answer, which realizations are rigid, which are flexible, and which will fall apart. This matters, in fields from architecture to robotics, in which metric properties of polyhedra play an indispensable role. (Most polyhedron models have definite angles and edge lengths; most buildings are built to precise architectural plans.) The theory of the rigidity of polyhedra and polyhedral frameworks has enjoyed a renaissance in recent decades.

The Polyhedrist's Bookshelf

This is the end of the guided tour of the Polyhedron Kingdom. But before you explore the frontier, take a few minutes to browse the Polyhedrist's Bookshelf. Mathematicians have been writing about polyhedra almost as long as they've been making them. Here are some of the highlights of the past 4,000 years, assembled with the help of Joseph Malkevitch and other contributors to this book. Today's theorists stand on the shoulders of these giants.

Antiquities: 1850 BCE–1599 CE

The Moscow Mathematical Papyrus, *circa* 1850 BCE, is held in the Pushkin State Museum of Fine Arts, Moscow.

The Rhind Papyrus, *circa* 1650 BCE, also known as Papyrus British Museum 10057, is held in the British Museum, London.

Archimedes, *The Sphere and the Cylinder*; the original codex on which this was written is lost, but see *The Works of Archimedes*, by Sir Thomas Heath, 1897; reprinted by Dover Publications, 2002.

Plato's *The Timaeus* is available in many languages and in many editions. Plato's contributions in this dialogue are invoked throughout this book, as are Euclid's:

Euclid's *Elements*. The original is lost, but Sir Thomas Heath's classic translation is not: *The Thirteen Books of Euclid's Elements,* 1909. Reprinted by Cambridge University Press in 1925 and by Dover Publications in 1956.

For the works of Pappus, see *La Collection Mathématique* translated by P. Van Eecke and published in Paris by De Brouwer/Blanchard in 1933.

Luca Pacioli's *Divina Proportione* was published in Milan in 1509; reprinted by Fontes Ambrosioni, also Milan, in 1956.

Albrecht Dürer's famous *Unterweysung der Messung mit dem Zyrkel und Rychtscheyd* was published in Nürnberg in 1525, and in English translation by Abaris Books in 1977.

Classics: 1600 CE–1899 CE

Johannes Kepler's *Harmony of the World—Harmonices Mundi*—was first published in Linz in 1619.

Descartes' manuscript was lost at sea, but P. J. Federico restored it through excellent detective work: see *Descartes on Polyhedra*. New York: Springer-Verlag, 1982.

For Euler's work, see *Novi Commentarii Academiae Scientiarum Petropolitanae 4*, 1752–53. Or see http://www.leonhard-euler.ch/.

Louis Poinsot described the four star polyhedra in "Mémoire sur les polygones et les polyèdres" in *Journal de l'École Polytechnique 10* in 1810.

For Cauchy's two important papers on polyhedra, see the *Journal de l'École Imp. Polytechnique* (1813).

M. E. Catalan's "Mémoire sur la théorie des polyèdres" was also published in that journal, but in 1865.

Modernists: 1900–Present

M. Brückner's *Vielecke und Vielfläche* was published by Teubner in Leipzig in 1900.

Max Dehn's solution to Hilbert's Third Problem appeared in the *Mathematische Annalen* in 1901.

Ernst Steinitz's "Polyeder und Raumeinteilungen" appeared in Volume 3 of the *Encyklopädie der mathematischen Wissenschaften* which Teubner in Leipzig published between 1914–1931.

A. D. Alexandrov's *Vupyklue Mnogogranniki* (1950) was republished as *Convex Polyhedra* by Springer-Verlag, Berlin, in 2005.

H. S. M. Coxeter's *Regular Polytopes* was first published in 1948; today it is available in several editions, including a paperback by Dover Publications.

Branko Grünbaum's *Lectures on Lost Mathematics*, though available only in a mimeographed edition (1978), sparked research bringing tiling theory and rigidity theory back to life after a long hiatus, synergetically timed for the computer revolution.

Branko Grünbaum's *Convex Polytopes* first appeared in 1967; a second edition was published by Springer in 2003.

14

Paneled and Molecular Polyhedra: How Stable Are They?

Ileana Streinu

Polyhedral models can be physically constructed in a variety of ways. The inexpensive methods described in Chapter 2 include sticks connected at endpoints, or paper, creased and glued with tape along edges. But the resulting structures are not always sufficiently *stable*: sticks may slip off their connecting joints, paper bends, and even when sturdy carton is used, not having enough tape may lead to loose paper ends or a flexible polyhedral model.

Popular building kits use *rigid* parts, and fall roughly into three categories:

Bar-and-joint frameworks. These models are made of rigid bars connected via fully rotatable (so-called *universal*) joints. In some popular building kits *(Magz, Roger's Connection)*, magnetized rods of various sizes and steel bearings are used.

Molecular structures. Ball-and-stick kits are routinely used by chemists to build molecular models, and polyhedra made with *ZomeTool* have a similar flavor. The spherical atoms, made of sturdy plastic, have holes drilled through them, in which rigid sticks ("bonds") are inserted for connectivity.

Paneled polyhedra. Rigid triangles, squares and other polygonal shapes made from some sturdy material are connected along their edges through hinges. Some popular kits such as *Poly-*

dron have rigid plastic shapes hinged with small attachments along edges.

Before setting off to build a polyhedron with one of these kits, we may ask: will the model hold? For instance, is a bar-and-joint buckeyball (see Chapter 9) stable? What if one of the rods is dropped: will it still hold? One "feels" that the molecular models are more stable than the bar-and-joint ones, but why? Which models are easier, or harder to build—and why? Will they hold if bonds or hinges are broken? What about the paneled models, are they more stable? What if some of the hinges are broken? If our supply of good parts diminishes, for instance when some of the polydron pieces lose their connecting hinges, is it possible to arrange the parts in such a way that a certain polyhedron can still be built as a stable shape? Or, if we just build the old-fashioned paper-and-tape models, and run out of tape, we may want to know where we shall safely leave the pieces unhinged, but still have a stable (rigid) polyhedral model. These questions refer to the rigidity, flexibility and minimality of these structures. Let us briefly introduce these mathematical concepts, and take the first steps towards formalizing some of these questions.

Rigidity. If we built a structure from rigid parts, connected according to the rules of the model (joints, hinges), can it be continuously deformed while preserving the rigidity of the parts and the connectivity? If not possible, the structure is said to be *rigid*, otherwise it is *flexible*.

Flexibility. If a structure is flexible, how can it be deformed? For instance, how many and which

I. Streinu
Department of Computer Science, Smith College,
Northampton, MA 01063, USA,
e-mail: streinu@cs.smith.edu

M. Senechal (ed.), *Shaping Space*, DOI 10.1007/978-0-387-92714-5_14,
© Marjorie Senechal 2013

bonds can be rotated independently to change the molecules' shape? A flexible polyhedron made from polydron pieces with broken hinges can be turned into a rigid one if some hinges, perhaps not all, are fixed: how do we find which ones?

Minimality. If the shape is rigid, could we have achieved it with fewer constraints? For instance, could we have used less tape to glue together a polyhedron model made from paper folded along creases? How reliable is the structure: which of the hinges of a polydron model can be broken, without destabilizing the structure?

Redundancy. This last question is related to the *independence* and *redundancy* of building elements. A bar (in a bar-and-joint framework) is redundant if it can be removed without changing the rigidity. Otherwise, it is *independent*. Similarly, some bonds (in a molecular structure), or faces and hinges (in a paneled polyhedron) can be removed without changing the rigidity, but in this context, the concepts of redundancy and independence turn out to be more subtle. How do we detect redundant and independent parts?

Building physical models of polyhedra is complemented nowadays with building electronic models, using Computer Aided Design (CAD) systems such as SolidWorks. The geometric constraints defining them (lengths of bars, rigidity of faces, incidences between faces along hinges, etc.) induce systems of polynomial equations that must be solved by the software. Insufficient constraints, which induce flexible structures, are reflected in under-constrained systems of equations. Redundant constraints lead to over-constrained algebraic systems, which are typically rejected by most CAD system engines. Understanding which sets of constraints *minimally* define the structure can be used to guide the user in the design process, rather than rejecting the construction without any explanation. These practical concerns motivate our study, in this chapter, of the following problems:

- **Minimal rigidity and independence.** We will see that when all the constraints imposed by the geometry of the polyhedron are present, the bar-and-joint structures are minimally

rigid, but the other two frameworks are in general heavily dependent. In the molecular model, the minimally rigid ones can be characterized in terms of an underlying graph.

- **Eliminating dependencies.** When the structure is dependent, how do we eliminate the dependencies? For molecular structures, we give an efficient algorithm based on the pebble game paradigm.

- **Structural properties.** In the process of eliminating dependencies, the polyhedron may become flexible. We investigate structural properties related to degrees of freedom and rigid component distribution.

We won't be able to answer all the questions we set up to, so we will conclude with some open problems and a conjecture for a rigidity characterization of panel-and-hinge polyhedra.

Bar-and-Joint Frameworks

The rigidity and flexibility of arbitrary bar-and-joint frameworks in 3D space is notoriously difficult to assess, but not for polyhedra: several theorems, some going back to the early 1800s, lay down the foundations of one of the classical results in rigidity theory. We will use this very well understood case to introduce the concepts and methodology required for evaluating (in the next sections) the rigidity of polyhedra built as other types of mechanical structures.

The starting point is Cauchy's famous theorem (1813), the oldest result in Rigidity Theory:

Theorem 14.1. *Let P and P' be two convex 3D polyhedra, with the same combinatorial structure and congruent corresponding faces. Then P and P' are congruent.*

This uniqueness theorem implies rigidity for *triangulated convex polyhedra*, viewed as spatial objects made from rigid bars connected by universal joints via a sequence of theorems and implications due to Dehn, Weyl and A.D. Alexandrov in the early 1900s.

In a different direction, these bar-and-joint structures are the 1-skeleta of convex polyhedra, and they are characterized by Steinitz's theorem as 3-connected planar graphs. A *realization* of such a graph $G = (V, E, F)$ as a polyhedral surface is a mapping $p : V \mapsto R^3$ of the vertices V to points in space, such that the edges $e \in E$ are mapped to line segments, and for all faces $f \in F$, the vertices incident to f are mapped to coplanar points. A realization need not be convex, and may even have self-intersections. When the faces are triangles, the coplanarity condition is automatically satisfied. A stronger form of rigidity, applicable to an entire class of combinatorial objects (graphs, in this case) rather than specific polyhedral realizations (such as the convex ones) is captured by a later theorem due to Gluck:

Theorem 14.2. *3D realizations of 3-connected triangulated planar graphs as bar-and-joint frameworks are generically infinitesimally rigid. This means that all but a measure zero set of the realizations are rigid.*

Infinitesimal rigidity is a stronger (and more technical) condition, but for our purpose here we do not have to define it formally. It suffices to know that it is easier to check mathematically and computationally, and that it implies rigidity.

When a combinatorial structure is infinitesimally rigid in all but a measure zero set of realizations, we say that it is *generically rigid*; in other words, 3-connected triangulated planar graphs, viewed as bar-and-joint frameworks, are generically rigid.

From the point of view of modeling polyhedra as bar-and-joint structures, this property says that for *most* of the possible lengths that we may choose for the bars, a polyhedron will be rigid. Sometimes, very rarely, it may be infinitesimally flexible, and—even more rarely—it may even be flexible. But if it is convex, it is guaranteed to be rigid. For simplicity, from now on we drop *infinitesimal* and simply refer to polyhedra as being *rigid* rather than *infinitesimally rigid*.

Finally, we focus on the *minimality* property exhibited by this classical rigidity result. When all the faces are triangles, Euler's theorem implies that such a planar graph on $|V| = n$ vertices spans exactly $|E| = 3n - 6$ edges and $|F| = 2n - 4$ faces; this is the maximal number of edges, respectively faces, that a planar graph may have.

Viewed as bar-and-joint frameworks, the triangulated planar graphs are *minimally rigid*: if any of the edges of the skeleton is removed, the structure becomes flexible. In other words, if we consider arbitrary 3-connected planar graphs, not just the triangulated ones, and fix the lengths of the edges (but do not impose any other constraints, such as planarity of the faces), then the rigidity results no longer hold: the polyhedra are flexible. This is why the buckeyball constructed from magnetic bars with ball joints will fall apart. One would have to add extra bars (e.g., in such a way that the faces get triangulated) to obtain a rigid structure. But once the minimally rigid threshold has been attained, additional bars become redundant: not only do they violate the planarity of the underlying graph, but also they are not needed to keep the structure rigid. A bar which can be removed from a framework without affecting its rigidity properties is called *dependent* or *redundant*. A structure containing no redundant bars is called independent.

Now let's look at molecular frameworks and panel-and-hinge polyhedra, which are much less understood than bar-and-joint frameworks.

Molecular Frameworks

A (hinged) *molecular framework* is a collection of atoms connected by covalent bonds, in such a way that the angles between pairs of incident bonds are fixed. We note from the outset that the fixed angle assumption, while predominant in most molecular models, does not apply to all atoms in nature. We work henceforth with abstract molecular frameworks, with no mention of the nature of the atoms. We first turn them into mechanical structures made from rigid parts interconnected via flexible joints to arrive at an abstract model of a *body-and-hinge structure*.

To assess the rigidity of bar-and-joint polyhedral frameworks, generically, we introduced an abstract, combinatorial structure: a planar graph. The characterization theorem for the generically rigid polyhedral frameworks stated that this graph must be triangulated and 3-connected. Similarly, to a body-and-hinge framework we associate a multi-graph. A theorem, due to Tay and Whiteley, characterizes minimally rigid generic body-and-hinge frameworks in terms of a sparsity condition of this graph.

Molecular Frameworks as Body-and-Hinge Frameworks

A finite set of rigid bodies, together with a finite set of rotatable hinges, rigidly attached to pairs of bodies, forms a body-and-hinge framework. The simplest example consists of two bodies connected with one hinge, and has one internal degree of freedom (dof).

Rigidity of generic body-and-hinge frameworks.

To a body-and-hinge framework, we associate a graph G, whose vertices correspond to the bodies and edges to the hinges. The multi-graph $5G$, obtained from G by taking each edge with multiplicity 5, is called the *Tay graph* associated to the framework.

A multi-graph on n vertices is $(6, 6)$-sparse if every subset of $n' \leq n$ of its vertices spans at most $6n' - 6$ edges. It is $(6, 6)$-tight if, in addition, it has exactly $6n - 6$ edges. We also say that the graph satisfies the $6n - 6$ counts *hereditarily*. If a graph contains a spanning $(6, 6)$-tight subgraph, we say that it is spanning, or combinatorially-rigid; otherwise, we say it is combinatorially-flexible. A graph which is not sparse is called combinatorially-dependent.

Tay and Whiteley proved:

Theorem 14.3. *A body-and-hinge framework is minimally rigid if and only if its Tay graph is $(6, 6)$-tight, and rigid if and only if its Tay graph contains a $(6, 6)$-tight spanning subgraph.*

Moreover, if the Tay graph is $(6, 6)$-sparse, the framework is flexible and contains no redundant hinges.

A body whose incident hinges are coplanar can be viewed as a *panel* spanning the plane of its hinges. A panel-and-hinge framework is a special type of body-and-hinge framework, where all the bodies are panels. Until recently, it was not known whether Tay and Whiteley's theorem also holds for panel-and-hinge structures.

Another specialization arises when the hinges incident to a body are all concurrent in one vertex. This situation occurs in the modeling of molecular frameworks.

A (hinged) *molecular framework* is a collection of atoms connected by covalent bonds in such a way that the angles between pairs of incident bonds are fixed. To a molecular framework, we will associate a body-and-hinge structure, as follows.

An atom incident to more than two other ones (coordination number at least 2), together with the set of all the bonds and atoms incident to it, is mapped to a 3D rigid *body*. We say that the body is *centered* at the atom. This is an accurate model, because of the assumption that the bonds incident to each atom are rigidly attached to it. The bond between two adjacent atoms, when viewed as connecting the two bodies corresponding to these two atoms, behaves like a hinge: it allows the rotation of one body around the other.

A characteristic of molecular frameworks is that the hinges incident to one atom are *concurrent*. Concurrency and coplanarity of lines are dual concepts in projective geometry, which turns statements about coplanar frameworks (panel-and-hinges) to statements about frameworks with concurrent hinges (molecular). Tay and Whiteley conjectured that their theorem correctly characterizes generic panel-and-hinge and molecular frameworks (respectively). Their guess was proved true by Katoh and Tanigawa.

Theorem 14.4 (Rigidity of Generic Panel-and-Hinge and of Molecular Frameworks). *A molecular body-and-hinge framework, or a panel-and-hinge framework is generically rigid if and only if its associated Tay multi-graph spans six edge-disjoint spanning tree (or, equivalently, a $(6, 6)$-tight graph). It is independent if and only if its Tay multigraph is $(6, 6)$-sparse.*

In the rest of this chapter we look at consequences of this breakthrough result for the generic minimal rigidity of molecular and panel-and-hinge polyhedral models.

Independent Frameworks by Edge Subdivisions

In a molecular framework G with no leaves (vertices of degree-1), all the edges become hinges. Tay's conditions for independence, respectively minimal rigidity, state that $5m \leq 6(n-1)$, where m is the number of edges in G. This implies that the independent frameworks come from very sparse graphs: the number of edges m is at most $\frac{6}{5}(n-1)$, hereditarily. Frameworks with vertices of degree at least 3 have at least $\frac{3}{2}n$ edges. Hence, such graphs lead to over-constrained molecular frameworks, and the degree-of-dependency (at least $\frac{3}{2}n - \frac{6}{5}(n-1) \approx \frac{3}{10}n$) is large (linear in n).

Independent molecular frameworks must therefore contain vertices of degree 2. To eliminate dependencies we need to introduce such vertices. A *subdivision* of a bond (an edge) consists of *splitting* it in two by the placement of an atom.

Proposition 14.5. *Subdividing a bond either decreases by one the degrees-of-dependency, or increases by one the degrees-of-freedom of a molecular framework.*

Proof 1 Let G be the graph associated to the molecular framework; $5G$ is its associated Tay graph. Notice that each bond of the molecular framework induces an edge in G, except when one of its incident atoms has coordination number 1 (it is incident to exactly one other atom, or is a *leaf*). Subdividing a bond incident to a leaf atom amounts to adding a vertex of degree-1 to G; this obviously increases the dofs by one. By the structure theorem for (k, k)-sparse graphs, applied for $k = 6$ to $5G$, an edge either lies in a rigid component or is a free edge joining two vertices lying in two distinct rigid components. But such an edge comes in groups of 5 parallel edges in $5G$ and corresponds to an edge in G, and to a bond in the original framework. If the edge (bond) chosen for subdivision lies in a rigid component (of m' edges on n' vertices) of the molecular framework, then in $5G$ we have $5m' \geq 6n' - 6$, with $5m' - (6n' - 6)$ being its degree-of-dependency. The subdivision increases by one

both the number of vertices and edges in G (and by 5 the number of edges in $5G$), thus decreasing by one the degree-of-dependency (if non-zero), or increasing by one the dofs (if the dod is zero). If the edge is not in a rigid component, then the subdivision creates a new component (consisting of just one body, centered at the new atom) and a new bond, i.e., adds 5 new edges to $5G$. This increases by one the dofs.

Given a molecular framework, define a 2-chain as a maximal sequence of degree 2 vertices (i.e., the endpoints have degree at least 3). The length of a 2-chain is the number of vertices of degree 2 contained in it. The edge subdivision operation creates 2-chains. Proposition 14.6 below captures a simple relationship between the length of the 2-chains and a Tay-graph type of modeling for the original framework (with minimum degree at least 3), described next.

Let G be a graph with minimum degree at least 3 and G' a subdivision of it, with no 2-chain longer than 4. Let $5G'$ be the Tay graph associated to G', and G'' be a multi-graph associated to G' as follows: if two vertices of degree at least 3 in G' are joined by a 2-chain of length d, for some $0 \leq d \leq 5$, then G'' contains $5 - d$ parallel edges between the corresponding vertices.

Degrees-of-freedom (dof) and degrees-of-dependency (dod). Let G be a $(6, 6)$ sparse graph with n vertices and $m \leq 6n - 6$ edges. The difference $6n - 6 - m$ counts its *degrees-of-freedom (dof)*, by analogy with the special case (defined above) when the counting is applied to evaluate rigidity, resp. flexibility of mechanical frameworks. If the graph is rigid, but not sparse, it has more than $6n - 6$ edges; the surplus $m - (6n - 6)$ counts its *degrees-of-dependency (dod)*. An arbitrary graph, however, may be both flexible and dependent. To understand that, we must look at induced subgraphs.

A rigid subset of vertices $V' \subset V$ (spanning a tight-subgraph), and maximal (as a set of vertices) with this property, is called a *(rigid) component*. Any graph G which is not rigid can be decomposed into components. If a rigid component is replaced by a tight one (spanning the same vertex subset), the resulting graph G' is sparse. The degrees-of-freedom of G are, by definition, the degrees of freedom of G'. The degrees-of-dependency of G are obtained by summing the dods of all the rigid components. We note that in this case, the components are vertex-disjoint.

Proposition 14.6. *There is a one-to-one correspondence between the rigid component decompositions of the two Tay graphs $5G'$ and G'', as follows. The rigid components of $5G'$ and G'' contain the same degree ≥ 3 vertices, joined by either 2-chains (in $5G'$), or by the corresponding multi-edges in G''. The free edges in G'' correspond to 2-chains in G' that do not fall into rigid components.*

The proof is a direct consequence of Proposition 14.5. This proposition will be the basis of simple algorithms described below. Another direct consequence of the dof counts is the following corollary.

Corollary 14.7. *Any 2-chain of length ≥ 6 belongs to a flexible molecular framework.*

To obtain independent subdivided frameworks with a prescribed number k of degrees-of-freedom, the splitting operation must be applied in a specific manner (i.e., not all subdivisions will work). Moreover, the subdivision is not unique. Here is a simple algorithm for accomplishing this task. It relies on the pebble game algorithm of Lee and Streinu for $(6, 6)$-sparsity; for completeness, a brief description of the pebble game appears below.

Algorithm 14.8. (Eliminating dependencies by edge subdivisions)
***Input** a molecular framework G of min degree ≥ 3, and an integer k*
***Output** a subdivision G' of G, with no dependent edges and k dofs*
1. Run the $(6, 6)$-pebble game on $5G$.
2. For all edges ij of G: if d edges have been rejected in $5G$ between the pair of vertices i and j, insert, in the output framework, a 2-chain of length d between the nodes i and j.
3. Arbitrarily subdivide in two k edges of the graph obtained so far.

The correctness is a direct consequence of Proposition 14.6.

Algorithm 14.9. (The $(6, 6)$-pebble game)
***Input** a multi-graph G*
***Output** a maximal $(6, 6)$-sparse subgraph, and a set of rejected edges*

The algorithm maintains pebbles on the vertices and a collection of already accepted edges that are given an orientation and induce a directed graph.

1. As initialization, 6 pebbles are placed on each vertex of the graph.
2. The edges are considered one after another, in an arbitrary order, and each one is either accepted, if 7 pebbles can be gathered on its two endpoints, or rejected otherwise.
3. Once an edge is accepted, it is oriented in an arbitrary way (tail to head), and a pebble is removed from its tail vertex.
4. The gathering of pebbles is performed by searching for them, using a standard depth-first search technique, in the directed graph of all currently accepted edges.
5. When a pebble is found along a search path, the orientation of the edges on the path is reversed, and the pebble is "brought" to the starting point of the path.

You can play the pebble game (and learn much more about it) at http://linkage.cs.umass.edu/pg/.

Structural properties. A variation of the pebble game algorithm, further described in Lee and Streinu, maintains subsets of vertices that span $(6, 6)$-tight components and achieves $O(n^2)$ time complexity. These tight components correspond directly, via Tay's Theorem, to *rigid components* in molecular frameworks. This leads to further refinements in the analysis of edge subdivisions and points to structural properties of flexible molecular frameworks. Each time an edge belonging to a rigid component is subdivided, it may either eliminate a degree of dependency and maintain rigidity, or create a degree of freedom. In the latter case, the edge is called *critical*. Simple applications of the definitions lead to pebble game algorithms to detect structural features such as critical edges and rigid components in molecular frameworks.

Polyhedral Molecular Frameworks

A *molecular polyhedron* is a molecular framework whose incidence structure is that of a 3D convex polyhedron or, more generally, a planar graph that is not necessarily 3-connected. In a molecular polyhedron there are no leaves and all vertices have degree at least 3.

Example. Probably the best known (and certainly the most famous) example of a polyhedral molecule is the C_{60} *fullerene*, or *buckyball*, see Chapter 9. A feature of the buckyball is that all atoms have coordination number 3, i.e., are incident with exactly 3 others. A simple calculation of the number of edges shows that all molecular frameworks with coordination number at least 3 are over-constrained. Below we show how to compute their degree-of-dependency. Note that when viewed just as bar-and-joint structures, they are here highly flexible; the high degree of dependency comes from the angle constraints between the edges incident to the vertices.

Consider a polyhedral framework $G = (V, E, F)$, with no multiple edges but possibly with degree 2 vertices, and let $G^* = (V^* = F, E^* = E, F^* = V)$ be the dual planar graph, which will contain no degree 2 vertices but may have multiple edges. We present now a simple yet surprising connection between frameworks associated to such a pair of dual planar graphs. Let G be treated as a molecular framework, with edges corresponding to hinges, and let G^* (which may contain multiple edges) be treated as a body-and-bar graph, with edges interpreted as bars (not hinges).

Theorem 14.10. *Let G be a subdivision of a planar framework with no multiple edges, and let G^* be the dual planar multi-graph. Then:*

1. *$5G$ is $(6, 6)$-sparse with a dofs if and only if G^* is $(6, 6)$-rigid, with a dods, and vice-versa.*
2. *$5G$ is $(6, 6)$-tight if and only if G^* is $(6, 6)$-tight.*

Moreover, the result holds for any (k, k)-sparsity (not just $k = 6$).

Proof 2 Let v, e and f be the number of vertices, edges and faces of G, and v^*, e^* and f^* their counterparts in the dual graph G^*. We first prove (2), the tight case. Since $(k-1)G$ is (k, k)-sparse, we have $(k-1)e \leq k(v-1)$, and the inequality holds hereditarily for subsets of $v' \leq v$ vertices. Then the number of faces is: $f = e - v + 2 \leq \frac{k}{k-1}(v-1) - v + 2 = \frac{v-1}{k-1} + 1$. We have $f - 1 \leq \frac{v-1}{k-1}$, and hence $v - 1 \geq (k-1)(f-1)$. The dual multigraph has $e^* \geq \frac{k}{k-1}(v-1) = k(f^* - 1)$. Equality holds exactly when $5G$ is (k, k)-tight, and thus G^* is also tight. Extending this argument to rigid subgraphs with d dofs of $5G$, we obtain part (1).

Paneled Polyhedra

We switch now to our third polyhedral model and address questions about the rigidity of *paneled polyhedra*.

A polydron piece (named so after the popular model building set Polydron — see http://www.polydron.co.uk/) is a rigid flat piece of material (a *panel*) in the shape of a convex polygon. Along each edge, there is a hinge. This allows the panels to snap together and create polyhedral surfaces, which we call *polydron structures* or surfaces. The hinges allow for the rotation of the incident panels, and in general these surfaces are flexible. When they close up into a convex polyhedron, Cauchy's Theorem applies and leads to rigid structures. However, not all hinges are necessary to keep the polyhedron together: for instance, breaking the hinge along one edge (any one would do) will not change the rigidity of the polyhedron. This indicates that the structure is over-constrained. More generally, we conjecture:

Problem 14.11. *Show that if a paneled polyhedron is cut along edges that form a matching in the graph given by the 1-skeleton of the polytope, then it remains rigid.*

Breaking two of the hinges of a face introduces flexibility, if the face is triangular, but not when it has more than four edges. More generally, breaking two hinges incident to a vertex leads to a flexible structure. This follows generically from a variation on Cauchy's rigidity theorem, since the breaking of such a pair of hinges creates, in the underlying planar graph, an additional (combinatorial) face which is a quadrilateral, and none of the metrical constraints implies the rigidity of this face.

A polydron structure is a special case of a panel-and-hinge structure, but it is even more specialized, because not only are the hinges incident to a face coplanar, but also the hinges incident to a vertex are concurrent. The molecular theorem of Katoh and Tanigawa does not cover this situation. Indeed, we can see right away that there are several differences. A *ring* of panels connected by hinges has the same rigidity as a body-and-hinge ring: for n hinges, the ring is minimally rigid iff $n = 6$. For $n \le 5$, it has $6 - n$ dependencies, and for $n \ge 7$, it has $n - 6$ degrees of freedom. However, when all the hinges are concurrent, the situation is different.

Lemma 14.12. *The collection of panels and hinges incident to a vertex of a polyhedron is a ring of panels with concurrent hinges (called a vertex-ring) with 3 more dofs, respectively 3 less dods, than predicted by the molecular theorem for panel-and-hinges.*

Proof 3 The framework described here appears under the name of a single-vertex origami in a paper by Streinu and Whiteley, where it is shown that its generic rigidity properties coincide with those of a planar bar-and-joint polygon. This implies that a ring of $n \ge 3$ panels with all hinges concurrent has $n - 3$ degrees-of-freedom.

However, if we *cut* the vertex ring to obtain an open chain of panels incident at a vertex, the degrees of freedom are $n - 1$ for n panels, and coincide with what is predicted for an arbitrary chain of panels, even though the hinges are all concurrent.

If we eliminate the hinge between two panels, we say that we *cut* the polyhedron along that edge. Notice that cutting an edge creates two cut vertex-rings. If we cut edges in such a way that every vertex is incident to one cut edge, then we do not anticipate any change in the rigidity of the polyhedron, but this will allow us to reduce to a panel-and-hinge framework on which the Katoh–Tanigawa theorem applies. However, it is not always possible to apply this cutting to all the vertices, since this is equivalent to the existence of a perfect matching in the 1-skeleton of the polyhedron, which is not always possible (e.g., when the number of vertices is odd). Moreover, even when this is possible, we may still be left with dependent structures. For instance, cutting along two non-adjacent edges of a tetrahedron creates a 4-ring of panels, which is generically dependent. Cutting any further leads to a flexible structure. Putting back the two cut hinges leads to more over-constraints. More generally, we would like to eliminate dependencies (all of them, if possible) by cutting edges, while maintaining the rigidity of the framework. The tetrahedron shows that this may not always be possible. This example can be easily generalized.

Lemma 14.13. *A rigid paneled polyhedron (with all hinges present) is dependent. Cutting the edges, while maintaining rigidity, may not always lead to an independent framework.*

A *vertex-cut* polyhedron has the property that each vertex is incident to at least one cut edge. This allows for the application of the panel-and-hinge model to count degrees-of-freedom. The following is a direct consequence of the discussion in the previous section.

Lemma 14.14. *A vertex-cut polyhedron is minimally rigid if and only if it is minimally rigid as a panel-and-hinge framework.*

As we have seen, not all polyhedra can be turned into minimally rigid ones after a series of cuts. We leave the following as an open problem:

Problem 14.15. *Characterize the polyhedra which become minimally rigid after a series of vertex-cuts.*

Even so, we would like to find a maximal, resp. maximum set of cuts which leave a paneled polyhedron still rigid (albeit perhaps over-constrained).

Problem 14.16. *Give a polynomial time algorithm to find a maximal, resp. maximum set of vertex-cuts that maintains the polyhedron's rigidity.*

Assume now that the cuts are given. Can we characterize the flexibility, resp. dependency of the polyhedron? If all the vertices are cut, we can simply apply Tay's counts, because this case is covered by the Katoh–Tanigawa proof of the molecular conjecture.

Problem 14.17. *Characterize   the   rigidity parameters (degrees of freedom and degrees of dependency) of cut polyhedra where not all vertices are incident to a cut edge.*

We do not have yet a precise answer, but we offer now a simple necessary condition.

In a 3-connected planar graph, any simple cycle is separating, and induces two disk-like polyhedral surfaces. Let us focus on one of them. Assume that it is not a face, hence it contains inside at least one other vertex. For exactly one vertex, its rigidity is captured by Lemma 14.12.

In either case, we must account for the adjustment required by the un-cut vertex rings.

We propose the following necessary conditions for counting rigidity parameters (dofs and dods) of cut polyhedra.

Sparsity counts for cut polyhedra:

Let G^* be the dual graph of the original, uncut polyhedron with 1-skeleton G. For each cut hinge, eliminate its edge from the dual graph. Mark in G^* the cut and the uncut vertices. Then the following are necessary conditions for the cut polyhedron to be minimally rigid: consider all simple cycles, and the corresponding disks which are not faces of the polyhedron. If the disk contains m panels, h (uncut) hinges, and k uncut vertices, then the following condition must be satisfied:

$$5h \leq 6(m-1) + 3k.$$

For illustrations of all this and more, see http://linkage.cs.umass.edu/shape/.

15

Duality of Polyhedra

Branko Grünbaum and G.C. Shephard

The expression "mathematical folklore" refers to the results that most mathematicians take for granted, but which may never have been proved to be true. (Indeed, some of them are not true.) It is widely believed that the first person to call attention to this phenomenon was the French mathematician Jean Dieudonné.

An author who wishes to use material from mathematical folklore faces two unpleasant alternatives: either to quote the result (qualified by a phrase such as "it is well known that") or to prove it. The latter course may lead a referee or reviewer to ridicule the effort, and possibly identify it with an ancient result from the *Upper Slobbovian Journal of Recreational Mathematics,* or some other equally obscure source. Usually the situation is even worse because much folklore is imprecisely formulated (if, indeed, one can say that it is formulated at all) and quite frequently it is definitely wrong. The purpose of this chapter is to show that many of the "well-known facts" about *duality of polyhedra* are of latter kind, and it is well worth some effort to clarify the situation and arrive at the truth.

So far as we can ascertain, some if not all of the following "facts" are generally accepted among working mathematicians:

1. For every polyhedron P there exists a dual polyhedron P^*, and the dual of P^* is equal to P, or at least is similar to it.
2. If a polyhedron P has any convexity, symmetry, transitivity, or regularity properties, then the same is true, possibly in an appropriately modified form, for P^*.
3. The dual of a polyhedron can always be obtained by reciprocation with respect to a suitable sphere or more general quadric.
4. Duality between polyhedra is consistent with the combinatorial duality of their boundary complexes, that is, the cell complexes whose cells are the proper faces of the polyhedron. Moreover, in the particular case of polyhedra in three-dimensional Euclidean space, duality is consistent with the duality of planar graphs.
5. With appropriate interpretations, projective duality, duality in algebra and duality (conjugacy) in functional analysis are consistent with duality for polyhedra.

B. Grünbaum
Department of Mathematics, University of Washington, Box 354350, Seattle, WA 98195, USA,
e-mail: grunbaum@math.washington.edu

G.C. Shephard
School of Mathematics, University of East Anglia, Norwich, NR4 7TJ, UK,
e-mail: g.c.shephard@uea.ac.uk

Before we examine these statements in detail, it is necessary to define some of the terms that we shall use. For simplicity, we restrict attention to polyhedra in Euclidean space of three dimensions, that is, to compact 2-manifolds in E^3 which have no boundary and can be expressed as a finite union of plane polygonal regions.

M. Senechal (ed.), *Shaping Space*, DOI 10.1007/978-0-387-92714-5_15,

If these regions are such that no two adjacent ones are coplanar, then they are called the *faces* of the polyhedron; the *edges* and *vertices* of the polyhedron are the edges and vertices of its faces. We shall sometimes use words like "convex" and "star-shaped" to describe a polyhedron P though, strictly speaking, these terms apply to the polyhedral solid bounded by P. A vertex or an edge of P is said to be *convex* if the intersection of the polyhedral solid bounded by P with a sufficiently small spherical ball, centered at the vertex or at an interior point of the edge, is a convex set.

By the *elements* of a polyhedron P we mean the family consisting of all the faces, edges, and vertices of P. Two polyhedra P_1 and P_2 are *isomorphic* (or *combinatorially equivalent,* or *of the same type*) if there exists a one-to-one correspondence (bijection) between the elements of P_1 and the elements of P_2 which preserves the relation of inclusion between the elements. In a similiar manner, P_1 and P_2 are called *combinatorial duals* of each other if there exists a bijection that is inclusion-reversing; this is the concept referred to in statement 4. This extends in a natural way to topological complexes more general than polyhedra (in particular, to planar graphs or maps, and to maps on other 2-manifolds). The notions have their roots in the eighteenth-century works of Euler and Meister. Dual polyhedra are called *reciprocals* with respect to a sphere S (see statement 3) if each face of one is the polar with respect to S of the dually corresponding vertex of the other.

Now let us examine statements 1–5. The first is formulated in a misleading way since it seems that, in general, it is impossible to define in any useful or canonical way a *unique* polyhedron P^* as *the dual* to a given polyhedron P. In other words, although polyhedra that are combinatorial duals of P may exist, there appears to be no reasonable way in which one of them can be singled out and called *the dual* polyhedron to P. At best, we must therefore think of duality as a relation between isomorphism classes of polyhedra rather than between individual polyhedra. In this generalized sense, the second part of statement

1 is true, but most parts of statement 2 become vacuous.

If we consider only convex polyhedra—which may be thought of as a very simple special case— reciprocation can always be applied; as the center of the reciprocating sphere S we may take any point in the interior of the polyhedral solid enclosed by the polyhedron. In this way we can construct a convex polyhedron P^* dual to P, but even this does not lead to a unique dual since there is arbitrariness in the choice of the center and the radius of the sphere S. In fact, all the duals obtained in this way are projectively equivalent to each other, so the same difficulties as before still arise except that projective equivalence classes, rather than isomorphism classes, need to be considered.

In this connection we remark that reciprocation is the *only* known method of actually *constructing* a polyhedron P^* dual to a given polyhedron P. In some special cases which we shall now examine, reciprocation can lead to an essentially unique dual polyhedron, and then statements 1–4 become true for this restricted meaning of duality. It seems likely that the existence of these special cases, and the emphasis on them by many authors has led to the misconception that these statements are true more generally.

The first special case is when P is one of the five regular polyhedra. Such polyhedra have a natural center O which is the circumcenter, incenter, and centroid of P. If we choose S as any sphere centered at O, then the reciprocal is a dual polyhedron P^* (defined within a similarity, its size depending upon the radius of S) which is also regular. The reciprocal of P^* with respect to the same sphere S is P, and assertion 1 is true. In particular, if S is chosen so that the edges of P are tangent to it, then the edges of P^* have the same property and intersect those of P at right angles. This leads many authors to describe a certain regular octahedron as *the* dual of a given cube, a certain regular icosahedron as *the* dual of a given regular dodecahedron, and a regular tetrahedron as self-dual. Similar is the situation concerning the Archimedean (uniform) polyhedra and the polyhedra reciprocal to them.

There are more general cases in which a natural center of P exists. Suppose P is *isogonal*; that is, the symmetry group of P is transitive on its vertices. Then all the vertices of P lie on a sphere S which may be used in the process of reciprocation. To each vertex of P there corresponds a face of the reciprocal P^*, and the planes of these faces are the tangent planes to S at the vertices of P; moreover, P^* is *isohedral*; that is, its symmetry group is transitive on its faces. Similar considerations apply if P is isohedral and then P^* is isogonal, or if P is *isotoxal* (the symmetry group of P is transitive on the edges of P, which are therefore tangent to a sphere) and then P^* is isotoxal as well. In a similiar way if P has an axis of rotational symmetry R, or a plane of reflective symmetry E, then reciprocation with respect to a sphere centered at a point of R, or of E, will lead to a dual polyhedron P^* which has the same symmetry as P. Therefore in all these cases statements 1–4 are true.

It should be carefully noted that the discussion in the previous two paragraphs depended essentially on the fact that only convex polyhedra are under consideration. All the situations in which statement 5 can be sucessfully applied also deal with such polyhedra only. If we drop the convexity restriction, then despite some encouraging signs, things go sadly awry.

These encouraging signs appear when we consider examples such as the following, which are culled from the rather meager literature on nonconvex polyhedra. The icosahedron in Figure 15.1a and the dodecahedron in Figure 15.1b are duals of each other and have the same group of symmetries. The first is isogonal and the second is isohedral; nonconvex edges of both correspond to each other. Another dual pair consists of the well-known Császár polyhedron and the less well-known but remarkable Szilassi polyhedron. The Császár polyhedron is a triangulation of the torus with 7 vertices, 21 edges, and 14 triangular faces and the Szilassi polyhedron is toroidal with 7 hexagonal faces, 21 edges, and 14 vertices. These polyhedra are not only duals of each other but also have analogous symmetry and convexity properties. Such examples may seem to vindicate

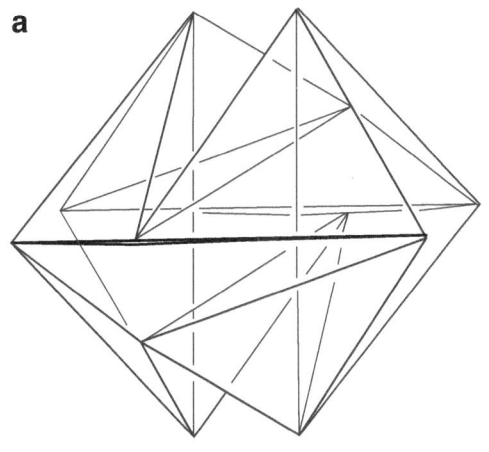

a

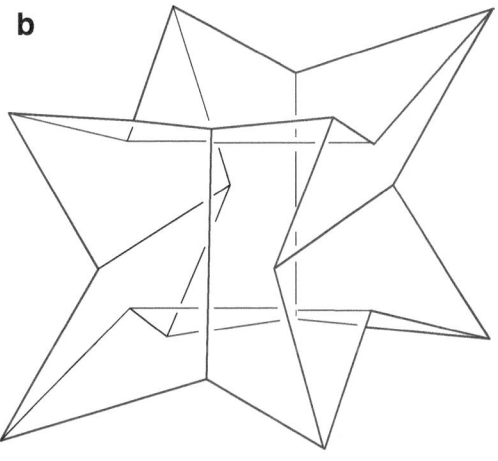

b

Figure 15.1. A dual pair of nonconvex polyhedra: (**a**) an isogonal icosahedron; (**b**) an isohedral dodecahedron.

the folklore and imply that it may be possible to prove the statements listed at the beginning of this chapter once we have learned how to deal with nonconvex polyhedra and, in particular, how to construct their duals.

This last is, in a sense, the nub of the problem: there is no difficulty finding topological complexes that are duals of any given polyhedron, but finding a *dual polyhedron* is a much more elusive goal. An indication that this goal may be unattainable is implied by our results concerning isohedral and isogonal polyhedra: isohedral polyhedra are always star-shaped and have star-shaped faces, whereas isogonal polyhedra have convex faces but need not even be simply

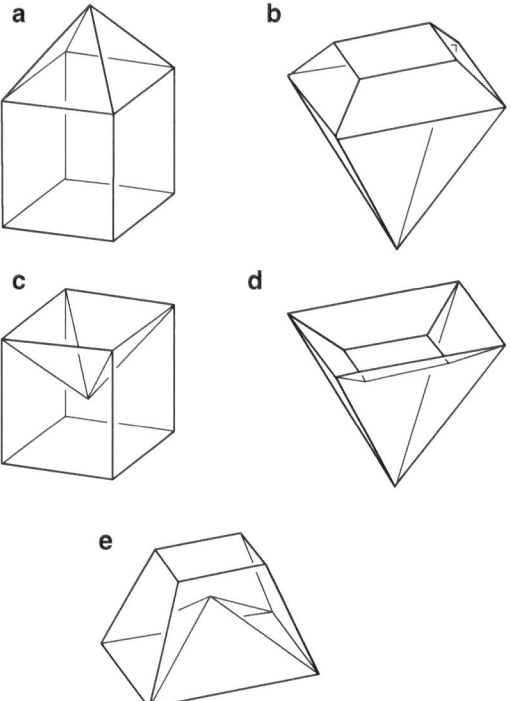

Figure 15.2. An illustration of the problem of duality for nonconvex polyhedra. The convex polyhedra illustrated in (**a**) and (**b**) are duals of each other, but the isomorphic nonconvex polyhedra in (**c**), (**d**), and (**e**) have no duals that preserve convexity properties.

connected. Hence these kinds cannot be related by duality.

To illustrate some of the difficulties, we shall consider a very simple example. In Figure 15.2a we show a polyhedron P which may be described as a cube with a four-sided pyramid adjoined to one of its faces. Reciprocation with respect to a suitable sphere (for example, the circumsphere of the cube) leads to the truncated octahedron of Figure 15.2b, which is therefore a dual P^* of P. In this particular case it happens that P and P^* are isomorphic, so P is self-dual, although this fact is only incidental to the following discussion. Now consider the nonconvex polyhedron P_1 of Figure 15.2c. Since this is isomorphic to P, every dual of P_1 will be isomorphic to P^*. However it is not hard to see that no such dual has corresponding convexity properties: P_1 has four nonconvex edges meeting at a vertex, so a dual *ought* to have four nonconvex edges bounding

a quadrangular face. Further, the four congruent nonconvex vertices of P_1 ought to correspond to four congruent nonconvex quadrangular faces of its dual. Examination of the various possibilities, such as those shown in Figure 15.2d, e, shows that no such dual exists. Hence, even in this very simple case, convexity properties cannot be preserved in duality.

It may seem that there is a very simple way out of this impasse. Let us start with any polyhedron P and then construct Q in the following way. After choosing a sphere S centered at an interior point of P, we define the vertices of V_j^* of Q as the polars with respect to S of the faces F_j of P. If two faces F_1 and F_2 of P meet in an edge, then the join of corresponding vertices V_1^* and V_2^* is defined to be an edge of Q. If $F_1, F_2, \ldots, F_r$ is a circuit of adjacent faces around a vertex of P, then the corresponding vertices of $V_1^*, V_2^*, \ldots, V_r^*$ are coplanar and so may be used to define a face of Q. Proceeding in this way, all the elements of Q may be defined, and so Q is completely determined. (Alternatively, and equivalently, the construction may be reversed: consider the set of planes that are the polars of the vertices of P, and then the edges and vertices of Q are defined as intersections of suitable subsets of these planes.) It is easy to verify that this procedure can be carried out for all polyhedra P, and that Q has the same symmetry and convexity properties as P. So it appears that Q is an obvious candidate for the dual of P. Unfortunately this is not the case since in general (and, in particular, if P is not convex) Q *will not even be a polyhedron as we have defined the word*. The union of the faces of Q *will not form a manifold either because they are not polygons or because they are mutually intersecting.*

We illustrate these assertions by an example that is chosen so as to be computationally and graphically easy to follow. It is not, in any sense, unique as the reader will discover by using the same procedure to find "duals" of the polyhedra in Figure 15.2c–e, or of the toroidal isogonal polyhedra described in our paper "Polyhedra with transitivity properties."

Consider the polyhedron P in Figure 15.3a. It is obtained from the octagonal prism in

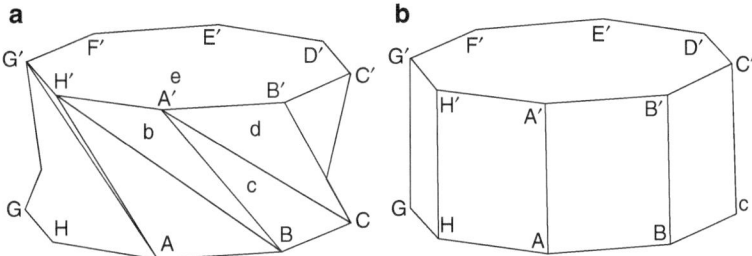

Figure 15.3. A nonconvex isogonal antiprism (**a**), with the same vertices as a prism (**b**). There exists no polyhedron dual to the antiprism, if "polyhedron" is understood in the usual sense.

Figure 15.3b by replacing its mantle of 8 regular faces by one of 16 triangles. The polyhedron P has convex faces and is isogonal, but since some of its edges and all of its vertices are nonconvex, so a dual P^* (if it exists) should have convex vertices, nonconvex faces, and be isohedral. Let us attempt to find such a dual by applying the construction for Q described in the previous paragraph. Take S (the reciprocating sphere) as the sphere that passes through the vertices of P. The set of planes tangent to S at these vertices (their polars) is easily visualized; it is the set of planes determined by the faces of a regular octagonal bipyramid. These will be the face planes of Q.

To determine the edges and vertices of Q we proceed as described above. There is an immediate simplification: the fact that P is isogonal implies that Q will be isohedral, so it is only necessary to determine *one* face of Q. The other faces will then arise by applying the symmetries of P to this face. Let the plane T be tangent to S at the vertex A' of P. Apart from the tangent plane to S at the point E, which is parallel to T, each of the other 14 face planes will meet T in a line. In Figure 15.4 we show the arrangement of 14 lines in T obtained in this way. Each line is marked with a letter indicating the vertex of P at which the corresponding plane touches S. The four faces b, c, d, and e of P contain a vertex A', so the corresponding points b^*, c^*, d^*, and e^* (defined as intersections of appropriate sets of lines in T) are the vertices of Q that lie in T. Since b, c, d, e form a circuit of faces at the vertex A' of P, the points b^*, c^*, d^*, e^* (in this order) should form a circuit of vertices

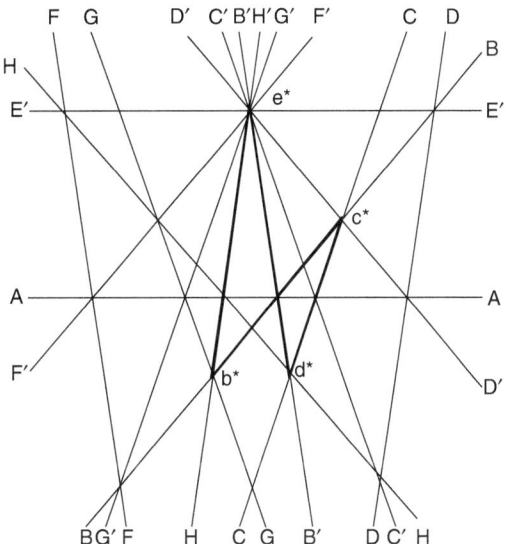

Figure 15.4. The arrangement of 14 lines in a plane, which can be used to prove the nonexistence of a polyhedron dual to the antiprism in Figure 15.3a.

around the corresponding face of Q. This is the nonconvex quadrangular face that we have been seeking. Unfortunately, it is not only nonconvex, but it is also self-intersecting. Hence it is not a polygon and so it is not acceptable as a face of *any* polyhedron. This Q is not a polyhedron, and our attempt to find a dual of P has failed.

We can summarize the situation by saying that in trying to establish a dual of each three-dimensional polyhedron we can *either* prescribe the character of the polyhedron as a 2-manifold, *or* we can adopt a wider definition of "polyhedron" admitting mutual intersections and self-intersections of the polygonal faces. In the latter case, as we have seen, duals can always be constructed by polarity. Then, if interpreted in a suitable manner, statement 2 at the beginning of this chapter will be true. It is interesting to note that this generalization of the concept of a polyhedron was adopted in part some 120 years ago by Möbius, and has often been applied (without any explicit definitions) in the study of regular, uniform, or other very special polyhedra. However it seems that no systematic study of "polyhedra" in this sense has ever been carried out, although there seems to be no reason why a completely satisfactory theory could not be developed.

So it appears that the various statements concerning duality made at the beginning of this chapter are mutually incompatible, and that the folklore is only a superstition! It may well be that this is the case; the brunt of our thesis is that, with the approaches to duality followed so far, this incompatibility is inescapable. It is not impossible, of course, that with some suitable generalizations, a theory might be formulated in which all the different aspects will fall into place so that folklore will be vindicated. We would like to suggest that the chief need at present is a theory of topological 2-manifolds in which faces may mutually intersect and even self-intersecting polygons are admissible as faces. We know of no attempts at such a theory, but it seems that it would be worth the effort to develop it, especially as it could have far-reaching applications and implications for other branches of mathematics.

16

Combinatorial Prototiles

Egon Schulte

Tiling problems have been investigated throughout the history of mathematics, leading to a vast literature on the subject. Our present knowledge of tilings of the plane is quite good, although there are of course many open problems even in the comparatively elementary and easily accessible levels.

As soon as we raise the dimension of the space from two to three or higher, our knowledge about tilings becomes comparatively poor. This is probably due to the fact that it is much harder to visualize the situation. In particular, considerations about the local structure of tilings are needed.

Before turning to the subject of this chapter, let us recall some definitions and notations. Although most of the results I will discuss can be extended to higher dimensions, I will restrict my considerations to ordinary three-dimensional space. Thus the underlying space for our tilings will be Euclidean 3-space, E^3, and we will tile this space by convex 3-polytopes, that is, with bounded convex polyhedra. A tiling of Euclidean 3-space is a family of convex 3-polytopes, called the tiles of the tiling, which cover the space without gaps or overlaps. This means that every point in space is contained in a tile, and no two tiles have common interior points.

To avoid pathological situations, we will always assume that our tiling is locally finite. By a *locally finite* tiling, we mean a tiling that has the following property: every point in space has a neighborhood that meets only finitely many tiles. Of natural interest are those tilings which respect the facial structure of the tiles or, more precisely, the facial structure of the boundaries of the tiles. These are exactly the face-to-face tilings. A tiling is called *face-to-face* if the intersection of any two tiles is either empty or a face of each; this means that two tiles may share a vertex, an edge, or a facet. Note that this definition of a face is slightly different from the usual one. Here a face can be 0-, 1-, or 2-dimensional. I will use the word "facet" to mean a two-dimensional face of a 3-polytope. (More generally, a facet of a $(d + 1)$-polytope will be a d-dimensional face.)

A tiling of E^d is called *normal* if its tiles are uniformly bounded, that is, if there are two positive real numbers r_1 and r_2 such that each tile contains a ball of radius r_1 and is contained in a ball of radius r_2. Obviously, normal tilings are necessarily locally finite. If each tile in a tiling happens to be congruent to one of the tiles in a finite family of k polytopes, then we say that the tiling has k isometric prototiles. Clearly such a tiling must be normal.

Finally, we recall that two polytopes P and Q are isomorphic, or *combinatorially equivalent*, or of the same combinatorial type, if there is an inclusion preserving bijection between the set of faces of P and the set of faces of Q (for 3-polytopes, that means, between the set of

E. Schulte
Department of Mathematics, Northeastern University,
Boston, MA 02115, USA
e-mail: schulte@neu.edu; http://www.math.neu.edu/
people/profile/egon-schulte

M. Senechal (ed.), *Shaping Space*, DOI 10.1007/978-0-387-92714-5_16,
© Marjorie Senechal 2013

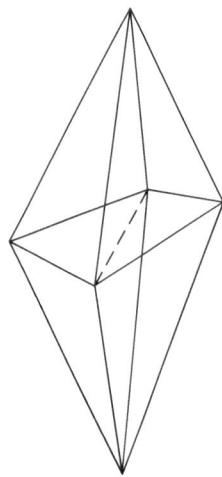

Figure 16.1. A combinatorial octahedron. The *dotted line* indicates that the four median vertices are not coplanar.

vertices, edges, and facets of P, and the set of vertices and edges and facets of Q). For example, the polytope in Figure 16.1 is combinatorially equivalent to the octahedron. Its faces are 3-gons and they fit together exactly the same way as the faces of the octahedron. Thus combinatorially these polytopes are the same, although they have totally different shapes. Notice, for instance, that the base of the polyhedron does not lie in a plane. If every tile of a tiling is combinatorially equivalent to a given 3-polytope P, we say that the tiling is *monotypic*. The type of the polytope is the *combinatorial prototile* of the tilings.

Nontiles

One of the main problems in the history of tilings of Euclidean 3-space is to characterize those convex polytopes, congruent copies of which tile space in a face-to-face manner. In other words, we are interested in finding those polytopes which play the role of triangles, quadrangles, pentagons, and hexagons in the plane. The answer to this extremely difficult problem is completely out of reach at the moment; we don't even know whether there are only finitely many combinatorial types of 3-polytopes that give such a tiling. Now

whenever people cannot solve a problem, they study related problems and hope that by doing so they will get additional information about the original one.

Thus we will discuss the three-dimensional analogue of the well-known fact, first observed by Schlegel in 1883, that for each $n \geq 3$, the Euclidean plane can be tiled by convex n-gons. Now the tiles in a tiling of the plane by n-gons, although they have the same number of edges, cannot be congruent if $n > 6$ (and indeed the tiling cannot be normal). Thus the tiles in such a tiling are combinatorially equivalent, but not congruent.

The combinatorial analogue of the tiling problem for higher dimensions was posed by Ludwig Danzer at a symposium on convexity in 1975. He suggested replacing the requirement of congruence for the tiles of a tiling of E^d by the much weaker requirement of the combinatorial equivalence of the tiles. This raises the following question:

Given a convex 3-polytope P, is there a locally finite tiling of space by convex polytopes isomorphic to P?

In other words, we would like to know whether every polytope is a combinatorial prototile of a monotypic tiling of three space. In particular, we would be interested to find face-to-face tilings that respect the facial structure.

Just as in the plane, this problem would be nonsense with combinatorial equivalence replaced by congruence of the tiles. But as long as we are only interested in combinatorial isomorphism, we have a great deal of freedom in the choice of the particular metrical shape of the tiles, so this is actually a reasonable problem. The general belief was that the answer to this problem should be positive, even in the strongest sense. That is, every convex three polytope was assumed to be the combinatorial prototile of a monotypic face-to-face tiling of three space. However, this is not true; in fact, the cuboctahedron, which is a very well known polyhedron (Figure 16.2), is not the combinatorial prototile of a face-to-face tiling: *There is no (locally finite, face-to-face) tiling of space by convex polytopes combinatorially equivalent to the cuboctahedron.*

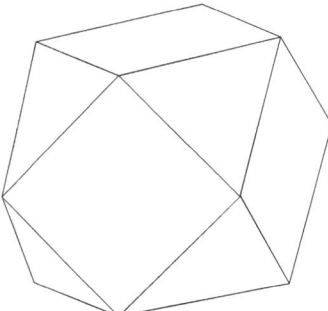

Figure 16.2. The cuboctahedron.

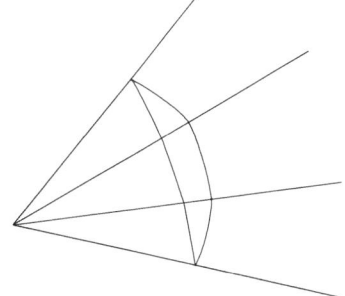

Figure 16.3. The facet of a spherical complex determined by a cuboctahedron with vertex X.

The proof is as follows. Let us look at the vertices of the cuboctahedron. They are all 4-valent and they are all surrounded by triangles and quadrangles in an alternating way: if we go around a vertex, then we meet a triangle, a quadrangle, a triangle, and a quadrangle. Thus it is natural to associate the vertices of the cuboctahedron the type [3,4,3,4]. It is important that it is of alternating type.

Now let us assume that there is a tiling of Euclidean space by polytopes that are combinatorial cuboctahedra and that fit together in a locally finite face-to-face manner. We fix one particular vertex of the tiling, that is, a vertex of one of the tiles, and call it X. This vertex is contained in many tiles. We choose a sufficiently small sphere centered at the vertex and consider the intersection of the sphere with the tiling. Now the tiling cuts out a spherical complex on our sphere. The facets of the spherical complex are just the intersections of the sphere with the tiles of the tiling which contain the fixed vertex X.

We will show that this spherical complex has quadrangular spherical facets and that each vertex is even-valent. This will immediately give us a contradiction, since Euler's theorem implies that a spherical complex without triangular facets must have a 3-valent vertex. (To see this, write Euler's formula in the form $2F - 2E + 2V = 4$ or $(2F - E) + (2V - E) = 4$. If the complex has no triangular facets, then counting the edges of the complex by going around each facet, and taking into account that each edge is shared by two facets, we have $2E \geq 4F$, so $2F - E \leq 0$. Similarly, if the spherical complex has no 4-valent vertices, then $2V - E \leq 0$. But these inequalities are not simultaneously compatible with Euler's formula.) This means that our spherical complex cannot exist and so the tiling cannot exist. So all we have to do is to show that our spherical complex has spherical quadrangles and even-valent vertices.

The first is easy. Why does the spherical complex have quadrangular facets? Recall that the facets are just the intersections of the sphere with the tiles that contain the fixed vertex X. Since the polytopes in our tiling are combinatorially cuboctahedra, the vertex X has valence 4 in each polytope that contains it. Thus the sphere cuts out a spherical quadrangle in each polytope, and so the spherical complex has spherical quadrangles as facets (Figure 16.3). Next, why are the vertices even-valent? The vertices of the spherical complex are just the intersections of the sphere with the edges of the polytopes that contain the fixed vertex X. What are the edges of the spherical complex? They are just the intersections of the sphere with the facets of the tiles that contain X. The facets of our tiles are triangles and quadrangles because the tiles are all combinatorial cuboctahedra and so, since the tiling is face-to-face, we can assign to each edge of the spherical complex one of the labels 3 or 4, according to the number of vertices of the facet which defines the edge.

And now comes the main point: since the vertices of the cuboctahedron are surrounded by triangles and quadrangles in an alternating way, the vertices of the spherical complex must also be surrounded in an alternating way. That means that the edges which come together in the vertex X can be labeled 3,4,3,4,...in an alternating way. But this means that the complex has even-valent vertices and, as we have seen, quadrangular facets. This spherical complex cannot exist, and so the tiling cannot exist.

In fact, this is a very strange result, because one would expect that such a nice symmetric polytope should give such a tiling. Indeed, the tetrahedron, octahedron, icosahedron, and dodecahedron, which are not isometric spacefillers, are combinatorial prototiles of monotypic tilings. The cube is even an isometric space-filler. Our counterexample reveals a very strange and very interesting aspect of the theory of tilings by convex polytopes: there seems to be no intrinsic relation between the regularity or symmetry of properties of the polytope and its tiling properties. In fact, in higher dimensions (seven or higher) it turns out that the regular crosspolytope (the higher dimensional analogue of the octahedron) does not give a face-to-face tiling by combinatorially equivalent polytopes.

Once we have a counterexample to our problem, we can ask to what extent we *can* expect positive results. The ideal situation would clearly be to give a characterization of all those polytopes which give a tiling, but this seems to be rather hopeless. The next best thing would be to try to determine certain classes of polytopes which do or do not give such tilings. For example, with the techniques used in the proof above we can prove this generalization:

Theorem 16.1. *Let P be a convex 3-polytope and $x_1, x_2, \ldots, x_k$ the vertices of P, all of even valence. Assume that it is possible to assign to each $x_i (i = 1, 2, \ldots, k)$ its type $[p_{i,1}, p_{i,2}, \ldots, p_{i,2m}]$ in such a way that*

$$\left\{ \bigcup_{i=1}^{k} p_{i,1}, p_{i,3}, \ldots, p_{i,2m-1} \right\}$$

$$\cap \left\{ \bigcup_{i=1}^{k} \{p_{i,2}, p_{i,4}, \ldots, p_{i,2m}\} \right\} = \emptyset$$

Then P is a nontile.

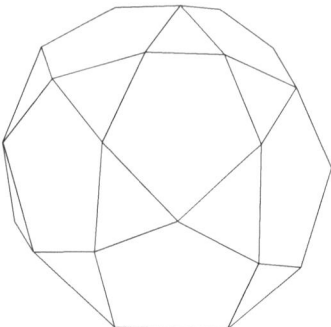

Figure 16.4. The icosidodecahedron.

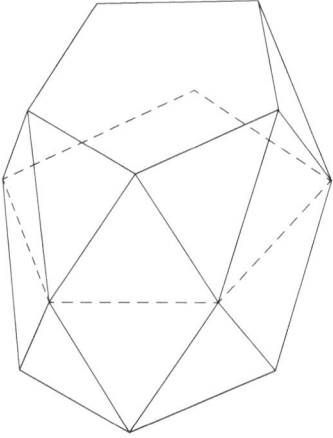

Figure 16.5. The polytope P_5. The *dotted lines* indicate a median cross section (they are not hidden edges).

That is, P does not give a locally finite face-to-face tiling.

With the help of this theorem, it is easy to construct many nontiles. If we start with a simple 3-polytope without triangular facets and cut off its vertices up to the midpoints of the edges incident with them, we obtain a nontile. (This operation, due to Steinitz, is denoted $I(G)$ by Grünbaum.) For example, starting in this way from the octahedron and icosahedron we obtain the cuboctahedron and icosidodecahedron (see Figures 16.2 and 16.4). Also, an infinite sequence of nontiles P_n can be obtained by appling the operation $I(G)$ to prisms over n-gons, where $n \geq 4$; see Figure 16.5.

The resulting polytopes have $3n$ vertices and $3n + 2$ facets. All vertices are 4-valent and of type [3,4,3,n]. For $n = 4$ we get a polytope combinatorially equivalent to the cuboctahedron. One might conjecture that the cuboctahedron (with 12 vertices and 14 facets) is the "smallest" nontile in three dimensions. It is worth noting that no simple or cubical nontiles are known yet. (A 3-polytope is called simple or cubical if all its vertices are 3-valent or all its facets are quadrangles, respectively.) I conjecture here that simple nontiles exist, but I doubt the existence of cubical nontiles.

As an example of a class of 3-polytopes which *do* tile E^3, Branko Grünbaum, Peter Mani, and Geoffrey Shepard have shown that all simplicial polytopes give locally finite face-to-face tilings.

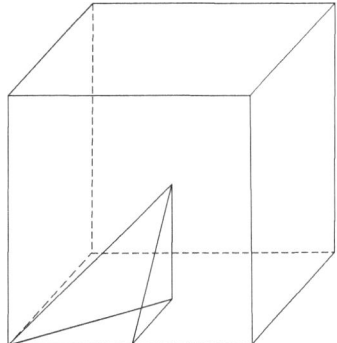

Figure 16.6. The fundamental region for the summetry group of the regular tessellation of E^3 by cubes.

Construction of Monotypic Tilings

One can also find very nice tilings of E^3 by projections of convex 4-polytopes. Intuitively we expect certain connections between the properties of 3-polytopes which are combinatorial prototiles of monotypic tilings of E^3 and 3-polytopes which are the 3-facet types of equifaceted 4-polytopes in E^4. (A $(d + 1)$-polytope is said to be equifaceted of type P if all its facets are isomorphic to a single d-polytope P. P is a nonfacet if it is not the facet-type of an equifaceted $(d + 1)$-polytope.)

Certainly a combinatorial prototile will not be a facet type in general, since this does not even hold in dimension 2; in fact, by Euler's theorem, a 3-polytope cannot have all its facets n-gons with $n > 5$, while on the other hand there are no restrictions on n for tilings of the plane by convex n-gons. However it has been proved that the reverse is true: every 3-polytope that is the facet-type of an equifaceted 4-polytope is also the combinatorial prototile of a locally finite face-to-face tiling of E^3. (It follows that all of the nontiles described above are also nonfacets!) The monotypic tiling of E^3 is obtained from the 4-polytope by an infinite sequence of projections. The construction works equally well in higher dimensions. Unfortunately this projection method produces non-normal monotypic tilings. However, in some instances another projection method provides normal face-to-face tilings with only finitely many isometric prototiles:

Theorem 16.2. *Let the convex 3-polytope P be realized as the facet-type of an equifaceted convex 4-polytope Q with at least one 4-valent vertex. Let m denote the number of facets of Q. Then P is the combinatorial prototile of a monotypic face-to-face tiling of E^3 with only $m - 4$ isometric prototiles.*

To prove this, let X be a 4-valent vertex of Q and T the 3-simplex whose vertices are the four neighboring vertices of X in the boundary complex of Q. By projecting Q centrally from X onto the affine hull of T, we get a face-to-face dissection of T into convex polytopes isomorphic to P. The 3-polytopes in the dissection are the images of facets of Q under the projection.

Next, we make use of the well-known fact that the fundamental region for the symmetry group of the regular tessellation of E^3 by cubes is a 3-simplex T' (see Figure 16.6). Mapping T affinely onto T' we turn the dissection of T into a dissection of T'. Then, if we apply all the symmetries of the tessellation, we obtain a tiling of the whole space, in which each tile is congruent to one of the 3-polytopes in the dissection of T'. Since the number of 3-polytopes in this dissection equals the number of facets of Q not containing X, that is, $m - 4$, the tiling has at most $m - 4$ isometric prototiles. In particular, the tiling is monotypic of type P, since P is the

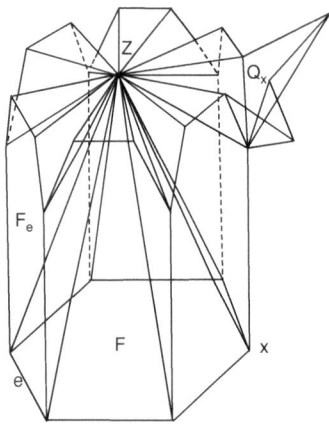

Figure 16.7. A face-to-face tiling of E^3 by hexagonal pyramids with three prototiles, derived from the usual hexagonal tessellation of the plane. To each hexagon F of the tessellation belong seven pyramids with a common apex Z. One has base F and is surrounded by six congruent pyramids, whose bases F_e share an edge e with F and lie in planes orthogonal to F. To each vertex X of the tessellation corresponds a nonconvex hexagonal pyramids with base Q_X. By taking suitable layers of this arrangement a tiling of E_3 by hexagonal pyramids arises.

facet-type of the equifaceted 4-polytope Q. The tiling is face-to-face because T' is a fundamental region for the symmetry group of the tessellation of E^3 by cubes, and that group is generated by the reflections in the planes bounding T'.

Another generalization of the tiling problem would be to relax the condition that the tiles be face-to-face. That is, we can ask: Is every polytope the combinatorial prototile of a tiling that is not necessarily face-to-face? And here we get the suprising result that the answer is definitely positive. But the construction of these tilings is too complicated to discuss it in detail here.

Finally, if we are willing to relax the conditions that our tiling must be convex, then we get some nice things. Figure 16.7 is a tiling by hexagonal pyramids, derived from the hexagonal tiling of the plane. It has only three types of pyramids; one of them is not convex.

Related Problems

So far our investigations of monotypic tilings were restricted to the case where the tiles of the tilings were topological balls, generally convex polytopes. But of course, we can as well consider combinatorial prototiles of other homeomorphism types, and we can ask if they admit locally finite tilings of the Euclidean space or not. Of particular interest is the case where the polytope P is a "polyhedron" in E^3 bounded by a closed polyhedral 2-manifold, for example a toroid. There are examples of space-filling toroids On the other hand, the existence of toroidal nontiles is almost trivial. In fact, if a toroid has only vertices of valence > 5, then isomorphic copies of it will not fit together to form a vertex-figure of a locally finite face-to-face tiling: this is an easy consequence of Euler's theorem applied to the spherical complex determined by the vertex-figure. Polyhedra with this property have recently been studied.

Another direction for research is offered by replacing the Euclidean space as the underlying space of the tilings by a topological 3-manifold M, and investigating monotypic tilings of M by topological polytopes or tiles of another homeomorphism type.

Polyhedra Analogues of the Platonic Solids

Jörg M. Wills

In this chapter we investigate polyhedra in Euclidean 3-space, E^3, without self-intersections and with some local and global properties related to those of the Platonic solids. A *polyhedron* is the geometric realization of a compact 2-manifold in E^3 such that its 2-faces are (not necessarily convex) plane polygons bounded by finitely many line segments. Adjacent faces and edges are not coplanar. A *flag* of a polyhedron P is any triple consisting of a vertex, an edge, and a face of P, all mutually incident.

Perhaps the most important property of the Platonic solids is that the set of their flags is transitive under the corresponding full Platonic symmetry group, consisting of all the rotations and reflections. There are no other compact (i.e., finite) and self-intersections-free polyhedra in E^3 with this property, so in order to carry the theory further one has to weaken this strict condition. A first step in this direction is to replace the global algebraic property of flag transitivity by the local property that all flags are combinatorially equivalent:

Definition 17.1. An equivelar manifold (i.e., with equal flags) is a polyhedron with the property: All faces are p-gons; all vertices are q-valent ($p \geq 3, q \geq 3$). Notation $\{p, q; g\}$, where g denotes the genus of the manifold.

For $g = 0$ (the sphere) one obtains the five Platonic solids, and for $g = 1$ (the torus) one obtains infinite series of tori, which were investigated long before. So in the following, all new polyhedra have genus $g > 1$.

In the definition it is not required that the faces are regular or congruent to each other and indeed all known equivelar manifolds contain at least one nonregular face. There exist infinitely many equivelar manifolds (and see the Problems at the end of this chapter). So equivelarity alone is too weak to yield close analogues to the Platonic solids and one has to find appropriate further conditions. It turns out that global algebraic conditions seem to be the most successful conditions, namely transitivity properties under certain symmetry and automorphism groups. (Symmetries are isometries of E^3 which map the polyhedron onto itself, and automorphisms are combinatorial isomorphisms.) Among the various possibilities we choose one in the following section which leads to nice polyhedra.

Platonohedra

Definition 17.2. A Platonohedron is an equivelar manifold such that a group isomorphic to its symmetry group acts transitively on its vertices or faces.

Simple combinatorial arguments show that there are only finitely many Platonohedra. Seven have been found so far: $\{3, 8; 3\}$, $\{3, 8; 5\}$,

J.M. Wills
Eichlingsborn 6, D 57076 Siegen, Germany
e-mail: wills@mathematik.uni-siegen.de

M. Senechal (ed.), *Shaping Space*, DOI 10.1007/978-0-387-92714-5_17,

$\{4, 5; 7\}$, $\{5, 4; 7\}$, $\{3, 9; 7\}$, $\{9, 3; 7\}$, and $\{3, 8; 11\}$. Figures. 17.1–17.4 show some of them. We let $f = (f_0, f_1, f_2)$ denote the number of vertices, edges, and faces. Because of their equivelarity the Platonohedra can be represented in a flag diagram (Figure 17.13), explained later. The Platonohedra have the same rotation group as the corresponding Platonic solids, whereas

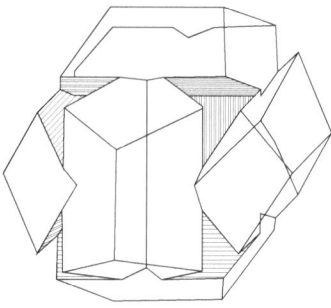

Figure 17.4. The Platonohedron $\{9, 3; 7\}$; $f = 12$ $(6, 9, 2)$.

the usual reflection in a plane is replaced, in the case of vertex-transitivity, by a reflection in a plane and a simultaneous inside-outside inversion. For the face-transitive Platonohedra the analogous reflection-inversion can be described for the normal vector.

Let us consider the simplest cases, $\{4, 5; 7\}$ and $\{5, 4; 7\}$ (see Figures 17.1 and 17.2). The 48 vertices of the $\{4, 5; 7\}$ lie pairwise on 24 rays which have their common endpoint at the rotation center. The inside-outside inversion interchanges the "inner" and "outer" vertices. The situation for the $\{5, 4; 7\}$ is analogous: its 48 faces fall into two classes of 24 "outer" and 24 "inner" faces and each "outer" face corresponds to one "inner" face. The inside-outside inversion interchanges the corresponding "inner" and "outer" faces. Clearly the inside-outside inversion is no isometry. Nevertheless it has a geometric meaning and is not a purely combinatorial automorphism. This corresponds to the fact that the usual reflection is an improper movement.

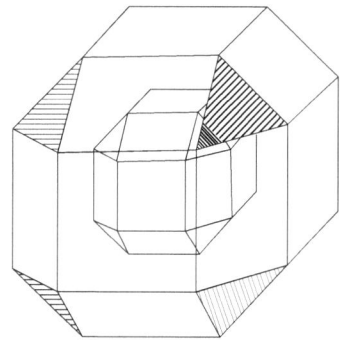

Figure 17.1. The Platonohedron $\{4, 5; 7\}$; $f = 12$ $(4, 10, 5)$.

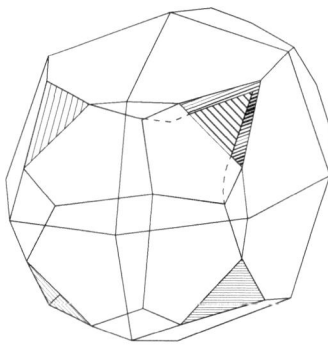

Figure 17.2. The Platonohedron $\{5, 4; 7\}$; $f = 12$ $(5, 10, 4)$.

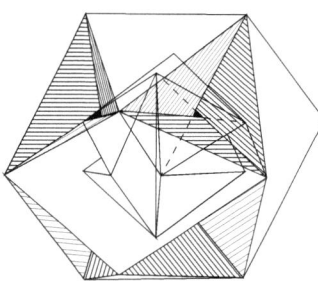

Figure 17.3. The Platonohedron $\{3, 9; 7\}$; $f = 12$ $(2, 9, 6)$.

If we require transitivities only under symmetries, then there exist no face-transitive polyhedra with $g > 0$, as Grünbaum and Shephard have shown. But they found three remarkable vertex-transitive Platonohedra $\{3, 8; g\}$, $g = 3, 5, 11$.

The first vertex-transitive polyhedron with $g > 0$ to be discovered was the flat torus ($g = 1$), which Brehm found in 1978; it is vertex-transitive under the dihedral group. It is three-dimensional; its name comes from the fact that the Gauss curvature in its vertices is zero. It was rediscovered by Grünbaum and Shephard, who found two more polyhedra with vertex-transitivity under symmetries; those polyhedra have two combinatorially distinct types of faces, and so they are not considered here.

In constrast with the three Platonohedra of genus 3,5, and 11, the four other (of genus 7) have two metrically different types of vertices and faces. This "disadvantage" corresponds to the "advantage" that they occur in dual pairs and that both their vertex-figures and their faces have one additional symmetry.

Construction of the Platonohedra

Figures 17.1–17.4 give an impression of four Platonohedra but clearly models of cardboard make it easier to understand them. Here is a brief description of their construction.

- $\{4,5;7\}$ is the simplest Platonohendron. It consists of 18 exterior squares of edge-length a, 18 interior squares of edge-length b, and 24 trapezoids (in the tunnels) of edge-length a,b and c. Suitable choice: $a = 9\,\mathrm{cm}$, $b = 6\,\mathrm{cm}$, $c = 4.2\,\mathrm{cm}$.
- $\{5,4;7\}$ consists of 24 outer and 24 inner pentagons. Their shape is shown in Figure 17.5. One should start with blocks of four outer and four inner pentagons and then fit the six blocks together.
- $\{3,9;7\}$ is easy to construct if one regards the following: the outer 12 vertices are those of a regular icosahedron; the 12 inner vertices are of a distorted icosahedron. Both types of vertices lie on two concentric cubes, where the exterior has, say, twice the edge-length of the interior one. From this it is easy to determine the coordinates of the vertices and so the five different edge-lengths of $\{3,9;7\}$.
- $\{9,3;7\}$ This "disdodecahedron" consists of 12 outer and 12 inner nonagons. It is much harder to construct than the three others, so we omit the construction.

The remaining Platonohedra belong to the family $\{3,8;g\}$, $g = 3,5,11$. Grünbaum and Shephard give a figure of $\{3,8;5\}$ from which a three-dimensional construction is possible; Egon Schulte described a precise construction of $\{3,8;3\}$.

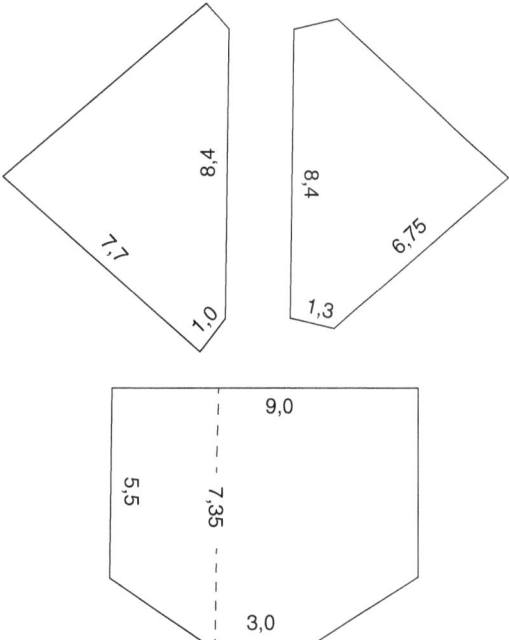

Figure 17.5. (*Above*) The two kinds of pentagons used in construction of the Platonohedron $\{5,4;7\}$. (*Below*) The nonregular hexagonal face used in the construction of the regular polyhedron $\{6,4;6\}$.

Regular Polyhedra

In this section we consider equivelar manifolds with flag transitivities under certain automorphism groups; this has been done by many authors in spaces other than E^3 and in abstract configurations. We use the same notation as above. The case $g = 0$ corresponds to the five Platonic solids and $g = 1$ to the regular toroidal polyhedra, which were found by Coxeter and Moser. Thus we consider $g > 1$; seven of them are shown in Figures 17.6–17.12. If one considers only polyhedra in E^3 and requires further that the polyhedron has, besides its automorphism group, a nontrivial symmetry group (for example, a Platonic group or a normal subgroup of it), then at least six of Coxeter's finite skew polyhedra can be realized in E^3 as equivelar manifolds, namely by suitable projections. If A and S denote the orders of their automorphism group

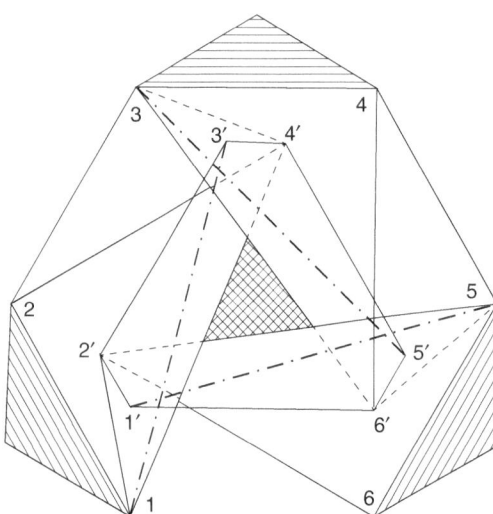

Figure 17.6. The regular polyhedron $\{3, 7; 3\}$; $f = 4(6, 21, 14)$.

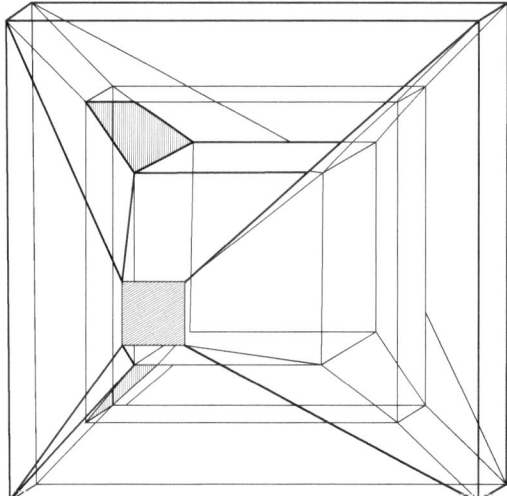

Figure 17.7. The regular polyhedron $\{4, 5; 5\}$; $f = 8$ $(4, 10, 5)$.

and of their symmetry group, respectively we find:

- $\{4, 5; 5\}$ and $\{5, 4; 5\}$, $A = 320$, $S = 8$.
- $\{4, 6; 6\}$ and $\{6, 4; 6\}$, $A = 240$, $S = 24$.
- $\{4, 8; 73\}$ and $\{8, 4; 73\}$, $A = 2{,}304$, $S = 48$.

Coxeter showed that the last four can be realized in E^4 and the first two in E^5 with the appropriate symmetry group. The geometric construction traces back to Alicia Boole-Stott in 1882; it has since been shown that they are projections of Coxeter's regular skew polyhedra.

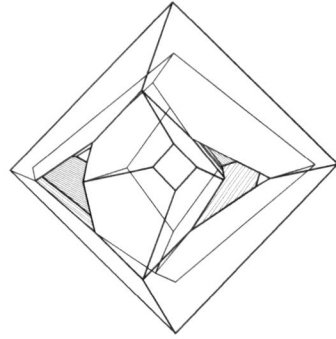

Figure 17.8. The regular polyhedron $\{5, 4; 5\}$; $f = 8$ $(5, 10, 4)$.

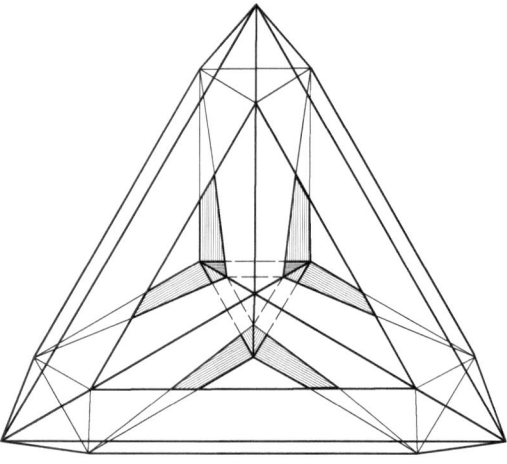

Figure 17.9. The regular polyhedron $\{4, 6; 6\}$; $f = 10(2, 6, 3)$.

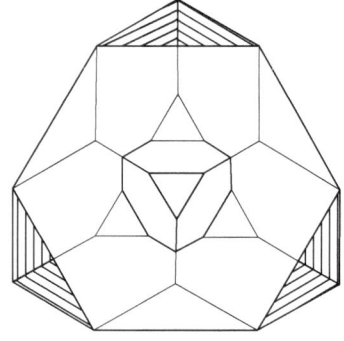

Figure 17.10. The regular polyhedron $\{6, 4; 6\}$; $f = 10(3, 6, 2)$.

The seventh and most spectacular example is the polyhedral realization of Felix Klein's famous quartic (a complex algebraic curve)

Figure 17.11. The regular polyhedron $\{4, 8; 73\}$; $f = 144(1, 4, 2)$.

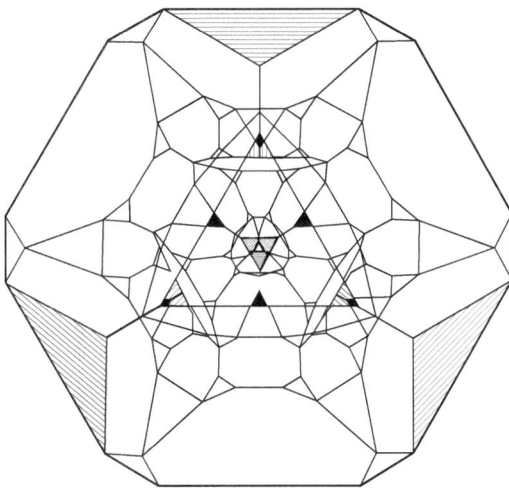

Figure 17.12. The regular polyhedron $\{8, 4; 73\}$; $f = 144(2, 4, 1)$.

as a polyhedron of genus 3. It is remarkable that this polyhedron has the same number of vertices, edges, and faces (and so the same genus) as our Platonohedron $\{3, 7; 3\}$, although the polyhedra differ combinatorially. This close coincidence was one motivation for finding the "Klein polyhedron." So far we know one infinite series and five single combinatorially regular polyhedra of genus $g > 1$. (See the Problems at the end of chapter).

We now describe the construction of the $\{6, 4; 6\}$ (Figure 17.10). This polyhedron has four regular hexagonal faces of edge-length a and four regular hexagonal faces of edge-length $3a$. Further it has 12 nonregular hexagonal faces which are all congruent to each other. In Figure 17.5b we show one of these faces (with $a = 3$). Three of these hexagons are fitted together along their edges of length 5.5 to make four tunnels. The four tunnels are first joined with the small regular hexagons (along the edges of length 3), and then the tunnels must be joined with the four large regular hexagons (see Figure 17.10).

The Flag Diagram

A survey of equivelar and regular polyhedra is given by the (p, q)-diagram or flag diagram (Figure 17.13). On the p-axis of the diagram the values p of the p-gons ($p \geq 3$) are plotted; on the q-axis are the valences q of the vertices ($q \geq 3$), the number of edges incident with the vertex. Thus, for example the tetrahedron can be found at $p = q = 3j$, the usual 3-cube can be found in $p = 4, q = 3$, and so forth. (Clearly, polyhedra with different types of faces or vertices—as, for example, the Archimedean solids—cannot be shown in the flag diagram.) The labels in the flag diagram denote the genera: thus $g = 0$ for the Platonic solids. In particular, the flag diagram shows in a very suggestive way the three types of polyhedral geometry, due to the values of p and q.

The hyperbola $1/p + 1/q = 1/2$ dissects the lattice points (p, q), $p \geq 3, q \geq 3$ of the diagram into three subsets:

- The elliptic case: $1/p = 1/q > 1/2$. Here we find the five Platonic solids and the two Kepler – Poinsot star polyhedra of genus 0 (which are isomorphic to the dodecahedron and the icosahedron, respectively).
- The parabolic or Euclidean case: $1/p = 1/q = 1/2$. Here we find the three regular tilings of the Euclidean plane, and the regular tori.

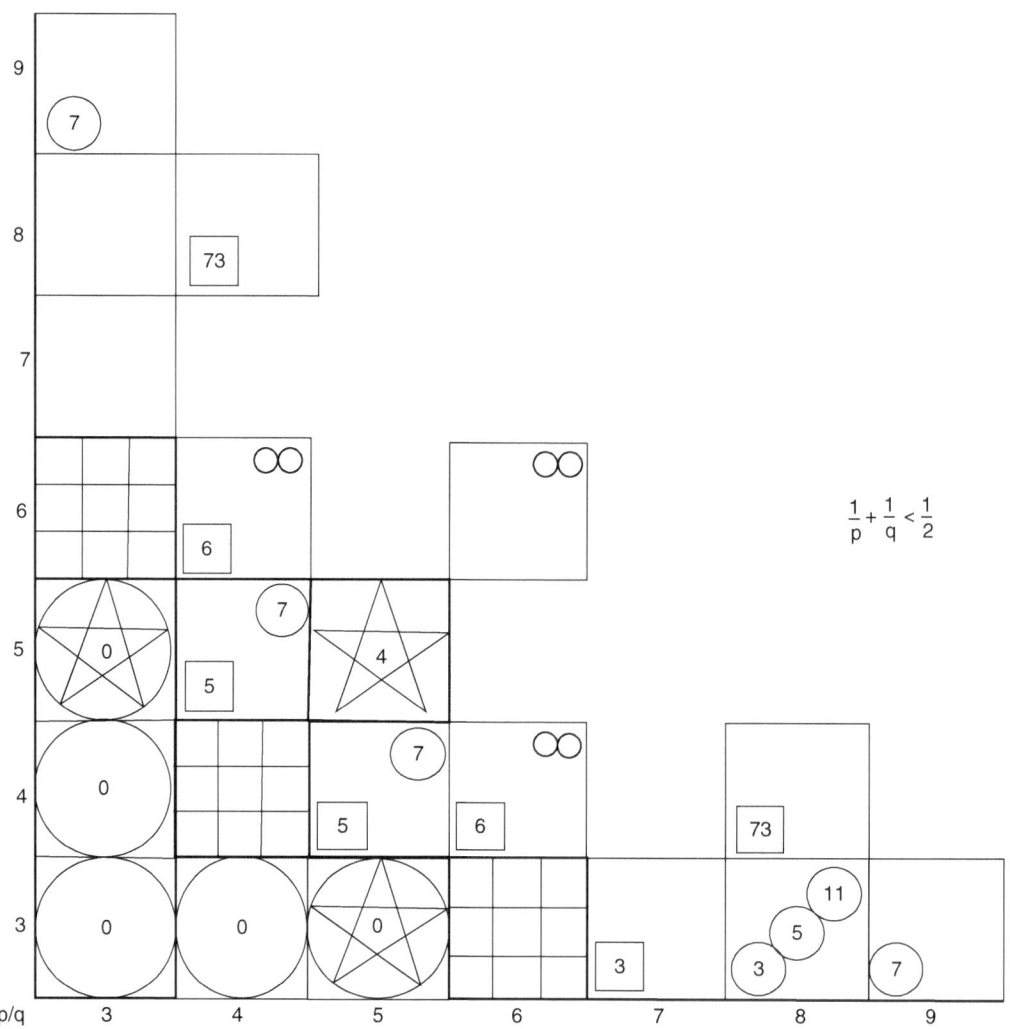

Figure 17.13. The flag diagram. The *numbers* denote the genera; *large circles* denote the five Platonic solids ($g = 0$), the *pentagrams* denote the four Kepler–Poinsot polyhedra, the *frames* denote the three regular tilings of the plane and tori ($g = 1$), *small circles* denote the seven known Platonohedra with $g > 1$, and *squares* denote seven of the known regular polyhedra with $g > 1$.

- The hyperbolic case: $1/p = 1/q < 1/2$. Here we find the two Kepler–Poinsot star polyhedra of genus 4 and the three infinite regular Petrie–Coxeter polyhedra. Further we find here the Platonohedra and the regular polyhedra of genus $g > 1$ mentioned previously.

Problems

I conclude this chapter with five open problems on equivelar manifolds and regular polyhedra. Although all of these problems are easy to understand, their solution seems not to be straightforward.

1. Do equivelar manifolds exist with $p \geq 5$ and $q \geq 5$?
2. Do equivelar manifolds exist with all faces being regular? (It has been shown that for $q = 4$ no such manifold exists.)
3. An equivelar manifold with $g = 577$ and number of vertices $f_0 = 576 < g$ has been constructed. Are these the smallest possible numbers? (In E^5 there is one with $g = 2$ and $f_0 = 19$.)

4. Are there more than seven Platonohedra of genus $g > 1$?

5. Are there other combinatorially regular polyhedra (in E^3 without self-intersection) for $g > 1$? In particular: Does the dual of Klein's quartic exist as an intersection-free polyhedron?

UPDATE 2010

- Problems 1 and 2. Coxeter's apeirohedron of type (6, 6), built up of regular hexagons is an infinite example for both problems. Finite examples are not known yet. And there is no proof that such finite examples do not exist.

- Problem 3. The infinite series of equivelar polyhedra give 0(g/log g) for the number of vertices (or faces), where g denotes the genus of the polyhedra. So the number of vertices (or faces) is significantly smaller than the genus. On the other hand this bound for geometric realizations is still far from Ringel and Young's bound for abstract realizations. Several authors have tried to narrow the gap between both bounds, but without success. The only progress is a nice and interesting proof by Joswig and Ziegler on the geometric bound via boundary complexes of 4-polytopes.

- Problem 4. No new results.

- Problem 5. Several new combinatorially regular polyhedra were discovered, in particular the dual of felix Klein's map of genus 3, built up of 24 nonconvex heptagons (and of course without self-intersections) This polyhedron was found by David McCooey in 2009. It has automorphism group of order 336 and the geometric symmetry group of the tetrahedron.

18

Convex Polyhedra, Dirichlet Tessellations, and Spider Webs

Walter Whiteley, Peter F. Ash, Ethan Bolker, and Henry Crapo

Plane pictures of three-dimensional convex polyhedra, plane sections of three-dimensional Dirichlet tessellations, and flat spider webs with tension in all the threads are essentially the same geometric object. At the root of this remarkable coincidence is a single geometric diagram that permits us to offer a unified image of the connections among these and other objects. Some hints of these connections are more than a century old, but others are very recent. We begin with an historical sketch.

In the nineteenth century, mathematicians and engineers investigated frameworks built from iron bars and pins to determine when they were rigid. (The technical term is *infinitesimally rigid*.) Their studies led them to consider static stresses: tensions and compressions in the bars in internal static equilibrium. In 1864, James Clerk Maxwell

W. Whiteley
Professor of Mathematics and Statistics, York University, Toronto, ON, Canada
e-mail: whiteley@mathstat.yorku.ca;
http://www.math.yorku.ca/~whiteley/

P.F. Ash
Senior Faculty, Cambridge College, USA
e-mail: peter.ash@gocambridgecollege.edu;
http://peterashmathedblog.blogspot.com

E. Bolker
Department of Mathematics and Computer Science, University of Massachusetts at Boston, Boston, MA, USA
eb@cs.umb.edu, www.cs.umb.edu/~eb

H. Crapo
Centre de recherche "Les Moutons matheux,"
34520 La Vacquerie, France
email: crapo@ehess.fr

discovered a geometric tool for studying the static equilibrium of forces on a plane framework with a planar graph: the *reciprocal figure*, a drawing of the dual planar graph with the dual edges perpendicular to the original edges and forces (see Figure 18.1). Maxwell built this reciprocal by patching together the polygons of forces expressing the vector equilibrium at each joint. He then observed that this construction yields a polyhedron in space which projects onto the framework. These results belong to the field of *graphical statics*, a branch of graphical and mechanical science that withered around the turn of the century, along with much of projective geometry.

Contemporary work on the statics of frameworks grows from these geometric roots. In particular, we now know that a convex polyhedron, projected from a point on one face onto a plane parallel to this face, corresponds to a *spider web:* a framework with no crossing edges and some edges going to infinity, which has an internal static equilibrium formed entirely with tension in the members. In the plane, the spider webs are frameworks with *convex reciprocals:* reciprocals in which the convex polygons have disjoint (that is, non-overlapping) interiors (Figure 18.2). Other recent work has extended some hints in the work of Maxwell and a conjecture by Janos Baracs, a modern structural engineer and geometer, to show that three-dimensional projections of convex 4-polytopes correspond to some, but not all, spider webs in 3-space.

In the 1970s, computer scientists sought algorithms to recognize and draw correct

M. Senechal (ed.), *Shaping Space*, DOI 10.1007/978-0-387-92714-5_18,

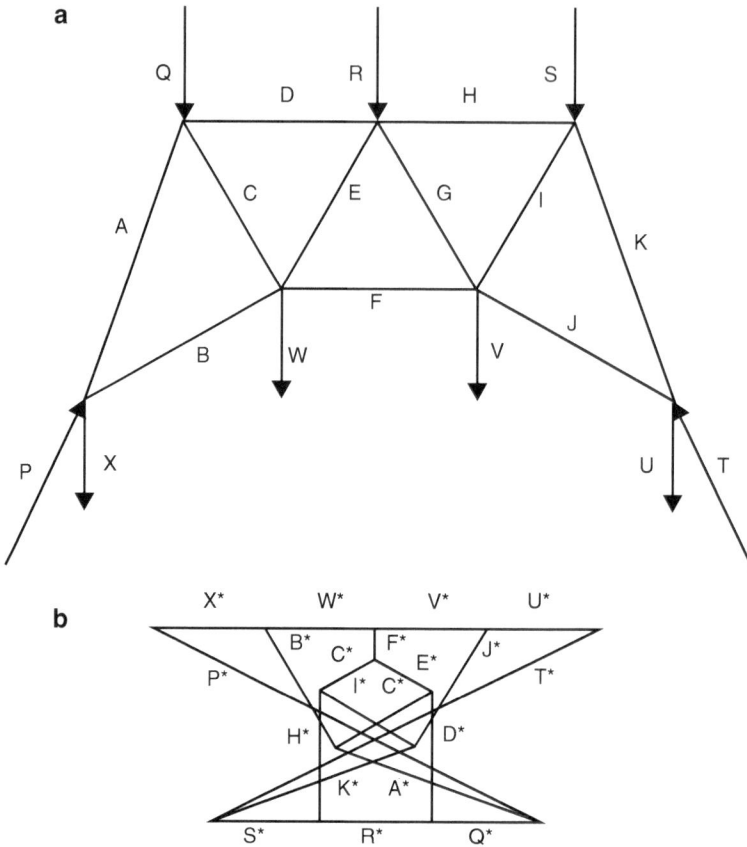

Figure 18.1. A framework (**a**) in static equilibrium with a set of external forces (*the arrows*), has a Maxwell reciprocal figure (**b**) with dual edge Z* perpendicular to the original edges, and a polygon dual to the edges at each vertex of the original.

pictures of objects in space. Several workers independently observed that the existence of a reciprocal figure was the natural geometric condition for a correct *picture of a polyhedron*, noting that the reciprocal figure records the normals to the faces (Figure 18.3). At first some critical topological details were not properly addressed, but this construction of plane reciprocals has been refined to give a necessary and sufficient condition for correct pictures of any oriented polyhedron.

At about the same time, computer scientists were studying *Dirichlet tessellations* (also known as Voronoi diagrams): subdivisions of the plane (and of *n*-space) into the polygonal (or polyhedral) regions of points closest to given centers (Figure 18.4). In 1979, Ezra Brown observed that a Dirichlet tessellation in the plane

corresponds to a convex polyhedron with all faces tangent to a sphere, projected from the point of tangency of one face onto a plane parallel to this face. He used this observation to develop efficient algorithms to compute the Dirichlet tessellation for a set of centers.

Peter Ash and Ethan Bolker observed that the diagram of centers forms a classical reciprocal figure for its Dirichlet tessellation. More generally, a *sectional Dirichlet tessellation*, a plane section of a Dirichlet tessellation in 3-space, has a plane reciprocal formed by the orthogonal projection of the spatial centers. At the 1984 Shaping Space Conference, Walter Whiteley and Bolker forged the last link in the proof that sectional Dirichlet tessellations and plane spider webs coincide. This says implicitly that convex polyhedra projected from one face

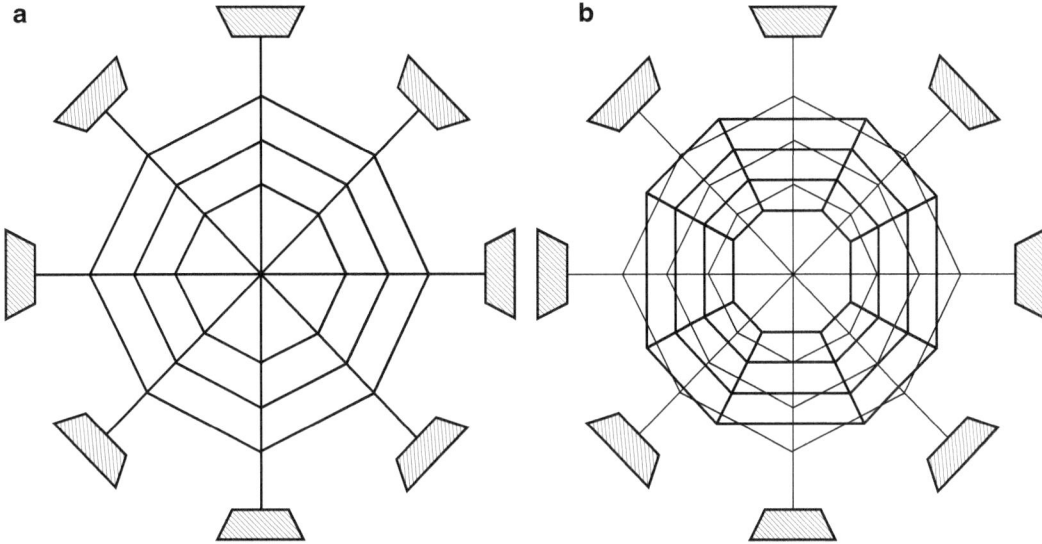

Figure 18.2. A plane spider web (**a**) has an internal static equilibirum with tension in all members, and a convex reciprocal figure derived from this equilibrium (**b**).

are just sectional Dirichlet tessellations, and conversely. Independently, Herbert Edelsbrunner and Raimund Seidel gave an explicit construction of a polyhedron which will correspond to a given sectional Dirichlet tessellation.

Thus many of the results we present are not new. However, the unified picture is, and some new results follow from this unification. We highlight the reciprocal figure as the central geometric construction and some direct geometric arguments replace previous, seemingly accidental coincidences. We will sketch proofs when they are simple or illuminating; otherwise we will refer the reader to the literature.

In the next section, we carefully describe the equivalence of the following finite geometric objects in the plane:

• A plane section of a Dirichlet tessellation of 3-space.
• A plane section of a furthest-point Dirichlet tessellation of 3-space.
• A projection of a convex polyhedron in 3-space from a point inside one of its faces onto a plane parallel to this face.
• A plane framework, without self-intersection, with a static equilibrium using tension in all members.

• A plane drawing of a planar graph, with a planar reciprocal figure of disjoint convex polygons.

We also describe the special correspondence between Dirichlet tessellations in the plane and convex polyhedra with all faces tangent to a sphere (or a paraboloid). Later we survey some infinite analogues and point out how most but not all equivalences remain true. Such infinite, but locally finite, structures occur in the study of circle packings of the entire plane and in both periodic and aperiodic tilings of the plane.

All of the questions we raise, and many of the answers, generalize to n-space, but in this chapter we limit ourselves to dimensions 1, 2, and 3.

Each of the fields we touch on has its own favorite questions, results, and unsolved problems. The connections we establish among these fields have important implications for those preoccupations. For example the question: "Which graphs can be realized (that is, constructed) as a Dirichlet tessellation in the plane?" now coincides with the classical problem: "Which spherical polyhedra can be realized with all faces tangent to an insphere?" The question: "Which graphs can be realized as a sectional Dirichlet tessellation in the plane?" is answered by

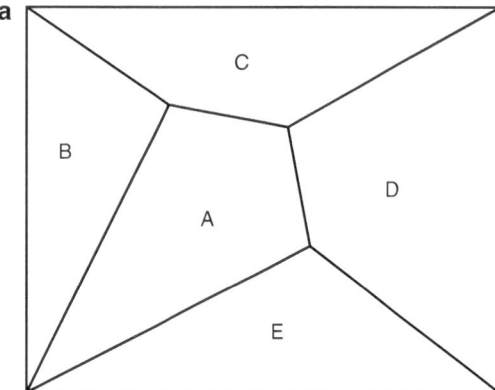

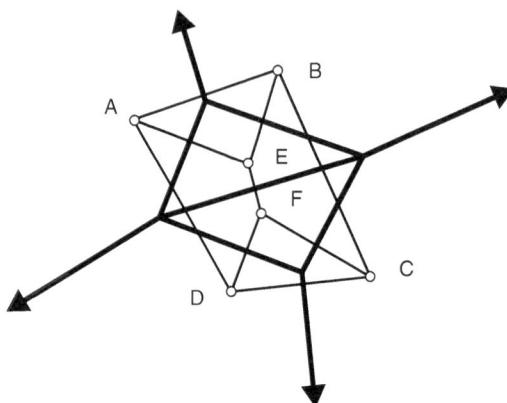

Figure 18.4. A set of centers (*the circles*) defines a Dirichlet tessellation (*heavy lines*) and a reciprocal diagram of centers (*lighter lines*).

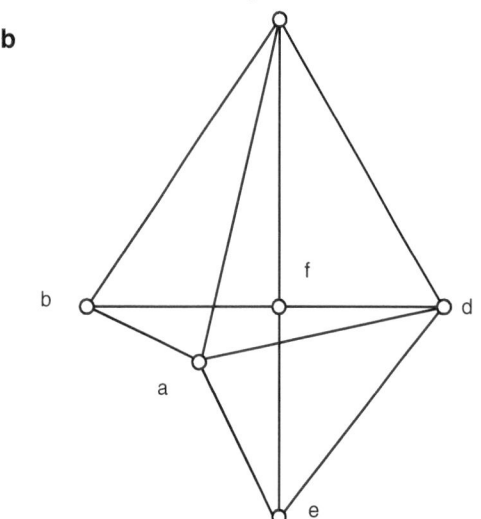

Figure 18.3. The projection of a polyhedron (**a**) has a reciprocal figure (**b**) which places the vertex dual to the face C at the point given by the gradient c of this plane.

Steinitz's theorem: adding the polygon at infinity must create a triply connected planar graph! We will draw out some of these implications as we proceed through the correspondences.

This study represents geometry in what is for us its best sense: recognizing the kinship among classes of tangible, visible objects that one can draw, build, and manipulate. We feel pleasure in seeing a spider weave a many-faceted diamond, excitement in discovering the geometric basis of common algorithms to recognize or create these patterns, horror in seeing a building shake because it contains a 4-polytope, and satisfaction

in knowing that a circle-packing is locally maximally dense because the graph of the centers is a rigid spider web.

Cell Decompositions and Reciprocal Figures

Imagine cutting the plane into a finite number of convex polygons and unbounded convex polygonal regions (see Figure 18.5a). A *proper cell decomposition* of the plane is a finite set of convex polygons and unbounded convex polygonal regions called the cells such that

- every point in the plane belongs to at least one cell;
- the cells have disjoint interiors;
- the decomposition is edge-to-edge; that is, every edge of a cell is a complete edge of a second cell.

For example, the *Dirichlet tessellations* described above are proper cell decompositions. Each proper cell decomposition D of the plane (henceforth in this section we shall omit "of the plane") has an abstract *dual graph* D*. The vertices of D* are the cells of D; two vertices of c and c' are joined by an edge just when the corresponding cells C and C' share an edge. It is clear that D* is always a planar graph, because

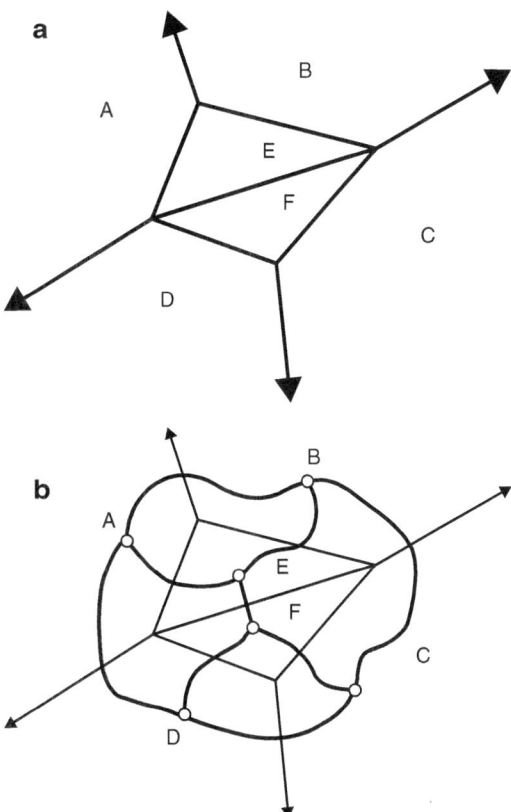

Figure 18.5. Any proper cell decomposition of the plane (**a**) has a dual graph (**b**).

So far our discussion of the reciprocal figure has concentrated on the graph: the edges and the vertices. The cell decomposition has vertices, edges, and cells, and we now restore this symmetry to the reciprocal. Around each vertex of the original decomposition we have a cycle of exiting edges and a corresponding polygon of orthogonal edges in D*. If all these dual edges have nonzero length, and the resulting polygons are convex and have disjoint interiors, we say we have a *convex reciprocal figure*. Figure 18.6a shows a convex reciprocal figure, while the cell decomposition in Figure 18.6b has no convex reciprocal because we have turned the edge between cells C and D. Figures 18.6c, d show a single cell decomposition with a convex reciprocal (C) and a nonconvex reciprocal, in which the convex polygons are not disjoint (D).

A set of parallel lines and the strips between them is a *trivial cell decomposiiton* which has no vertices (Figure 18.7). Such decompositions are just perpendicular translations of cell decompositions of the line. They have reciprocal figures in which all the dual vertices lie on some line perpendicular to the edges. For the purposes of our theorems and constructions we call such a trivial reciprocal figure convex if the order of the dual vertices along the line matches the order of the strips along the line. We note that any nontrivial proper cell decomposition in the plane has at least one vertex on each edge.

These trivial cell decompositions hint at the fact that our entire theory can be restated simply for figures on the line. A cell decomposition of the line is a set of line segments with disjoint interiors. A convex reciprocal is just a set of reciprocal points for these cells, which respect the order of the cells (see Figures 18.9 and 18.13). From time to time we shall use figures on the line to illustrate the concepts we are exploring for figures in the plane.

Returning to the example of a Dirichlet tessellation, we see that the diagram of centers is a convex reciprocal figure. If all the vertices are 3-valent, the reciprocal will be a triangulation known as the Delauney triangulation. This observation is our first theorem.

it can be drawn in the plane simply by choosing a point inside each of the cells and joining the points in cells which share an edge (see Figure 18.5b). For a Dirichlet tessellation, the centers are the vertices of a planar embedding of D* and the edge separating two cells C and C' is the perpendicular bisector of the segment cc' in this drawing of the dual (see Figures 18.4 and 18.8b).

This example suggests a way to try to draw the dual graph of a cell decomposition. A *reciprocal figure* for D is a plane drawing of D* in which the edges are straight-line segments which are (when extended) perpendicular to the (extended) edges of D. Figure 18.6 shows additional examples of cell decompositions with reciprocal figures. Note that we do *not* demand that the vertices of a reciprocal figure lie in the cells to which they correspond.

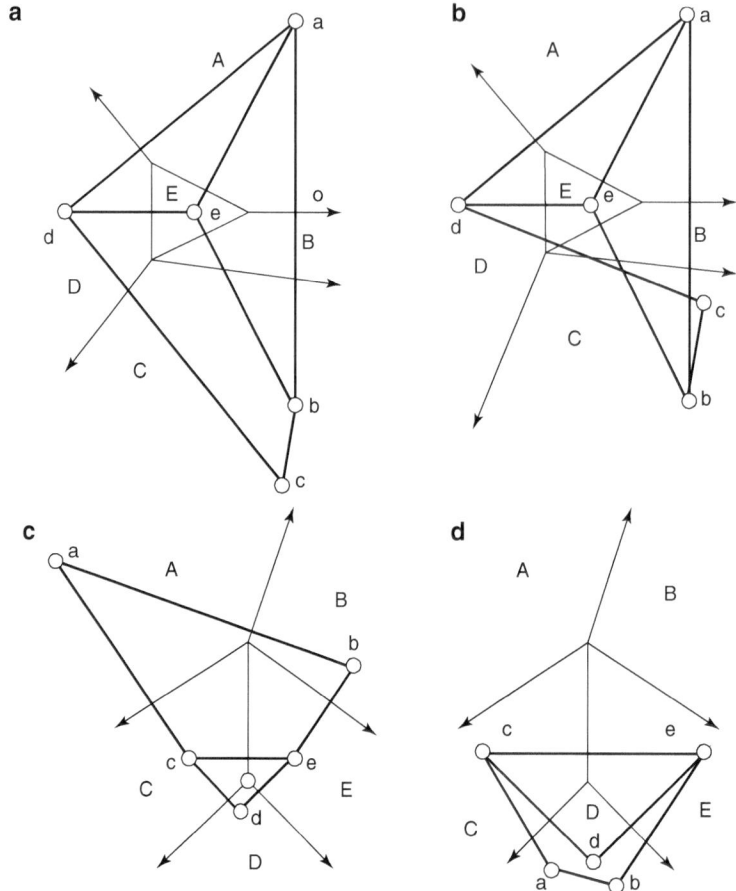

Figure 18.6. Some cell decompositions (*light lines*) with reciprocal figures (*heavy lines*). The cell decomposition in (**a**) has only convex reciprocals, but that in (**b**) has only nonconvex reciprocals. A single cell decomposition may have both convex (**c**) and nonconvex (**d**) reciprocals.

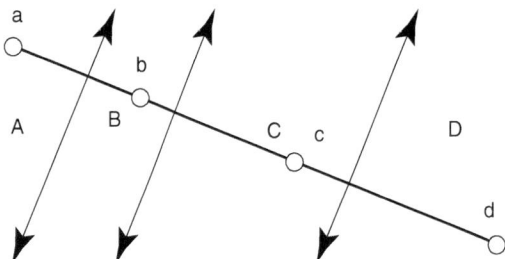

Figure 18.7. A trivial cell decomposition of the plane (*light lines*) has a convex reciprocal which lies on a perpendicular line (*heavy line*).

Theorem 18.1. *A Dirichlet tessellation has a convex reciprocal figure. The converse of this statement is false.*

Figure 18.8a shows a proper cell decomposition which has a convex reciprocal figure but is not a Dirichlet tessellation: there is no way to position the centers so that the edges of the original decomposition bisect the edges of the dual (as in Figure 18.8b). Experimental evidence is quite convincing; a proof can be found in Ash and Bolker's "Recognizing Dirichlet Tessellations."

To find a converse, we must broaden our search.

A *sectional Dirichlet tessellation* is a plane section of a Dirichlet tessellation of 3-space. (We define a Dirichlet tessellation of 3-space by replacing "the plane" by "space" in our previous definition.) If we throw away any centers in the space whose cells do not meet the slicing

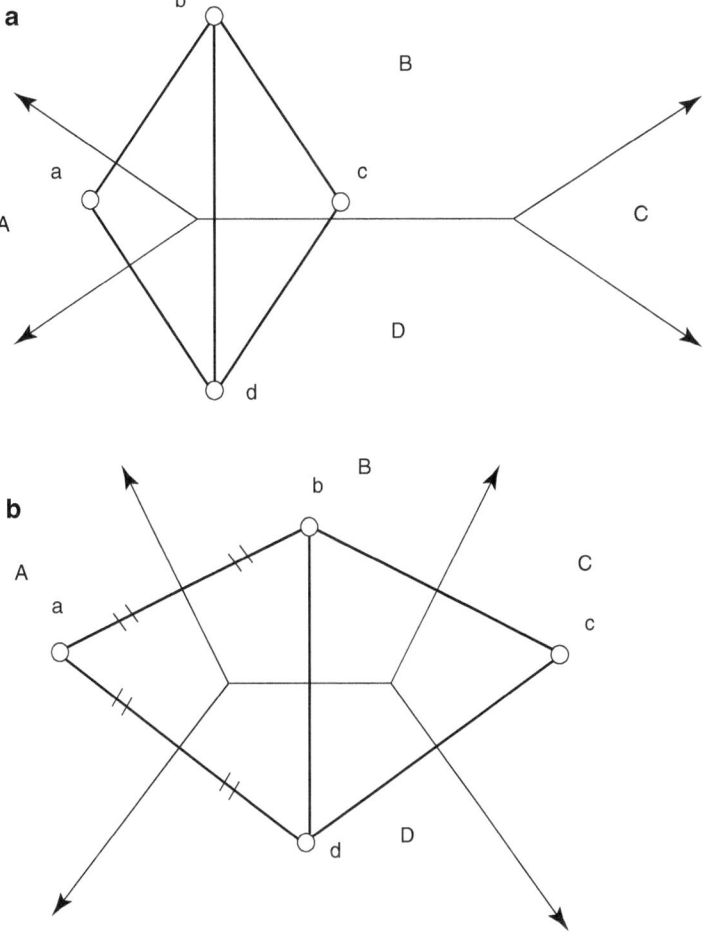

Figure 18.8. Some cell decompositions with a convex reciprocal cannot be a Dirichlet tessellation (**a**), while others with a similar structure are Dirichlet tessellations (**b**).

plane in a nonempty open set, the remaining centers are in one-to-one correspondence with the cells of the sectional Dirichlet tessellation. Their orthogonal projections onto the slicing plane form a convex reciprocal figure for the sectional Dirichlet tessellation (see Figure 18.9 for the analogue on the line). Ash and Bolker proved:

Theorem 18.2. *A sectional Dirichlet tessellation has a convex reciprocal figure.*

These sectional Dirichlet tessellations, also called power Voronoi diagrams or Voronoi diagrams in Laguerre geometry, model a simple

biological phenomenon. Suppose bacteria start to grow at center c at time t_c, with growth rate at the boundary inversely proportional to the distance from the center. If the colonies cannot overlap, the cells occupied by the colonies form the sectional Dirichlet tessellation on the plane $z = 0$ of the spatial tessellation with centers $(c, \sqrt{t_c})$. If all the bacteria start at the same time, we have a Dirichlet tessellation. In this model each cell contains its center; this need not always be true.

We shall prove the converse of Theorem 18.2 and discuss the furthest-point Dirichlet tessellations after we examine spider webs and projections of convex polyhedra.

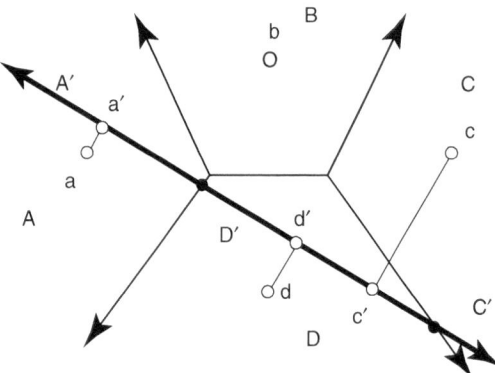

Figure 18.9. Any section of a plane Dirichlet tessellation creates a sectional Dirichlet tessellation on the line (*the black dots on the heavy line*), with a convex reciprocal (*the circles*) given by the orthogonal projection of the plane centers.

Spider Webs and Projections

A proper cell decomposition is a *spider web* if it supports a spider web stress: a set of nonzero tensions in the edges which leads to mechanical equilibrium at each vertex (Figure 18.10a). More specifically, a spider web stress is a nonzero force F_{VE} in each edge E at a vertex V, directed from V out along the edge, such that

- for a finite edge E joining V and V', the forces at the two ends are equal in size: $F_{VE} = -F_{V'E}$;
- for each vertex V, the vector sum of the forces on the edges leaving V is zero.

Spider webs are interesting and important. If they are built with cables and pinned to the ground on the infinite edges they are rigid in the plane. In fact they are the basic building blocks of all rigid cable structures in the plane. At the other extreme, if a plane bar-and-joint framework has the minimum number of bars needed to restrain $|V|$ joints ($|E| = 2|V| - 3$), then the appearance of a spider web signals that it is shaky. Finally, packing nonoverlapping circles of a fixed into a convex polygon cannot be made denser by a small jiggle if, and only if, the associated graph of centers and contact points is an infinitesimally rigid spider web.

If a cell decomposition has a spider web stress, then the vanishing of the vector sum of forces at each vertex says that these forces can be drawn as a closed convex polygon (Figure 18.10b). If we rotate each such polygon clockwise by 90°, then each edge is perpendicular to the edge of the original figure to which it corresponds. Since the forces at the two vertices of a finite edge are equal in size and opposite in direction, these polygons can be glued together to make a planar drawing of the dual graph D* (Figure 18.10c); we have constructed a convex reciprocal.

Conversely, assume that a cell decomposition has a convex reciprocal. We turn this diagram 90° counterclockwise and use the length of each dual edge to define the size of the tension in the corresponding edge of the decomposition. The convex polygons of the reciprocal imply the vector equilibrium of these tensions at each vertex. Thus we have proved:

Theorem 18.3. *A proper cell decomposition is a spider web if and only if it has a convex reciprocal figure. The convex reciprocal determines the spider web stress, and the spider web stress determines the convex reciprocal up to translation and rotation by 180°.*

The trivial cell decompositions satisfy this theorem in an appropriately trivial way. They all have convex reciprocals, and require no tensions for edges with no vertices!

Theorem 18.3 shows that if a proper cell decomposition has one convex reciprocal figure, it has many. Any translation of a reciprocal produces another reciprocal; clearly this translation has no effect on the tensions. If we turn the reciprocal by 180°, we also get a reciprocal. In the classical literature, this turn corresponds to a switch from tensions to compressions, but we have chosen to concentrate on the tensions. Any dilation converts a convex reciprocal to a new reciprocal; the stress corresponding to the new reciprocal is a scalar multiple of the one corresponding to the original.

For some figures, we can freely choose the lengths of several different reciprocal edges and still complete the reciprocal. For the example in Figure 18.11 the lengths of the reciprocal edges

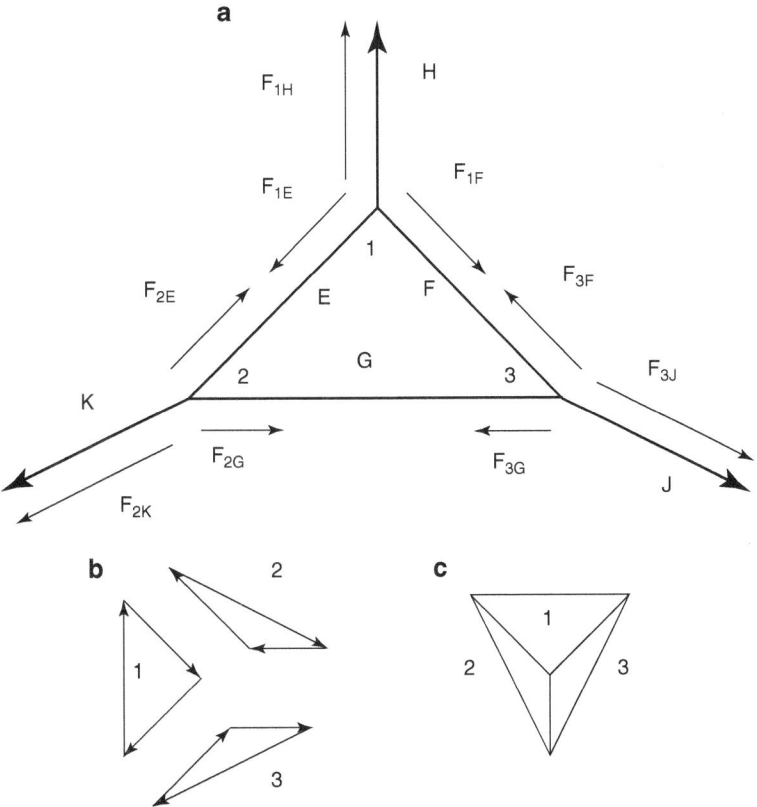

Figure 18.10. The *arrows* in (**a**) show the tensions of a spider web stress on a cell decomposition. The polygons of forces for the equilibria at the vertices (**b**) are pieced together and rotated 90° to form a convex reciprocal figure (**c**).

bc and ab are independent choices. The existence of such dissimilar convex reciprocals reflects the fact that the set of stresses, which is a vector space, has dimension greater than 1.

We now turn to study the projections of polyhedra. Consider the intersection of the upper half-spaces of a finite set of nonvertical planes. The faces, edges and vertices of this intersection form a *convex polyhedral bowl*. (Our choice of up-turned bowls is simply a convenient convention, as you will see below.) The vertical projection of a convex polyhedral bowl is a proper cell decomposition of the plane.

To construct a convex reciprocal figure for such a projection, suppose that the boundary planes P and P' which meet at an edge E have equations

$$Ax + By - z - C = 0,$$

$$A'x + B'y - z - C' = 0.$$

Then the line joining the points (A, B) and (A', B') in the plane is perpendicular to the vertical projection of the edge E, because those points are the intersections of the plane and the normals to P and P' drawn from the points $(0, 0, 1)$ (Figure 18.12). The set of points (A, B), one for each face of the bowl (and thus one for each cell of the projected cell decomposition) form a reciprocal figure for the projection (see Figure 18.13 for an example on the line).

The convexity of the polygons in the reciprocal follows from the convexity of the vertices in the polyhedral bowl. Finally, observe that this reciprocal is also the vertical projection of a dual object in 3-space. This *dual polyhedral bowl* has vertices (A, B, C) corresponding to the planes $Ax + By - z - C = 0$ of the original and boundary planes $Px + Qy - z - R = 0$, one

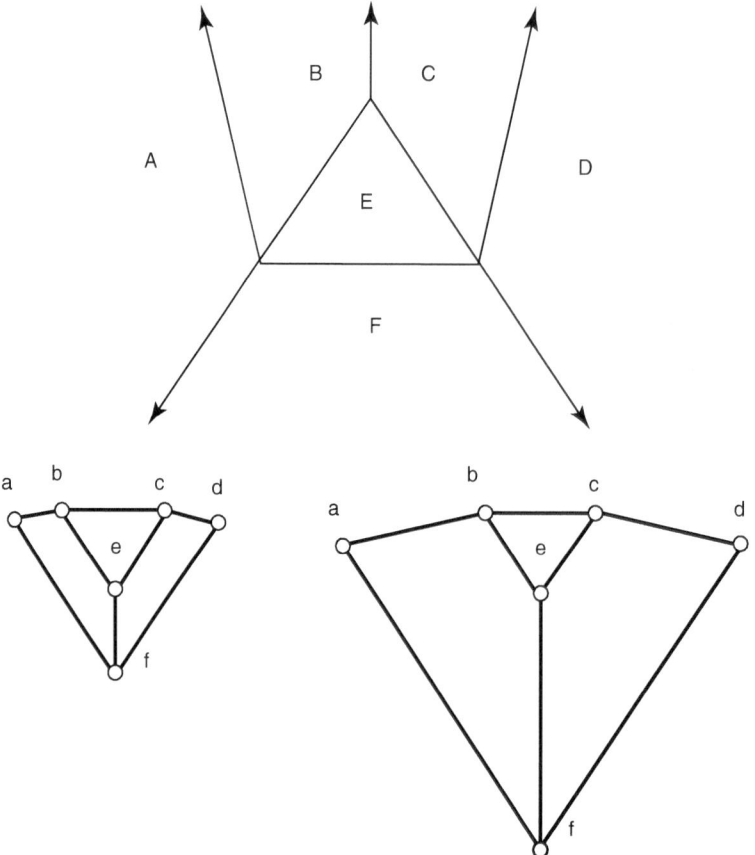

Figure 18.11. A cell decomposition (*light lines*) can have two dissimilar convex reciprocal figures (*heavy lines*) given by free choices for the lengths of some reciprocal edges.

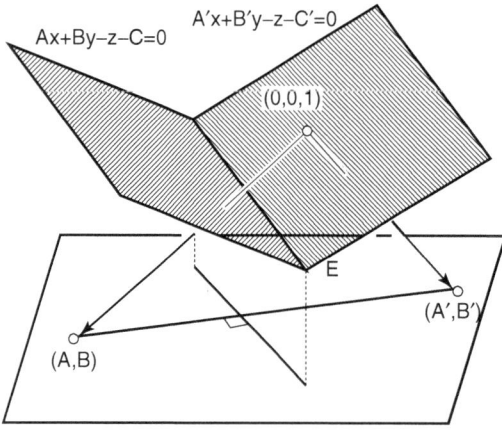

Figure 18.12. These normals to the face planes at a polyhedral edge section to the gradients, and to a reciprocal edge perpendicular to the projection of the polyhedral edge.

for each vertex (P, Q, R) of the original bowl. This dual bowl is created by a projective polarity about the *Maxwell paraboloid* $x^2 + y^2 - 2z = 0$. (Notice that a dual polyhedral bowl is also convex, but it has a cylinder of vertical planes dual to the points at infinity on the unbounded edges of the original bowl.)

The converse is also true. If a proper cell decomposition has a convex reciprocal, then there is a convex polyhedral bowl projecting to this decomposition, with the normals to the faces given by the reciprocal vertices (or by the reciprocal turned 180°). Any single boundary plane of the bowl can be chosen freely (perpendicular to its known normal). Then the positions of the remaining planes can be deduced. Therefore:

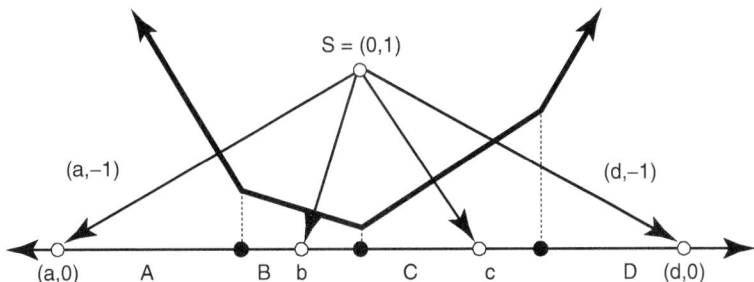

Figure 18.13. A polygonal bowl (*heavy lines*) projects to a cell decomposition of the line (given by the *black dots*), and the normals to the edges produce a convex reciprocal figure (*the circles*).

Theorem 18.4. *A proper cell decomposition has a convex reciprocal if and only if it is the vertical projection of a convex polyhedral bowl. The convex reciprocal can be reconstructed from the bowl by taking the normals to the faces, and the convex reciprocal determines the bowl, up to the vertical translation*

The trivial cell decompositions are projections of trivial convex polyhedral bowls formed by planes parallel to a line. The normals to the faces of such a bowl yield the trivial convex reciprocals as we defined them.

A vertical scaling of a convex polyhedral bowl projecting to our cell decomposition (that is, changing all the z-coordinates by a positive constant factor) corresponds to a similarity transformation of the reciprocal by the same factor. The translations of a reciprocal come from a more subtle "rolling" of the angles of the planes (see Figure 18.14a). If we reflect the bowl in the xy-plane, the reciprocal turns $180°$. We turn all our bowls up so that the reciprocals created match those for sectional Dirichlet tessellations, for which the dual edge cc' is oriented so that it crosses from cell C to cell C'.

While the reciprocal figure is given by an Euclidean construction, the existence of a reciprocal is an essentially affine geometric property: any affine transformation of a cell decomposition with a convex reciprocal figure extends to an affine transformation of the corresponding polyhedral bowl, which then gives the new reciprocal figure. (Alternatively, if the affine transformation of the cell decomposition is AX, then the reciprocal vertices are transformed

by $(A^\top)^{-1}X^*$.) In fact, the invariance has projective overtones as well. If we add the plane at infinity to a nontrivial convex bowl, we get a closed polyhedron in projective space, with the point of projection on this added face. We can make this a finite polyhedron by a projective transformation that brings the plane at infinity to the plane $z = 2$, and leaves the xy-plane unchanged (Figure 18.14b). The cell decomposition is now the projection of this polyhedron from a point on the face which is parallel to the xy-plane. This polyhedron is convex if the vertical direction is enclosed by the cone of normals to the faces of the bowl, or equivalently if the origin is in the convex hull of the vertices in the reciprocal. This can be arranged by a simple translation of the plane, so we have the following corollary:

Corollary 18.5. *A nontrivial proper cell decomposition has a convex reciprocal figure if and only if it is the projection of a convex polyhedron from a point inside a face that is parallel to the projection plane.*

The Main Result

We are now ready to work toward the converse of Theorem 18.2. Consider a proper cell decomposition with a convex reciprocal. This decomposition is the projection of a convex polyhedral bowl with faces of the form $Ax + By - z - C = 0$. Choose the centers to be the points

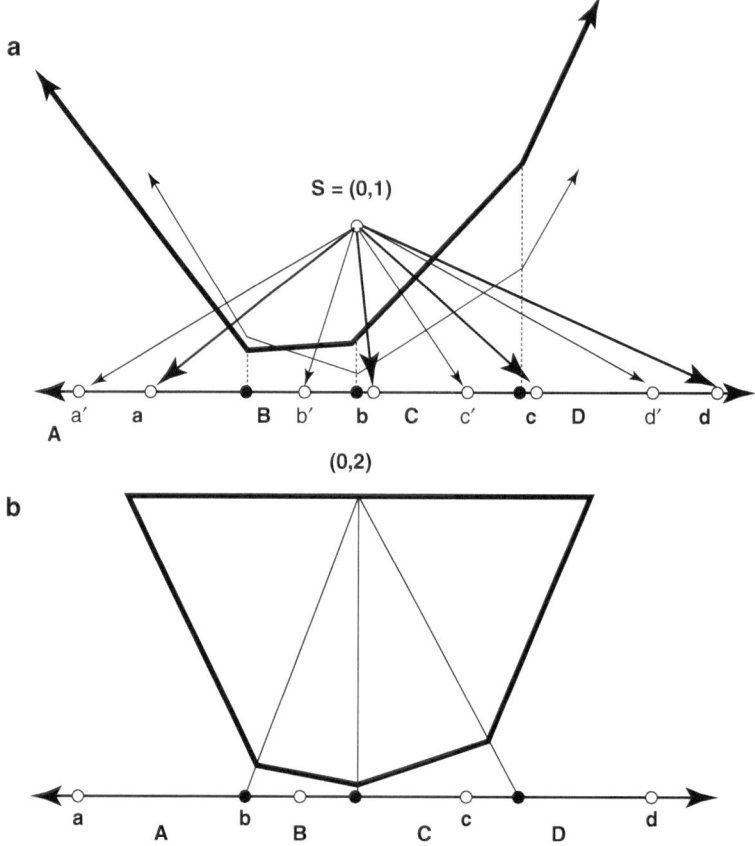

Figure 18.14. If a reciprocal figure to a projected polygonal bowl (*heavy lines* in (**a**)) is translated (b to b′) then the bowl is rolled (*light lines*) in (**a**). The same cell decomposition is also the projection of a closed polygon from a point on one edge (**b**).

$(A, B\sqrt{2C - A^2 - B^2})$; if necessary, the bowl can be lowered by adding a constant d to all the C to make all these z-coordinates well defined. It is a simple exercise to check that a point (x, y) is in the projection of a face if and only if it is closest to the corresponding center (see Figure 18.15), and therefore is in the cell of the sectional Dirichlet tessellation.

Theorem 18.6. *Each sectional Dirichlet tessellation corresponds to a convex polyhedral bowl, and each convex polyhedral bowl can be vertically translated to correspond to a sectional Dirichlet tessellation.*

If we translate the bowl by a constant d, a center at height h moves to one at height

$h' = \sqrt{2d + h^2}$ (provided that $2d + h^2$ is positive for all vertices). We are also free to choose the sign $\pm h$ for each center independently. This completes the converse to Theorem 18.2:

Corollary 18.7. *A cell decomposition has a convex reciprocal if and only if it is a sectional Dirichlet tessellation. The centers in space can be chosen over the vertices of the convex reciprocal, and these centers are unique up to a vertical scaling of the form $h' = \sqrt{2d + h^2}$.*

If we reflect the bowl through the xy-plane, and therefore turn the reciprocal by $180°$, then the point (x, y) is in the region corresponding to $(A', B') = (-A, -B)$ if the associated plane gives the minimum value of z. After rescaling all C' by a vertical translation, our construction

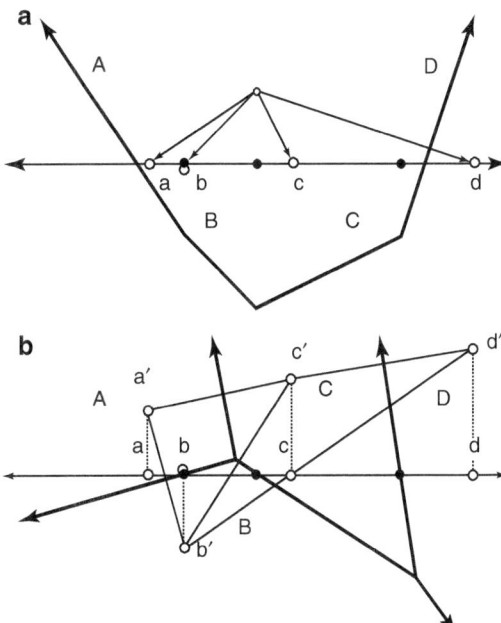

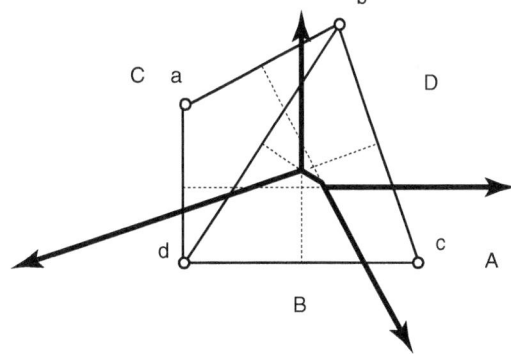

Figure 18.16. A set of centers (*the circles*) produces a furthest-point Dirichlet tessellation (*heavy lines*) with a reciprocal diagram of centers (*medium lines*).

Figure 18.15. A polygonal bowl projects to a cell decomposition with its reciprocal on the line (**a**) and the corresponding plane Dirichlet tessellation sections to the same cell decomposition with the same reciprocal (**b**).

(i) D has a convex reciprocal figure, (ii) D is a spider web. (iii) D is the vertical projection of a convex polyhedral bowl in 3-space. (iv) D is a sectional Dirichlet tessellation. (v) D is a sectional furthest-point Dirichlet tessellation. If D is nontrivial, these are also equivalent to (vi) D is the vertical projection of a convex polyhedron from a point inside one face into a plane parallel to this face.

shows that $(A', B', \sqrt{2C' - A'^2 - B'^2})$ gives the maximum distance from $(x, y, 0)$. Our cell decomposition is thus the section of the *furthest-point Dirichlet tessellation* of centers; that is, each point belongs to the cell of the center which is furthest from the point (Figure 18.16). Our entire theory of reciprocal diagrams applies to these *sectional furthest-point Dirichlet tessellations*. In particular, our inversion of the bowl gives the following result:

Theorem 18.8. *Given a proper cell decomposition D, the following are equivalent: (i) D is a sectional Dirichlet tessellation with projected centers P. (ii) D is the sectional furthest-point Dirichlet tessellation with projected centers* $-P$.

This completes our chain of equivalences. We summarize:

Theorem 18.9. *Given a proper cell decomposition D in the plane, the following are equivalent:*

These equivalences have some interesting consequences. For example, if we build a proper cell decomposition from rubber bands, pin down the edges that go to infinity, and let the tensions position the vertices at a mechanical equilibrium, we will have drawn a picture of a convex polyhedral bowl; spiders really draw such pictures. Moreover, any plane picture of a spider web in space is also a plane spider web, even if the spatial web is warped. Conversely, if we wish to design a rigid cable structure in the plane with a planar graph, we must build a spider web, so we can just use the picture of some convex polyhedron.

The plane Dirichlet tessellations we first studied have very special reciprocals: ones whose dual edges are bisected by the edges of the tessellation. We will see that the polyhedral bowl which projects to a Dirichlet tessellation is correspondingly special; its faces are tangent to the Maxwell paraboloid. Consider such a *Maxwell bowl*. The point of tangency on each face is

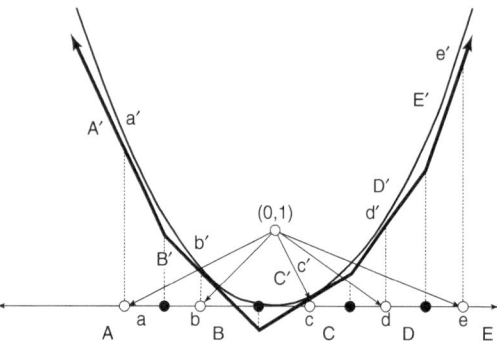

Figure 18.17. Each Dirichlet tessellation on the line is the projection of a polygonal bowl (*heavy lines*) circumscribed about the Maxwell parabola. The centers are the projections of the points of contact of the edges.

the spatial polar of this face (using the polarity described above for the dual bowl), so the plane $Ax + By - z - C = 0$ meets the paraboloid at the point $(A, B, z) = (A, B, 1/2(A^2 + B^2))$. Therefore, $C = 1/2(A^2 + B^2)$. Thus in the construction used in Theorem 18.6, the height $h = \sqrt{2C - A^2 + B^2} = 0$, and we have a Dirichlet tessellation (Figure 18.17). This argument and its converse prove:

Theorem 18.10. *A proper cell decomposition is a Dirichlet tessellation if and only if it is the vertical projection of a Maxwell bowl.*

As we noted before, the furthest-point tessellation corresponds to taking the smallest z value among the planes over the point (x, y), or equivalently, the intersections of the lower halfspaces of the planes. Thus a furthest-point tessellation is the projection of an inverse bowl all of whose faces are tangent to the Maxwell paraboloid: a *Maxwell inverse bowl* (Figure 18.18). (It is not the projection of the inverse of a Maxwell bowl, and is not a Dirichlet tessellation.)

Theorem 18.11. *A proper cell decomposition is a furthest-point Dirichlet tessellation if and only if it is the vertical projection of a Maxwell inverse bowl.*

The plane at infinity is also tangent to the Maxwell paraboloid at the infinite point of projection. In the construction of the corollary to Theorem 18.6 the projective transformation

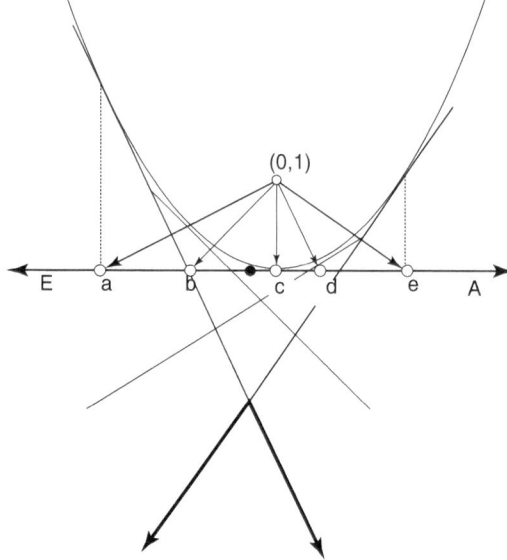

Figure 18.18. Each furthest-point tessellation on the line is the projection of an inverse polygonal bowl (*the heavy lines at the bottom*) with edges tangent to the Maxwell parabola. The centers are the projections of the points of contact.

converts this paraboloid into the sphere $x^2 + y^2 + (z - 1)^2 = 1$. After a suitable translation of the Dirichlet tessellation and its centers, this creates a convex polyhedron with all faces tangent to the sphere that projects onto the Dirichlet tessellation from the point of contact $(0, 0, 2)$ of a horizontal face (Figure 18.19). This gives

Corollary 18.12. *A nontrivial proper cell decomposition is a Dirichlet tessellation of the plane if and only if it is the central projection of a convex polyhedron circumscribed about a sphere, from the point of contact of one face, onto a plane parallel to this face.*

We note a curious consequence of this vision of Dirichlet tessellations. Given a convex polyhedron with all faces tangent to a sphere, we can turn the sphere so that any one face is parallel to the xy-plane, and project from the point of contact. This defines a new and unusual equivalence relation among Dirichlet tessellations: a tessellation with n cells is equivalent to n other tessellations.

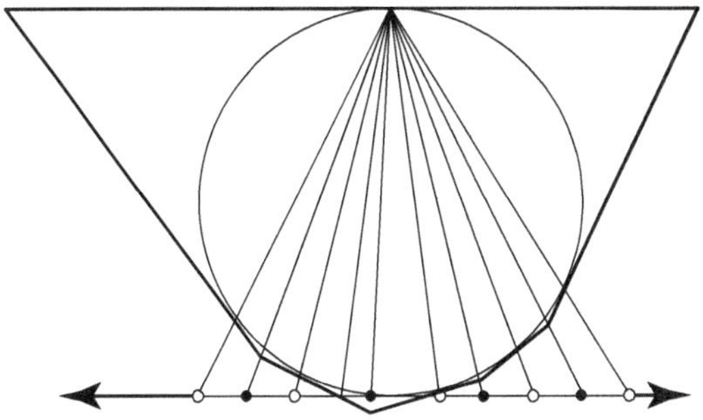

Figure 18.19. Each Dirichlet tessellation on the line is the projection of a convex polygon with an inscribed circle, from the point of contact of one edge. The centers are the projections of the other points of contact.

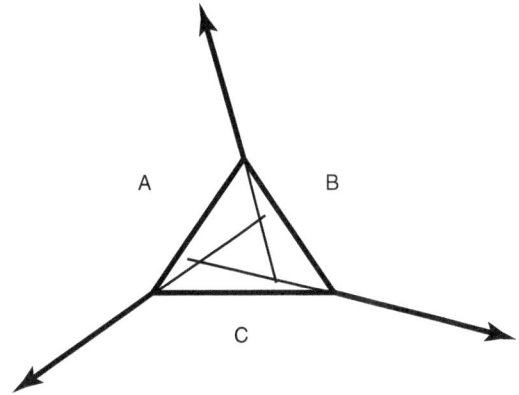

Figure 18.20. This cell decomposition has no stress and no reciprocal figure, because the three extended lines are not concurrent.

It is easy to construct examples of proper cell decompositions of the plane which are not spider webs, or, equivalently, are not projections of convex polyhedral bowls. Consider the cell decomposition of Figure 18.20. If this were the projection of a convex polyhedral bowl, the three planes over the cells A, B, and C would meet in a point. This point would be at the intersection of the three lines separating these cells; this intersection does not exist. Equivalently, if this were a spider web, the three forces in these three separating edges would be in equilibrium (since the forces on any cut set in a spider web will be in equilibrium) but three forces in the plane can reach equilibrium only if they are on concurrent lines. Finally, if this were a sectional Dirichlet tessellation, then the line of points equidistant from the spatial centers of the three exterior cells would lie in the three planes separating these cells and the intersection with the plane would be three concurrent lines.

Realizations of Abstract Graphs

Given a proper cell decomposition of the plane, we can consider the vertices, edges, and cells as a combinatorial structure G (whose precise definition will be given soon) and ask: "Which realizations of G as proper cell decompositions have convex reciprocals?" Work on plane stresses shows that answering this question is equivalent to finding the convex cone of entirely positive solutions to a homogeneous system of linear equations whose unknowns represent the positions of the vertices and the directions of the infinite edges. To be specific, we have a set of finite vertices, V, and infinite vertices V^o, one for each positive direction of an unbounded edge, as well as cyclic order for these infinite vertices. The finite and unbounded edges are given in the obvious way, with two unbounded edges sharing an infinite vertex if they go in the same direction. We take all realizations as proper cell decompositions which respect the order of the cycle of infinite vertices (or its reverse). From general results on planar graphs, this graph structure uniquely determines the possible cells. For such an abstract structure G there are five possibilities:

(i) No realization of G as a proper cell decomposition has a convex reciprocal.

(ii) Some realizations of G as proper cell decompositions have a convex reciprocal; almost all small changes in the position of at least one vertex destroy the spider web stress. These special realizations must satisfy a *geometric condition* expressed by nontrivial polynomial equations in the coordinates of the vertices.

(iii) Many realizations of G as proper cell decompositions have a convex reciprocal; all realizations near a given spider web will also be a spider web. These correct realizations meet a *qualitative condition* expressed by nontrivial polynomial inequalities in the coordinates of the vertices.

(iv) Almost every realization of G as a proper cell decomposition is a spider web; certain special positions are improper. These improper positions, expressed by polynomial equations in the coordinates of the vertices, have zero tension in some edges.

(v) Every realization of G as a proper cell decomposition is a spider web.

Case (i) cannot happen. It is easy to check that all graphs of nontrivial proper cell decompositions are also triply connected planar graphs if we add the polygon at infinity. Theorem 18.9 tells us that graphs of nontrivial spider webs are all constructed from convex polyhedra, by projecting from a point inside one face onto a plane parallel to this face. A classical theorem of Steinitz shows that any triply connected planar graph can be realized as a convex polyhedron.

Figure 18.20 illustrates case (ii). If a graph has $|V|$ finite vertices, and $|E|$ edges, then the general theory of plane frameworks guarantees that when $|E| \leq 2|V|$ there are geometric conditions that must be satisfied if there is to be a stress on all members. If the graph has all vertices 3-valent, the conditions for a spider web can be expressed as a set of equations, one for each finite cell in the graph.

Figure 18.6 illustrates case (iii). All realizations near Figure 18.6a will be spider webs, since the convexity of the reciprocal is preserved by small changes. Similarly, all realizations near Figure 18.6b will have only nonconvex reciprocals and will not be spider webs.

We conjecture that case (iv) cannot occur. As evidence, we mention a study of White and Whiteley in which the authors conjecture that such boundary events, where one tension must be zero, lead to nearby points where the sign of the stress in the edge switches. The set of realizations as proper cell decompositions forms an open convex cone, so both signs of stress must appear in proper cell decompositions near the boundary event, putting us back in case (iii).

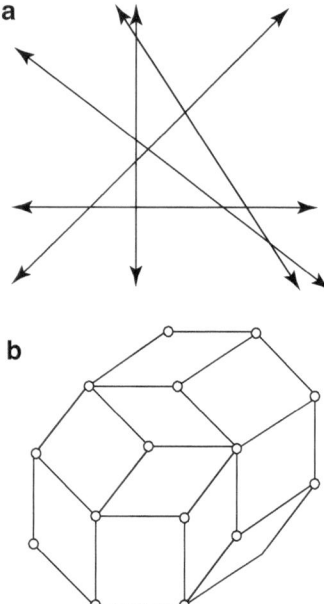

Figure 18.21. Any arrangement of lines in the plane gives a cell decomposition (**a**) with a zonohedral cap as a convex reciprocal figure (**b**).

Cell decompositions with no polygons illustrate case (v). Examples are the trivial decompositions and trees with all vertices are least 3-valent. The interested reader can check directly that such cell decompositions always have a convex reciprocal and thus are always the projection of convex polyhedral bowls. Some other graphs of cell decompositions share this property, but we have not been able to characterize them. A *line arrangement* yields a cell decomposition that is always a spider web. A finite set of lines in the plane creates a cell decomposition (Figure 18.21a). If for each line we choose a nonzero tension and assign this tension to all segments of the line, we have created a spider web stress. (At each vertex on the line, the two tensions in the line cancel; this local cancellations gives the equilibrium.) Therefore this cell decomposition has a convex reciprocal. The reader can check that this reciprocal will be a drawing of the zonohedral cap corresponding to the line arrangement (Figure 18.21b), and the dual bowl over this reciprocal will be a zonohedral cap.

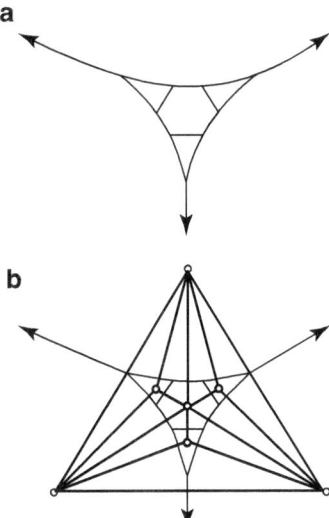

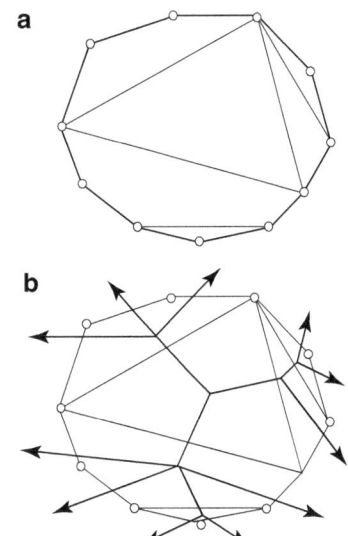

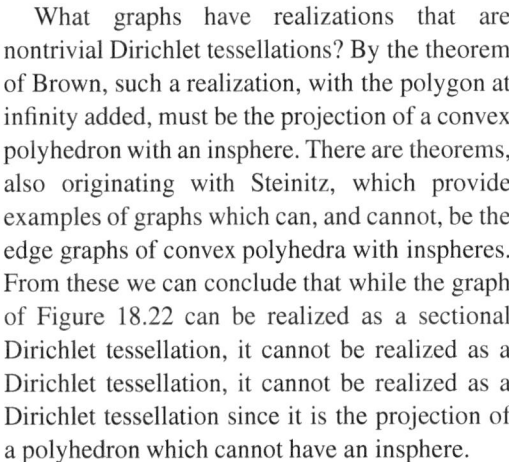

Figure 18.22. The cell decomposition in (**a**) cannot be a Dirichlet tessellation, by its graph, but it is a sectional Dirichlet tessellation, by the convex reciprocal in (**b**).

Figure 18.23. Any split polygon (**a**) is the dual of a tree (**b**) which is the graph of some furthest-point Dirichlet tessellation.

What graphs have realizations that are nontrivial Dirichlet tessellations? By the theorem of Brown, such a realization, with the polygon at infinity added, must be the projection of a convex polyhedron with an insphere. There are theorems, also originating with Steinitz, which provide examples of graphs which can, and cannot, be the edge graphs of convex polyhedra with inspheres. From these we can conclude that while the graph of Figure 18.22 can be realized as a sectional Dirichlet tessellation, it cannot be realized as a Dirichlet tessellation, it cannot be realized as a Dirichlet tessellation since it is the projection of a polyhedron which cannot have an insphere.

If a nontrivial graph can be realized as a Dirichlet tessellation, the realization must satisfy geometric conditions to be a Dirichlet tessellation; it must have a convex reciprocal with the dual edges bisected by the original edges. Ash and Bolker provide a geometric characterization of proper cell decompositions that are Dirichlet tessellations. We will not attempt here to connect their characterization with that of Theorem 18.13 or its corollary. There may be some nice geometry waiting for someone who wishes to explore the connection.

We can characterize those graphs that can appear as nontrivial furthest-point Dirichlet tessellations: they are the trees with all vertices at least 3-valent. To prove this, consider such a tessellation, its reciprocal figure, and the corresponding Maxwell inverse bowl. Look at the convex hull of points of contact with the Maxwell paraboloid. The upper surface of this hull projects to the reciprocal figure. since the points are on the paraboloid, this figure is a convex polygon with some interior edges forming a "split polygon" (Figure 18.23a) or it is a line segment. A split polygon is the dual of a tree (Figure 18.23b) and a line segment is the dual of the 2-cell trivial decomposition. Conversely, every embedded tree has a split polygon as its dual, and we can arrange the points on the paraboloid to realize any combinatorial split polygon as the projection of an upper convex hull. Of course, the realizations of a tree which forms furthest-point Dirichlet tessellations must satisfy additional geometric conditions, but we have not seen these explicitly worked out.

We close this section by remarking that all the results in it are special, convex cases of more general theorems. A planar graph that has a (possibly

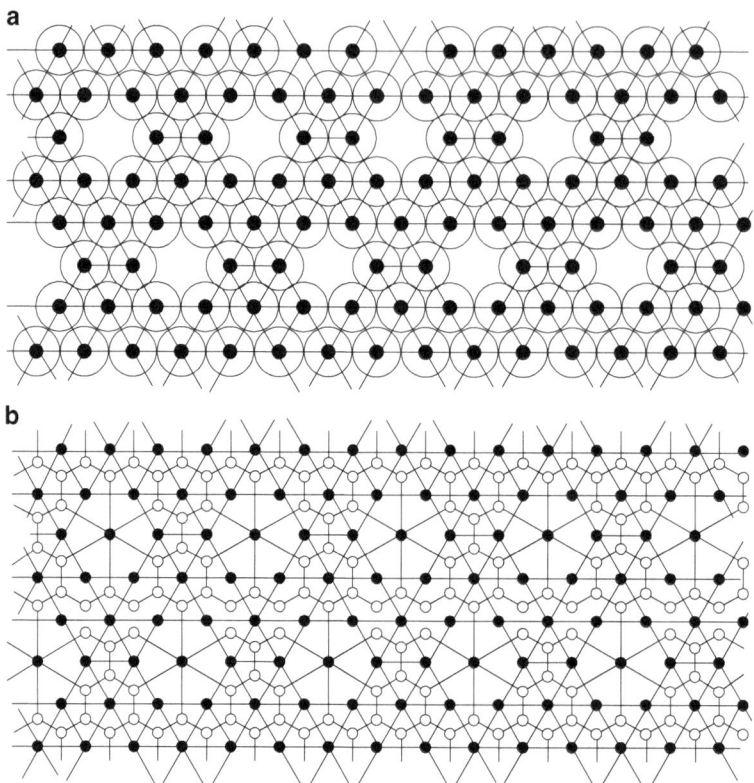

Figure 18.24. The graph of centers of this circle packing gives an infinite cell decomposition of the plane (**a**) which has a convex reciprocal figure (**b**) which is also an infinite cell decomposition of the plane.

nonconvex) reciprocal is the projection of a general spherical polyhedron in 3-space (possibly self-intersecting). The reciprocal corresponds to a set of nonzero tensions and compressions in the edges of the graph, in a static equilibrium at each vertex. This general case was the one first studied in graphical statics. These closing observations emphasize the important but often forgotten fact that statics and the equivalent theory of infinitesimal mechanics both truly belong to projective geometry. So too does the study of projected polyhedra and general reciprocal diagrams.

Infinite Plane Examples

In order to study packings of the plane by identical circles, a number of mathematicians have considered the locally finite graph in the plane which has the centers of the circles for vertices and an edge joining two vertices whenever the two circles touch (Figure 18.24a). Connelly discovered that there is an intimate connection between the problems: "Is this diagram a spider web?" and "Is this packing locally maximally dense?" He even uses convex reciprocal diagrams to study these examples (Figure 18.24b).

Another class of infinite but locally finite examples comes from the study of Dirichlet tessellations arising from periodic lattices in the plane. Surprisingly, there are also aperiodic sections of periodic tessellations of 3-space. (Figure 18.25 shows an example on the line.) To study these and other similar examples, we change the definition of a proper cell decomposition by allowing infinitely many cells, but we require local finiteness: no point belongs to infinitely many cells.

We then refer to the cell decompositions of the previous section as *finite decompositions.*

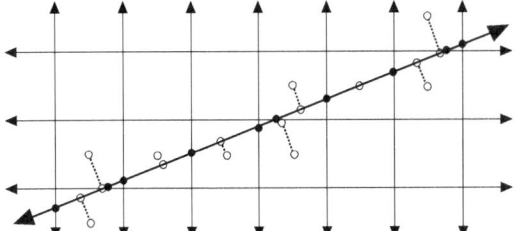

Figure 18.25. A periodic Dirichlet tessellation of the plane (*light lines*) can section to a nonperiodic cell decomposition of the line (*the black dots on the heavy line*).

Our new infinite proper cell decompositions still have abstract dual graphs. The local finiteness implies that the dual polygon for each original vertex is finite, so the definition of a *convex reciprocal figure* is unchanged. However, since an unbounded cell of the decomposition may have infinitely many edges, the dual and the reciprocal may not be locally finite. We still include the trivial examples formed by an infinite number of parallel strips and their reciprocals.

An infinite set of centers, with only a finite number of centers in any bounded set, defines an *infinite Dirichlet tessellation*. More generally, an infinite sectional Dirichlet tessellation is a plane section of an infinite Dirichlet tessellation of space. The argument used for finite tessellations still works to show the following.

Theorem 18.13. *An infinite sectional Dirichlet tessellation has a convex reciprocal figure.*

Since the definition of a spider web refers to tensions in equilibrium at each vertex, local finiteness guarantees that we have an immediate extensions of the definition and of Theorem 18.3:

Theorem 18.14. *An infinite proper cell decomposition is a spider web if and only if it has a convex reciprocal figure.*

An *infinite convex polyhedral bowl* is the intersection of the upper half spaces of an infinite set of nonvertical planes such that

- no finite region of space intersects an infinite number of the planes;
- this intersection includes points over all points (x, y).

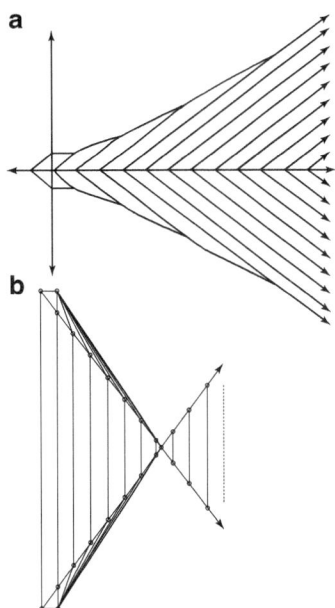

Figure 18.26. The cell decomposition in (**a**) has a local convex reciprocal for arbitrarily large finite sets of vertices, but all global reciprocals are nonconvex (**b**).

With this definition, which ensures that the projection of such a bowl is a proper infinite cell decomposition, Theorem 18.4 remains true.

Theorem 18.15. *An infinite proper cell decomposition has a convex reciprocal if and only if it is the vertical projection of an infinite convex polyhedral bowl.*

However, if we add the plane at infinity to an infinite Maxwell bowl, we do not create a closed polyhedron. There are no analogues of Corollary 18.5. Note that each element of an expanding sequence of finite subpieces of an infinite cell decomposition may have a convex reciprocal, while the entire structure has no convex reciprocal. (An example is shown in Figure 18.26.) Thus these finite pieces of an infinite cell decomposition may each be the projection of a convex polyhedral bowl, while the entire structure is not.

If our cell decomposition has only bounded polygons, then it must have infinitely many of them. If a convex reciprocal covers the plane, it is also an infinite cell decomposition; this reciprocal and the original decomposition form a reciprocal pair (Figure 18.24). Such a reciprocal pair corresponds to a full bowl for which the total curvature is 2π, since the normals to the faces cover a hemisphere. For example Robert Connelly has observed that any infinite cell decomposition of the plane using only regular convex polygons as cells corresponds to such a bowl. In fact, the centers of the polygons form a reciprocal: the Dirichlet tessellation for the vertices of the regular polygons, as illustrated in Figure 18.24b. How about the analogue of Theorem 18.6? For the spatial center corresponding to the plane $Ax + By - z - C = 0$ we took the point $(A, B, \sqrt{2C - A^2 - B^2})$. To make all these z-coordinates well defined, we translated the bowl by adding a sufficiently large constant d simultaneously to all values of C. With an infinite set of planes, this may be impossible, and there would be no Dirichlet tessellation to section. Consider the trivial example on the line formed by the plane polygon with vertices: $\ldots, (0, 0), (1, -2), (2, -6), (3, -12), \ldots,$ $(n, -n(n + 1)), \ldots$ The face gradients have A: $\ldots, 2, 4, 6, 8, \ldots, 2n, \ldots,$ and intercepts $C: 0, -2, -6, \ldots, -n(n - 1), \ldots.$ No constant d can make $2C + d - A^2 > 0$ for all n. This example with this reciprocal cannot correspond to any sectional Dirichlet tessellation. Each finite segment of the cell decomposition is a sectional Dirichlet tessellation with centers in the plane over the reciprocal vertices, but the entire object is not. (However, it is a Voronoi diagram in the Laguerre geometry, since this algebraic definition allows h^2 to be negative.)

The transformation from a sectional Dirichlet tessellation to a convex polyhedral bowl still applies.

Theorem 18.16. *Each infinite sectional Dirichlet tessellation corresponds to an infinite convex polyhedral bowl, determined up to vertical translation.*

Since it is not meaningful to talk about furthest-point Dirichlet tessellations for infinite sets of centers, we have completed our chain of analogues. The following summarizes the limited analogue of Theorem 18.8

Theorem 18.17. *Given an infinite proper cell decomposition D in the plane, the following are equivalent: (i) D has a convex reciprocal figure; (ii) D is a spider web; (iii) D is the vertical projection of an infinite convex polyhedral bowl in 3-space.*

For infinite Dirichlet tessellations, our construction did not require a vertical translation of the bowl and we have a complete analogue for Theorem 18.9.

Theorem 18.18. *An infinite proper cell decomposition is a Dirichlet tessellation if and only if it is the vertical projection of an infinite Maxwell bowl.*

We note that a periodic cell decomposition may have only nonperiodic reciprocals, and that even a periodic cell decomposition with a periodic reciprocal may induce no periodicity in the spatial polyhedron or in the spatial Dirichlet tessellaton it sections. Conversely, some of the interesting aperiodic tessellations of the plane are sections of periodic Dirichlet tessellations of some higher space, but neither the tessellation nor the reciprocal is periodic. For example, it has been observed that a version of the nonperiodic Penrose tiling of the plane, drawn with rhombi (Figure 18.27a), is a section of the regular cubic tessellation in R^5. The corresponding reciprocal figure for this tiling consists of a line arrangement of five families of parallel lines (Figure 18.27b), called a pentagrid; this pentagrid and the aperiodic rhombic tiling form a reciprocal pair.

Finally, we do not know an infinite analogue of Steinitz's theorem, so we cannot answer the question: "What infinite planar graphs can be realized as the edge skeleton of an infinite convex polyhedral bowl?" nor the equivalent question: "What infinite graphs can be represented as spider webs?" *We conjecture that any graph that can be realized as a proper infinite cell decomposition can also be realized as the edge skeleton of an infinite convex polyhedral bowl.*

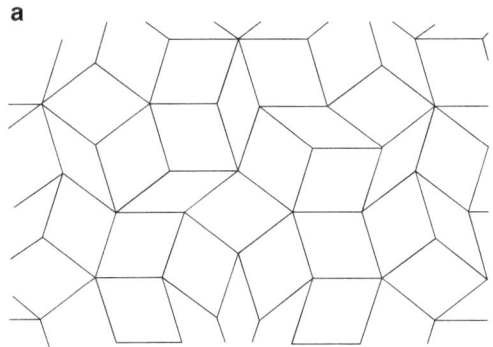

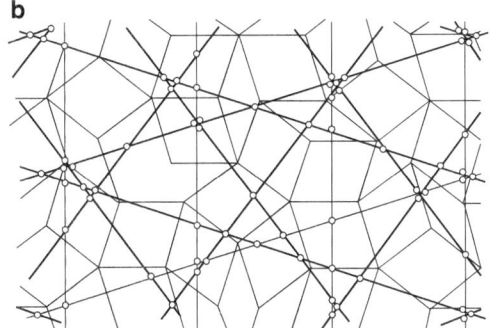

Figure 18.27. The aperiodic rhombic tiling (**a**) has a pentagrid of five families of parallel lines as a convex reciprocal (**b**).

UPDATE 2009

In the quarter century since this chapter was written, there has been significant interest in a number of related problems which can be viewed as extensions of these connections, as well as a number of relevant publications. For key references, see http://www.marjoriesenechal.com/. Here we briefly outline some of the current areas being explored.

There are a large number of publications on Voronoi Diagrams (the more common term for Dirichlet tessellation now, in all dimensions),

as fundamental structures for computational geometry, both in the plane and in higher dimensions, including continuing work on weighted tessellations. Many of these implicitly include the reciprocal diagrams and projections.

There have also been advances on reciprocal diagrams in the plane, including exploration of the vector spaces of self-stresses and the space of associated reciprocals. The reciprocals remain a subject of continuing research and application, including extensions such as reciprocal diagrams on the sphere—where all of the results here for finite configurations extend directly. In general, much of this work does not have restrictions for convexity, although there are results that continue to connect with Voronoi diagrams. One area where convexity continues to be central is the work on finite, zonohedral tilings (cubic partial cubes) which are implicitly connected to reciprocals of line arrangements. This has been extended to pseudo-line arrangements where the infinite ends of the lines are parallel and their direction is used to determine the (perpendicular) direction of all reciprocal edges of the pseudo-line. There is also interest in symmetries of the line-arrangements and the associated zonogonal rosettes. With growing interest in periodic structures, with periodic self-stresses, there is an opening for further exploration of the infinite examples and their reciprocals.

There is now a significant body of work on projections of higher dimensional cell complexes, their associated self-stresses, and higher dimensional reciprocals. This is an area of continuing research. All of the basic theory extends, though only some spider-webs are created this way. A sample area still ripe for explorations would be higher dimensional zonagonal tilings as reciprocals of higher dimensional hyperplane arrangements.

19

Uniform Polyhedra from Diophantine Equations

Barry Monson

A simple set of coordinates eases the study of metrical properties of uniform polyhedra. For instance, the six vertices of the regular octahedron $\{3, 4\}$ have Cartesian coordinates $(\pm 1, 0, 0)$, etc. where "etc." means "permute the coordinates in all possible ways." I find it pleasing in such examples that the coordinates are given by systematic choices. Observe further that the coordinates provide all *integral* solutions to the Diophantine equation

$$x^2 + y^2 + z^2 = N, \qquad (19.1)$$

when $N = 1$. If instead $N = 3$, we obtain the eight vertices $(\pm 1, \pm 1, \pm 1)$ of the cube $\{4, 3\}$. Less obviously, we get vertices for the cuboctahedron $\{^3_4\}$ (here I use Coxeter's notation $\{^p_q\}$ for a quasiregular polyhedron in which p-gons and q-gons alternate at each vertex) when $N = 2$ and the truncated octahedron $t\{3, 4\}$ when $N = 5$. Clearly, Equation 19.1 is unchanged by the eight possible sign changes or six possible permutations of x, y, and z. Thus the $48 = 6 \times 8$ geometric symmetries in the octahedral group are represented as algebraic symmetries of (19.1). In fact, for any N, we may thus construct a polyhedron with octahedral symmetry although it may be uninteresting; there usually is no uniform way of defining its edges and faces.

B. Monson
Department of Mathematics and Statistics,
University of New Brunswick, PO Box 4400,
Fredericton, NB E3B 5A3, Canada,
e-mail: bmonson@unb.ca, http://www.math.unb.ca/~barry/

Interlude

The remaining uniform polyhedra with octahedral symmetry pose another problem. For example, the truncated cube $t\{4, 3\}$ has typical vertex $(1, 1, \sqrt{2} - 1)$, but the irrational $\sqrt{2} - 1$ is not the kind of integer required in (19.1). We tackled these cases with some (untidy) success using the ideas exploited in the next section. Also, (19.1) is invariant under the central inversion $(x, y, z) \rightarrow (-x, -y, -z)$; thus a homogeneous quadratic must be replaced by some other equation when describing polyhedra without central symmetry, such as the tetrahedron or the snub cube.

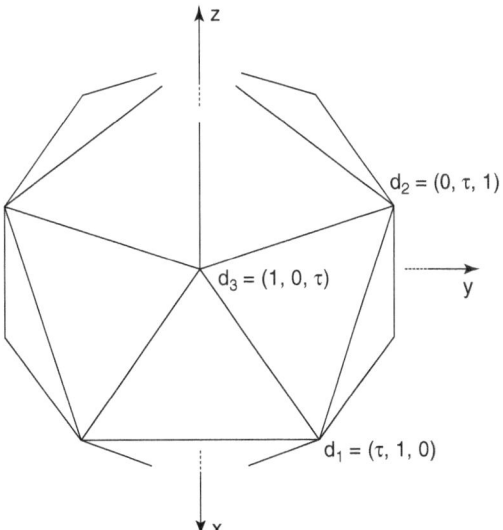

Figure 19.1. The icosahedron $\{3, 5\}$.

M. Senechal (ed.), *Shaping Space*, DOI 10.1007/978-0-387-92714-5_19,
© Marjorie Senechal 2013

Uniform Polyhedra with Icosahedral Symmetry

The icosahedron $\{3, 5\}$ and its relatives have fivefold rotational symmetry (see Figure 19.1). Hence our coordinates must somehow involve the number

$$\cos(2\pi/5) = 1/2\tau = (\sqrt{5} - 1)/4,$$

where the Golden Ratio $\tau = (1 + \sqrt{5})/2$ satisfies the condition

$$\tau^2 = \tau + 1.$$

To reconcile this irrational with the integral nature of our equations we replace the rational field Q and its ring of integers Z by the quadratic number field $Q(\sqrt{5})$ and its ring of algebraic integers $Z[\tau]$. The ring $Z[\tau]$ consists of all polynomials in τ with integral coefficients; any integer $x \in Z[\tau]$ is uniquely expressed as $x = x_1 + \tau x_2, (x_1, x_2 \in Z)$. (For the number-theoretic properties of the Euclidean domain $Z[\tau]$, see Hardy and Wright's *An Introduction to the Theory of Numbers*.) Note that the arbitrary magnitude of the units $\pm \tau^n$, $(n \in Z)$, complicates the solution of Diophantine equations over $Z[\tau]$.

The icosahedron has 12 vertices with Cartesian coordinates $(\pm \tau, \pm 1, 0)$ and its *cyclic* permutations only. Now let us solve (19.1) for

$$N = \tau^2 + 1^2 + 0^2 = \tau + 2.$$

Letting $x = x_1 + \tau x_2, y = y_1 + \tau y_2, z = z_1 + \tau z_2$, we split (19.1) into its rational and irrational parts, then solve two simultaneous ordinary Diophantine equations in six variables:

$$x_1^2 + x_2^2 + y_1^2 + y_2^2 + z_1^2 + z_2^2 = 2;$$

$$2x_1 x_2 + x_2^2 + 2y_1 `y_2 + y_2^2 + 2z_1 z_2 + z_2^2 = 1.$$

Dissappointingly, we find 24 solutions, namely *all* permutations of $(\pm \tau, \pm 1, 0)$. But then we recall that our solution must have 48 octahedral symmetries, whereas the icosahedrron has 120 symmetries. Since $120/48 = 2.5$ we should have expected this incompatibility; the convex hull of the 24 points is a nonuniform truncation of $\{3, 4\}$ with two naturally inscribed $\{3, 5\}$'s.

Thus we must abandon Cartesian coordinates in favor of some system of oblique coordinates referred to a basis $d_1 d_2 d_3$. Much effort leads to the obvious choice of three vertices of a triangular face of $\{3, 5\}$, say $d_1 = (\tau, 1, 0), d_2 = (0, \tau, 1), d_3 = (1, 0, \tau)$ in Cartesian coordinates (Figure 19.1).

A typical point $u = x d_1 + y d_2 + z d_3$ thus has squared length $u \cdot u = N$, that is

Table 19.1. Solution of Equation 19.2

Polyhedron	Symbol	Vertices	N	Number of solutions to 19.2
Icosahedron	$\{3, 5\}$	12	$2 + \tau$	12
Icosidodecahedron	$\{^3_5\}$	30	4	30
Dodecahedron	$\{5, 3\}$	20	3	20
Truncated icosahedron	$t\{3, 5\}$	60	$10 + 9\tau$	60
Rhombicosidodecahedron	$r\{^3_5\}$	60	$6 + \tau$	60
Truncated dodecahedron	$t\{5, 3\}$	60	$7 + 4\tau$	60
Truncated icosidodecahedron	$t\{^3_5\}$	120	$14 + 5\tau$	180

The theory of regular (and other, less symmetric) polytopes has exploded these last 28 years. Do have a look at *Abstract Regular Polytopes*, P. McMullen and E. Schulte, Encyclopedia of Mathematics and its Applications **92**, (Cambridge: Cambridge University Press, 2002), particularly chapters 1 and 5. Recently, Egon Schulte and I have exploited the arithmetic of $Z[\tau]$ in a much more significant way in "Modular Reduction in Abstract Polytopes", *Canadian Mathematical Bulletin*, **52** (2009), pp. 435–450. There, in §5, we give a natural construction for the 11-cell and 57-cell, two abstract regular 4-polytopes discovered by Grünbaum and Coxeter in the 1970's.

$$(x^2 + y^2 + z^2)(2 + \tau) + 2\tau(xy + xz + yz) = N. \tag{19.2}$$

This equation has built-in icosahedral symmetry, since each of the 120 symmetries preserves points with coordinates in the ring $Z[\tau]$. We solve it by splitting it into rational and irrational parts; some results are tabulated below. In the last case we find 60 superfluous solutions.

It is unclear what is merely fortuitous in the last example. A more insightful account may appear elsewhere. Perhaps, however, the reader has enjoyed yet another duet played by geometry and number theory.

20

Torus Decompostions of Regular Polytopes in 4-space

Thomas F. Banchoff

When a regular polyhedron in ordinary 3-space is inscribed in a sphere, then a decomposition of the sphere into bands perpendicular to an axis of symmetry of the polyhedron determines a corresponding decomposition of the polyhedron. For example, a cube with two horizontal faces can be described as a union of two horizontal squares and a band of four vertical squares, and an octahedron with a horizontal face is a union of two horizontal triangles and a band formed by the six remaining triangles.

We may approach the study of regular figures in 4-space in a similar way. The corresponding statement one dimension higher says that if a regular polytope in 4-space has its vertices on a hypersphere such that a symmetry axis coincides with the axis perpendicular to ordinary 3-space, then the polytope can be described as a union of polyhedra arranged in "spherical shells." For example, a 4-cube with one cubical face parallel to 3-space can be described as a union of two cubes and a shell made from the remaining six cubes.

Similar shell decompositions have become a standard means of describing the way various three-dimensional faces fit together in 4-space to form regular polytopes. In this chapter we examine an alternative way of describing regular

figures in 4-space, presenting them as unions not of spherical shells but of rings of polyhedra known as *solid tori*. Such *torus decompositions* are especially convenient for studying symmetries of these figures and for investigating their topological properties. A valuable tool in this project is a remarkable mapping discovered by Heinz Hopf which relates the geometry of circles on the hypersphere in 4-space to the geometry of points on the ordinary sphere in 3-space. One of the aims of this chapter is to give additional geometric insight into the *Hopf mapping* by describing its relationship to torus decompositions of regular polytopes.

Decompositions

Decompositions of objects are often easier to visualize when we project them into lower-dimensional spaces. For a regular polyhedron inscribed in a 2-sphere centered at the origin, if we use central projection from the North Pole to the horizontal plane which passes through the origin, the images of the vertices and edges of the polyhedron form a *Schlegel diagram* of the polyhedron. In such a diagram we may identify the convex cells in a decomposition of the polyhedron corresponding to the decomposition of the 2-sphere into horizontal bands (see Figure 20.1).

If we follow the same procedure one dimension higher, we project centrally from a point on the hypersphere to our three-dimensional space

T.F. Banchoff
Mathematics Department, Brown University, Royce
Family Professor in Teaching Excellence 2005-8,
Providence, RI 02912, USA
e-mail: thomas_banchoff@brown.edu; http://www.math.
brown.edu/~banchoff

M. Senechal (ed.), *Shaping Space*, DOI 10.1007/978-0-387-92714-5_20,

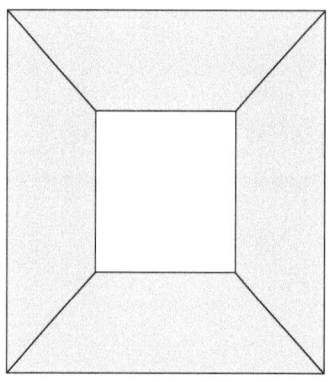
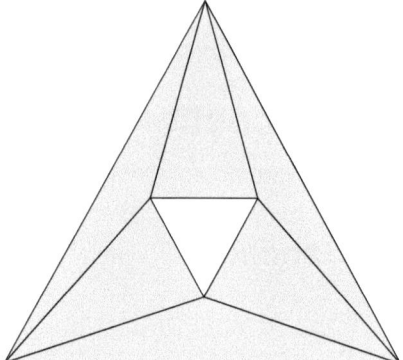

Figure 20.1. Band decompositions of the cube and octahedron.

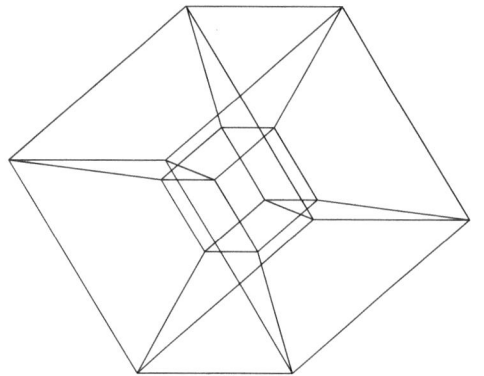

Figure 20.2. Cube-within-a-cube projection of the hypercube.

and the images of the vertices and edges of a regular figure determine its Schlegel diagram. Just as the central projection of a cube to the plane leads to a "square-within-a-square," the central projection of a hypercube may appear as a "cube-within-a-cube" with corresponding vertices connected in each case. The annular band in the plane formed by four trapezoids separating two squares corresponds to the region separating two cubes which is decomposed into six congruent truncated square pyramids (Figure 20.2).

The four vertical truncated pyramids in the central projection of the hypercube fit together to form a solid torus which is the image of a ring of four cubes on the hypercube in 4-space. The remaining four cubes also form a ring and the common boundary of these two rings is the surface of a torus formed by 16 squares in the

hypercube. This is the prototype of a torus decomposition, and it is this kind of analysis we wish to carry out with respect to other regular polytopes in 4-space. In this chapter we pay special attention to the 24-cell, a polytope formed from 24 regular octahedra. This polytope is complicated enough to exhibit most of the interesting phenomena of torus decompositions and it is still relatively easy to visualize, especially when we use the techniques of computer graphics.

The Cube and Its Associated Polyhedra

Associated with the cube are other polyhedra with vertices at the centers of faces or edges of the cube. If we take the six centers of square faces, we obtain the regular octahedron inscribed in the cube. If, on the other hand, we take the midpoints of the 12 edges of the cube, we obtain a cuboctahedron, a semiregular polyhedron with faces of two types: squares determined by the midpoints of the edges of the cube's square faces and triangles determined by the midpoints of the three edges coming from each vertex of the cube.

If we project the cube into the plane by central projection, we can identify the projections of the octahedron and the cuboctahedron by joining images of centers of faces and edges of the cube (Figure 20.1). The cuboctahedron in particular can be expressed as a union of a large square surrounding a small square, with the region between them subdivided into four squares

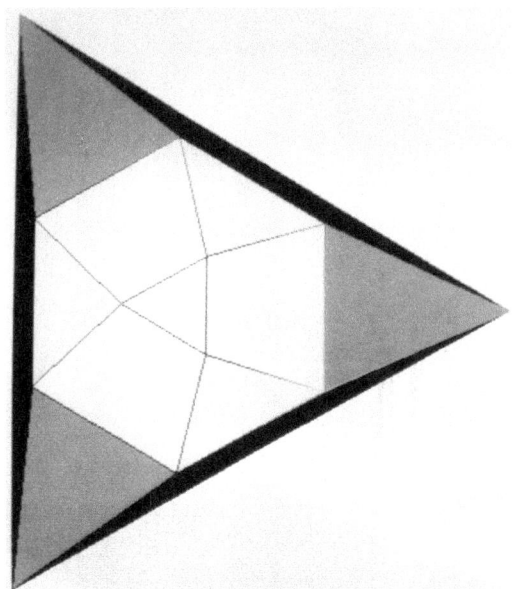

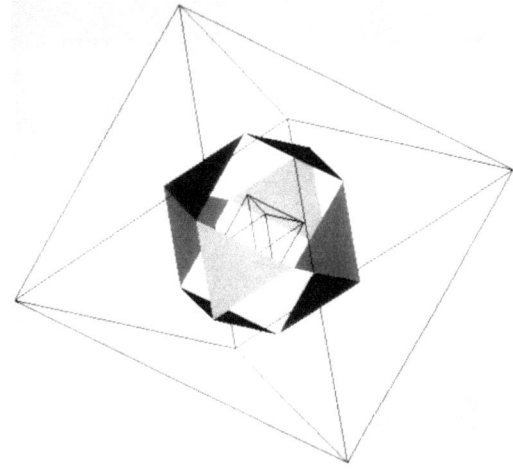

Figure 20.4. Projection of the 24-cell.

Figure 20.3. Band decomposition of the cuboctahedron.

and eight triangles (Figure 20.3). This gives a "band decomposition" corresponding to the decomposition of the cube itself into two horizontal squares and a band of four vertical squares.

The Hypercube and Its Associated Polytopes

In a similar manner we may identify regular and semiregular polytopes associated with a hypercube by taking the midpoints of the faces of certain dimensions. If we take the midpoints of the eight cubical faces, we obtain the vertices of a cube-dual determining the 16 tetrahedral faces of the 16-cell. If we take the midpoints of all 32 edges, we obtain a semiregular polytope with 24 cells: 16 tetrahedra connecting the midpoints of quadruples of edges emanating from each of the vertices of the hypercube, and eight cuboctahedra determined by the midpoints of edges of each of the cubical faces of the hypercube.

On the other hand, if we take the centers of all square faces of the hypercube we obtain a polytope which is regular. Each vertex of the hypercube is a vertex of six squares, each with a pair of sides chosen from among the four edges emanating from the vertex. The midpoints of these six squares will be vertices of an octahedron. Moreover, the midpoints of the six square faces of a cube in the hypercube also determine an octahedron. We thus obtain a polytope with 24 octahedral faces, 16 corresponding to vertices of the hypercube and eight corresponding to its dual polytope. This polytope is called the *24-cell*, and it is the main object of our study in this chapter.

As in the lower-dimensional situation, we may identify the projections of these semiregular and regular polytopes by referring to the central "cube-within-a-cube" projection of the hypercube (Figure 20.2). In particular, the 24-cell may be presented as a large octahedron surrounding a small octahedron, with the region between them decomposed into six octahedra meeting vertex-to-vertex and 16 octahedra each meeting either the large octahedron or the small one along a triangular face. We may think of this polytope as consisting of one octahedron on each face of the larger one. Each octahedron of the first set shares a triangle with one octahedron of the other set. The gaps left between these 16 octahedra determine the places for the remaining six octahedra (Figure 20.4).

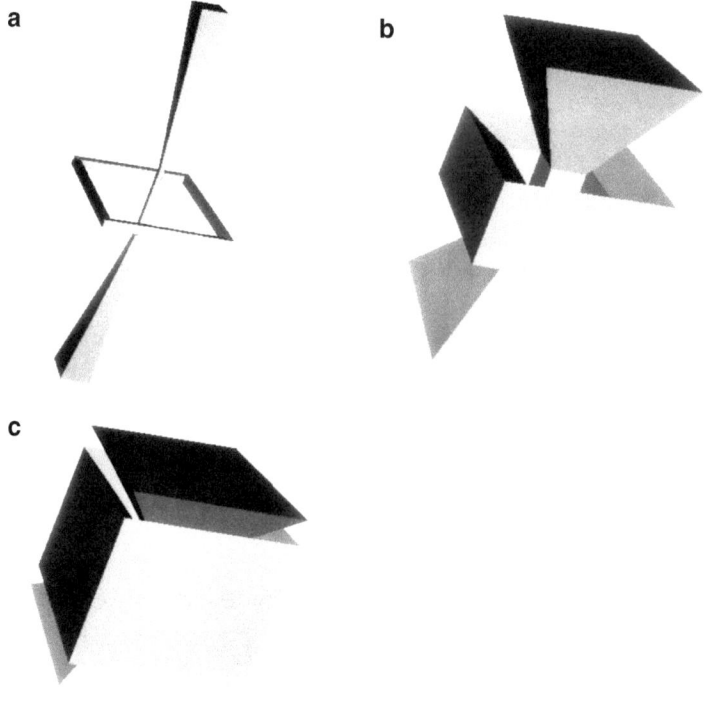

Figure 20.5. Torus decompositions of the hypercube.

The hypercube may be decomposed into two solid torus rings, each a cycle of four cubes meeting along square faces. In 4-space the centers of the cubes in one of these rings will form the vertices of a square. In the cube-within-a-cube projection, the centers of four of these cubes lie on a vertical straight line and the other four centers are vertices of a horizontal square. The vertical line meets the horizontal square disc in exactly one point, and this fact implies that the line and the square are linked (Figure 20.5).

Analogously we may express the 24-cell as a union of four solid tori, each a cycle of six octahedra meeting along triangular faces. The centers of the six octahedra in a ring will form a planar hexagon, with four vertices lying in two parallel edges of the hypercube and two opposite vertices from the dual 16-cell. Any two of these hexagons are linked, so that any hexagon meets the disc bounded by any other hexagon in exactly one point. We can identify such hexagonal cycles in the central projection of a hypercube and its dual polytope. One of the hexagons includes a

vertex at infinity, so it is a straight line. The other three hexagons are arranged symmetrically about this line (Figure 20.6).

Fold-Out Decomposition of the Hypercube and 24-Cell

The decomposition of a hypercube into two solid tori with a common polyhedral boundary can be described in a different way by "folding the figure out into 3-space." We may express a cube folded out into the plane by giving two squares together with a strip of four squares. The ends of the strip are to be identified to form a cylinder with two boundary square polygons, which will match up with the boundaries of the remaining two squares. The analogous decomposition of the hypercube starts with two solid stacks of four cubes. The ends of the stacks are to be identified by folding up in 4-space to obtain the two solid tori. The common boundary of these solid tori is a polyhedral torus which

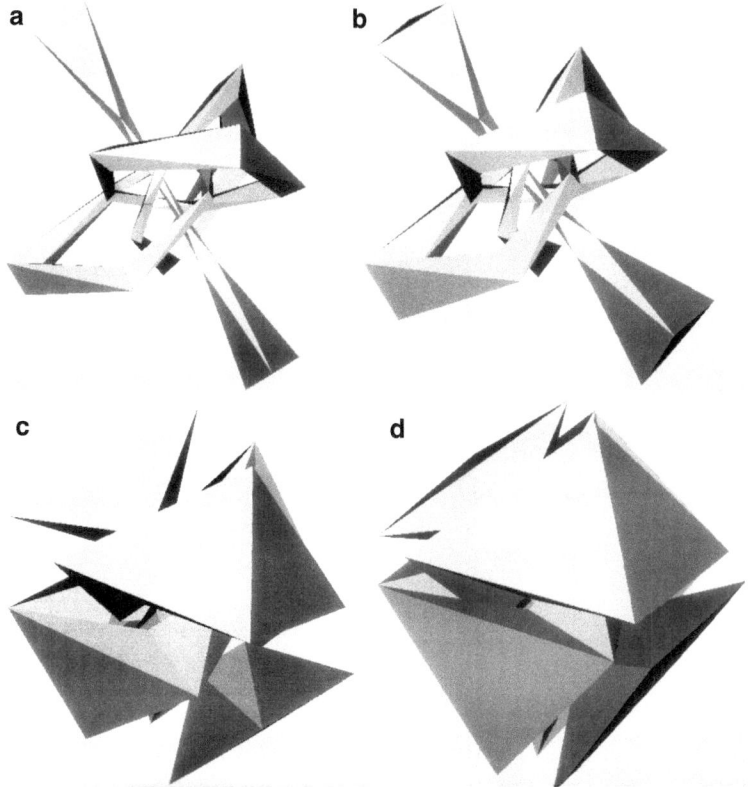

Figure 20.6. Torus decompositions of the 24-cell.

can be expressed as a square subdivided into 16 squares, with its left and right edges identified and its top and bottom edges identified (Figure 20.7).

The corresponding fold-out description of the 24-cell starts with a solid stack formed by six octahedra. The top and bottom triangles of the stack can be identified by folding up into 4-space to obtain a solid torus. The boundary of this solid torus can be expressed as a union of 36 triangles arranged in a polygonal region in the plane, to be folded up in 4-space so that its left and right edges are identified and its top and bottom edges are identified (Figure 20.8a, b).

The remaining 18 octahedra in the 24-cell can be arranged into three other stacks, each with six octahedra. We can place the four stacks together in 3-space to indicate the way the four solid tori will be linked when the stacks are folded together in 4-space (Figure 20.9).

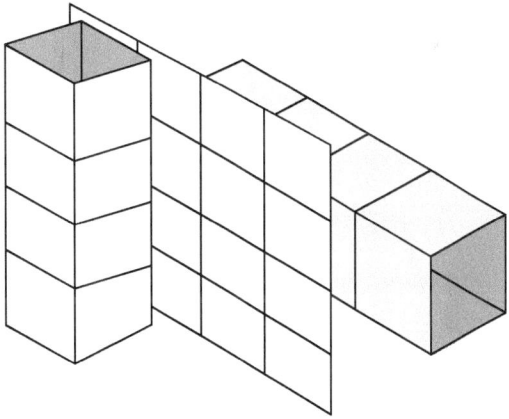

Figure 20.7. Fold-out decompostion of the hypercube.

Cartesian and Torus Coordinates

To describe polyhedra in 3-space in coordinates, it is most convenient to parametrize the unit

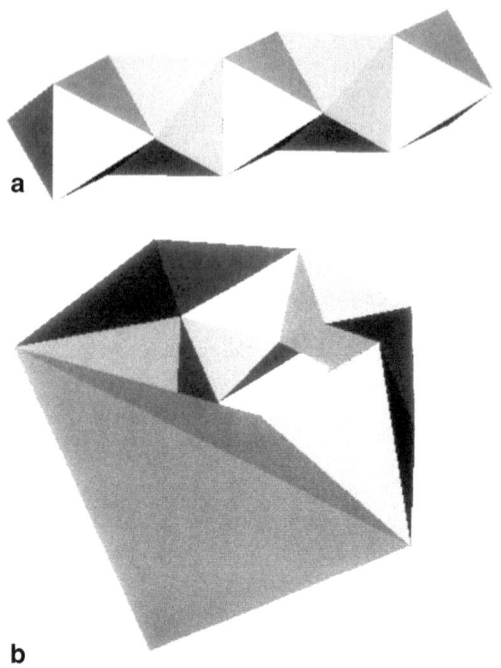

a

b

Figure 20.8. A stack of six octahedra, (**a**) unfolded, (**b**) folded.

Figure 20.9. Four stacks of octahedra.

sphere by longitude and co-latitude (measured down from the North Pole instead up from the Equator). A point on the unit sphere then has Cartesian coordinates

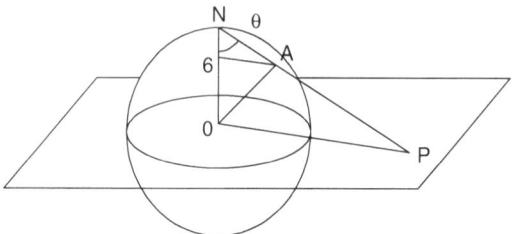

Figure 20.10. Stereographic projection of the 2-sphere.

$$(\cos(\theta)\sin(\varphi), \sin(\theta)\sin(\varphi), \cos(\varphi)).$$

Stereographic projection from the North Pole to the horizontal plane which passes through the origin sends a point to the intersection of the line through the North Pole and the point with the horizontal plane. The point whose coordinates are given above is then sent to

$$(\cos(\theta)\sin(\varphi)/(1 - \cos(\varphi)), \sin(\theta)\sin(\varphi)/$$
$$(1 - \cos(\varphi)), 0) = (\cos(\theta)\cot(\varphi/2),$$
$$\sin(\theta)\cot(\varphi/2), 0)$$

(Figure 20.10). Circles of latitude are sent to circles centered at the origin and semicircles of longitude are sent to straight lines passing through the origin. A rotation of reflection of the sphere about the axis in 3-space corresponds to a rotation or reflection in the plane. A regular polyhedron with a vertical axis of symmetry will have its vertices on certain parallels of latitude and the symmetries of the polyhedron preserving the axis will lead to symmetries of its Schlegel diagram in the plane.

In 4-space the sphere of points at unit distance from the origin can be parametrized in different ways. In this exposition, we concentrate on a coordinate system on the 3-sphere which has especially interesting geometric properties. In this *torus coordinate system* the coordinates of a point are

$$(x, y, u, v) = \{(\cos(\theta)\sin(\varphi), \sin(\theta)\sin(\varphi),$$
$$\cos(\psi)\cos(\varphi), \sin(\psi)\cos(\varphi)\}$$

where θ and ψ run from 0 to 2π and where $0 \leq \phi \leq \frac{\pi}{2}$. The points with $\varphi = 0$ give the unit circle in the (u, v) plane and if $\varphi = \frac{\pi}{2}$, the locus is the unit circle in the (x, y) plane.

Points on the 3-sphere corresponding to other values of φ give tori on the 3-sphere. For example, if $\varphi = \frac{\pi}{4}$ we get a symmetric torus,

$$(1/\sqrt{2})(\cos(\theta), \sin(\theta), \cos(\psi), \sin(\psi)).$$

This torus is the Cartesian product of two circles, one lying in the xy-plane and the other in the uv-plane.

As φ moves from 0 to $\frac{\pi}{2}$, these tori sweep out the region between the two linked circles. In particular, the entire 3-sphere is then displayed as a union of two tori, one corresponding to negative values of φ and the other corresponding to positive values.

If we position the vertices of a regular polytope symmetrically with respect to this coordinate system, we obtain a torus decomposition of the regular polytope. The torus coordinate system on the 3-sphere is particularly well-suited to the study of the Hopf mapping, the final topic of this chapter.

Coordinates for Polyhedra and Polytopes

In three-dimensional space we may describe the cube by the eight vertices $(\pm t, \pm t, \pm t)$ for a positive constant t. These points lie on the sphere of radius $\sqrt{3}t$. The centers of cubical faces will then be the points $(\pm t, 0, 0)$, $(0, \pm t, 0)$ and $(0, 0, \pm t)$, the vertices of a regular octahedron inscribed in a sphere of radius t. The midpoints of edges of the cube will have coordinates $(\pm t, \pm t, 0)$, $(0, \pm t, \pm t)$ and $(\pm t, 0, \pm t)$, forming the vertices of a cuboctahedron inscribed in a sphere of radius $\sqrt{2}t$.

For the hypercube we may choose 16 vertices $(\pm t, \pm t, \pm t, \pm t)$ situated on the hypersurface of a 3-sphere of radius $2t$. In torus coordinates, these 16 points

$$2t(\cos(\theta)\sin(\varphi), \sin(\theta)\sin(\varphi),$$

$$\cos(\psi)\cos(\varphi), \sin(\psi)\cos(\varphi))$$

all lie on the torus with $\varphi = \pi/4$. The coordinates are given by letting θ and ψ take on the values $\pi/4 + k\pi/2$ for $k = 0, 1, 2, 3$. Just as the 3-sphere is expressed as a union of two solid tori with a common boundary torus, the boundary of the hypercube is expressed as a union of two solid tori with a common boundary. The boundary polyhedral torus can be expressed as the Cartesian product of two square polygons. It includes all 16 vertices and all 32 edges of the hypercube as well as 16 of its squares.

The centers of the eight three-dimensional cubical faces of the hypercube have coordinates $(\pm t, 0, 0, 0)$, $(0, \pm t, 0, 0)$, $(0, 0, \pm t, 0)$ and $(0, 0, 0, \pm t)$, lying on a hypersphere of radius t. This gives coordinates for the regular 16-cell in 4-space. The midpoints of edges of the hypercube will be the 32 points $(0, \pm t, \pm t, \pm t)$, $(\pm t, 0, \pm t, \pm t)$, $(\pm t, \pm t, 0, \pm t)$ and $(\pm t, \pm t, \pm t, 0)$, lying on a hypersphere of radius $\sqrt{3}t$, giving the coordinates of the vertices of a semiregular polytope.

The vertices of the regular 24-cell in 4-space can be given by the midpoints of square faces of the hypercube, with coordinates $(\pm t, \pm t, 0, 0)$, $(0, \pm t, 0, \pm t)$, $(\pm t, 0, \pm t, 0)$, $(\pm t, 0, 0, \pm t)$, $(0, 0, \pm t, \pm t)$ and $(0, \pm t, \pm t, 0)$, lying on a hypersphere of radius $\sqrt{2}t$. Another set of coordinates for the 24-cell is given by taking the vertices of a hypercube $(\pm s, \pm s, \pm s, \pm s)$ together with the vertices of a dual 16-cell $(\pm 2s, 0, 0, 0)$, $(0, \pm 2s, 0, 0)$, $(0, 0, \pm 2s, 0)$ and $(0, 0, 0, \pm 2s)$. These 24 coordinates lie on a hypersphere of radius $2s$.

The Hopf Mapping

Torus coordinates are especially well suited for describing the Hopf mapping, a mapping from the 3-sphere to the 2-sphere for which every point of the 3-sphere lies on a circle that is the preimage of a point on the 2-sphere. One of the easiest ways to describe the Hopf mapping is to think of the 3-sphere as a collection of pairs of complex numbers with the squares of their lengths adding up to 1. We then have

$$S^3 = \{(x+iy, u+iv), x^2 + y^2 + u^2 + v^2 = 1\}$$

$$\equiv \{[z, w], z^2 + w^2 = 1\}.$$

Hereafter we shall adopt the convention of using square brackets to indicate the description of a point of S^3 by pairs of complex numbers and parentheses to indicate the usual description in terms of quadruples of real numbers.

To describe the Hopf mapping we send a pair of complex numbers to their quotient; that is, $h[z, w] = w/z$ if $z \neq 0$ and $h[0, w] = \infty$, the infinite point in the extended complex plane. If we write a point in S^3 in polar coordinates, then we have

$$[z, w] = [sin(\varphi)e^{i\theta}, \cos(\varphi)e^{i\psi}]$$

and

$$h[z, w] = \cot(\varphi)e^{i(\psi-\theta)}$$

if $\varphi \neq \frac{\pi}{2}$ and $h[0, w] = \infty$ as before.

To complete the description of the Hopf mapping, we use inverse stereographic projection to map the extended complex plane to the 2-sphere. The effect of this is to map the point $[z, w] = (x, y, u, v)$ to

$$h[z, w] = [2\bar{z}w, w\bar{w} - z\bar{z}] = (2xu + 2yv,$$

$$2xv - 2yu, -x^2 - y^2 + u^2 + v^2, 0).$$

Here $\bar{w} = u - iv$ indicates the complex conjugate. The coordinate $w\bar{w} - z\bar{z}$ is a real number and the image of each point of S^3 lies in three-dimensional space.

In torus coordinates, the Hopf mapping is given by

$$h[\sin(\varphi)e^{i\theta}, \cos(\varphi)e^{i\psi}]$$

$$= [\sin(2\varphi)e^{i(\psi-\theta)}, \cos(2\varphi)].$$

It is clear from this form that the image of each point lies on the 2-sphere of radius 1 in 3-space, so we write

$$h : S^3 \rightarrow S^2.$$

Under this mapping the unit circle in the xy-plane corresponding to $\psi = \frac{\pi}{2}$ is sent to the point $(0, 0, -1)$ and the unit circle in the uv-plane corresponding to $\varphi = 0$ is sent to point $(0, 0, 1)$. The middle torus $(1/\sqrt{2})[e^{i\theta}, e^{i\varphi}]$ corresponding to $\varphi = \frac{\pi}{4}$ is sent to the Equator $[e^{i(\psi-\theta)}, 0]$. The preimage of the point $(1, 0, 0)$ on the Equator is the set of points with $\varphi = \frac{\pi}{4}$ and $\theta = \psi$. This curve lies on the sphere and in the plane given by $x = u$ and $y = v$, so it must be a circle. Similarly, any point $[e^{i\beta}, 0]$ on the Equator will be a circle determined by the conditions $\alpha = \frac{\pi}{4}$ and $\varphi = \theta + \beta$.

More generally, for any point $[\sin(\gamma)e^{i\beta}, \cos(\gamma)]$ with $\sin(\gamma)$ positive on the 2-sphere, the preimage under the Hopf mapping will be a circle determined by the conditions $\varphi = \frac{\gamma}{2}$ and $\psi = \theta + \beta$. All of these circles will be great circles on the 3-sphere and no two of them will have a point in common. Since the discs bounded by these circles will meet only at the origin in 4-space, the circles will be linked.

The Hopf Decomposition of the Hypercube

We now consider the hypercube from the point of view of the Hopf mapping. In complex coordinates, the 16 vertices of the hypercube on the unit hypersphere may be given by $\frac{1}{2}[\pm 1 \pm i, \pm 1 \pm i]$. Under the Hopf mapping the images of these vertices will be four points on the Equator of S^2, namely $[\pm 1, 0]$ and $[\pm i, 0]$ (Figure 20.11).

We indicate on the unfolded torus diagram two of the four quadrilaterals containing the vertices of the hypercube. These four quadrilaterals are squares in 4-space which we may call "Hopf polygons." The edges of these Hopf polygons are diagonals in the square faces of the flat torus which is the preimage of the Equator under the

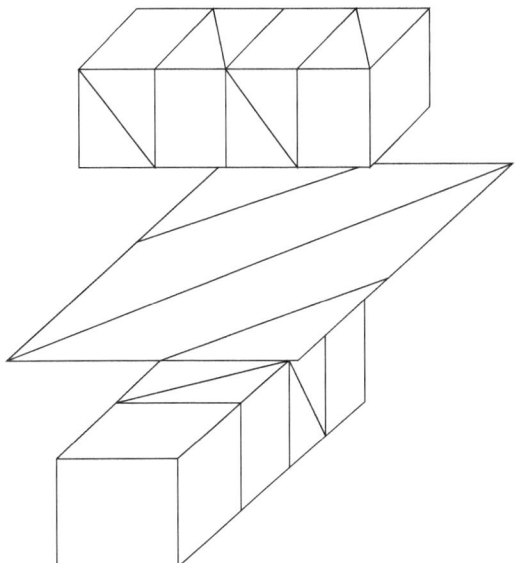

Figure 20.11. Hopf polygons on the hypercube.

Hopf mapping. One of the solid tori is the preimage of the upper hemisphere and the other is the preimage of the lower hemisphere.

Torus Decomposition of the 24-Cell

We may now attempt a similar decomposition of the 24-cell. In the fold-out version we may identify three hexagonal helices on each stack of six octahedra corresponding to the four quadrilateral helices on the stacks of four cubes in the hypercube. Unfortunately these hexagons are not preimages of points of S^2 under the Hopf mapping in either of the two coordinate systems we have described for the 24-cell. The coordinates which we have given for the hypercube are sent by the Hopf mapping to four points on the Equator and the eight vertices of the 16-cell are sent to the North and South Poles of S^2.

Thus the 24 coordinates of one coordinate system on the 24-cell are sent to six points of S^2 situated at the vertices of a regular octahedron. Similarly if we use the coordinates of the 24-cell obtained by taking midpoints of the square faces, again the images under the Hopf mapping are the same six vertices of an octahedron. In order

to obtain the decomposition of the 24-cell into four solid tori each with six octahedra, we need to reposition the 24-cell so that its vertices are sent to the four vertices of tetrahedron under the Hopf mapping.

To determine a rotation that will align the 24-cell so that it is situated well with respect to the Hopf mapping, we carry out a closer examination of the previously given coordinates for the 24-cell where we now consider the case $t = 1$ so the polytope lies on a sphere of radius equal to 2. If we project stereographically from the point $(0, 0, 0, \sqrt{2})$, then the image of a point (x, y, u, v) is $\{\sqrt{2}/(\sqrt{2} - v)\}(x, y, u)$. If $v = 0$, then the image of $(x, y, u, 0)$ is (x, y, u). Thus the images of the 12 vertices of the 24-cell of the forms $(\pm 1, \pm 1, 0, 0), (\pm 1, 0, \pm 1, 0)$ and $(0, \pm 1, \pm 1, 0)$ will be the vertices of a cuboctahedron in 3-space. The six vertices of the form $(\pm 1, 0, 0, 1)$, $(0, \pm 1, 0, 1)$, and $(0, 0, \pm 1, 1)$ will be sent to the vertices of large octahedron containing the cuboctahedron, and the six vertices with the fourth coordinate minus 1 will be sent to a small octahedron contained within the cuboctahedron.

The cuboctahedron has eight triangular faces, each only lying in one distorted octahedron with its opposite triangle on the small octahedron and another with an opposite triangle on the large octahedron. This accounts for 18 of the octahedra in the 24-cell. The remaining six each have four of the vertices on the square faces of the cuboctahedron, and one vertex on the large and one on the small octahedron.

We may then identify one of the four tori in the toroidal decomposition by taking the small octahedron and two adjacent octahedra with their opposite triangles on the semiregular polyhedron. The octahedron opposite these three complete a cycle of six octahedra on the 24-cell. We could for example take the large octahedron, with center $(0, 0, 0, 1)$, the small one with center $(0, 0, 0, -1)$, two others adjacent to the small one with centers $1/2(-1, -1, -1, -1)$ and $1/2(1, 1, 1, -1)$, and their opposite octahedra with center $1/2(1, 1, 1, 1)$ and $1/2(-1, -1, -1, 1)$. These six vertices may be arranged in a hexagon so that

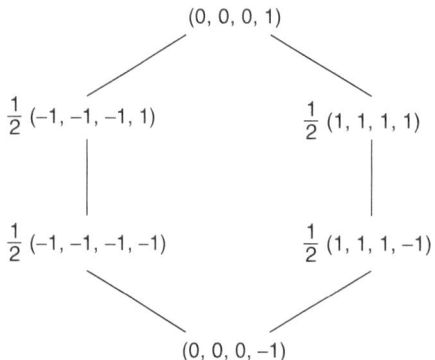

Figure 20.12. A hexagon of centers of octahedra in a Hopf cycle.

the angle between any two adjacent vertices is 120°, as shown in Figure 20.12.

We may label the vertices of the projected 24-cell so that this cycle of six octahedra appears as five octahedra in a vertical stack, together with the large octahedron. The remaining solid tori are obtained by taking a pair of octahedra meeting the central polyhedron in opposite square faces and then connecting them by two other pairs each with one octahedron inside and one outside the central polyhedron. Once we have one of these we obtain the other two by rotating about the vertical line by angles of 120° and 240°.

Fortunately it is possible to find a rotation of 4-space which realigns the vertices so that their images are situated at the vertices of a regular tetrahedron in 3-space! To find one such rotation, we look for a linear transformation T which sends the hexagon to a regular hexagon in the uv-plane in 4-space. We can let T fix the vector $(0,0,0,1)$ and let T send $1/2(1,1,1,1)$ to $(0,0,3/2,1/2)$ and $1/2(-1,-1,-1,1)$ to $(0,0,-\sqrt{3/2},1/2)$. It follows that $T(1/\sqrt{3},1/\sqrt{3},1\sqrt{3},0) = (0,0,1,0)$. The vectors sent to $(1,0,0,0)$ and $(0,1,0,0)$ by T must be a pair of mutually orthogonal unit vectors which are perpendicular to $(0,0,0,1)$ and $(1/\sqrt{3},1/\sqrt{3},1/\sqrt{3},0)$ and we may choose

these preimages to be $(1/\sqrt{2},-1/\sqrt{2},0,0)$ and $(1/\sqrt{6},1/\sqrt{6},2/\sqrt{6},0)$. This completely determines the matrix of T, and we may then check that after this rotation, the vertices of the 24-cell do indeed lie in four (planar) regular hexagons in 4-space which are mapped to the vertices of a regular tetrahedron inscribed in the unit 2-sphere in 3-space under the Hopf mapping.

The preimages of the four triangular faces of this spherical tetrahedra correspond to the four cycles of six octahedra described in the previous paragraph. To see how these four rings of octahedra fit together to fill out the 24-cell, we may shrink each ring toward the hexagon, with vertices at the centers of the six triangles between adjacent octahedra. We may interpolate linearly between the 24-cell and this union of four hexagons, constantly projecting the vertices centrally to the hypersurface of the 3-sphere. Illustrations of several stages of its deformation are shown in Figures 20.8–20.11.

Note that the 24 centers of four hexagons may be obtained from the centers of the 24 octahedra by a rotation in 4-space which moves each Hopf circle along itself by 60°. The comparable treatment of the hypercube shrinks the two rings of four cubes to the quadrilaterals determined by the centers of the eight squares where adjacent cubes meet in the two rings, as shown in Figures 20.5–20.7. These eight points may also be obtained from the centers of the cubes in the two rings by a rotation in the 4-space moving each Hopf circle along itself by 45°.

To conclude: we have seen that the familiar central projection of the hypercube suggests a decomposition of the hypersphere into solid tori, and this decomposition carries over to other polytopes as well, in particular the 24-cell. This investigation gives additional geometric insight into the properties of these polytopes and at the same time it elucidates some of the geometry of the Hopf mapping.

21

Tensegrities and Global Rigidity

Robert Connelly

In 1947 a young artist named Kenneth Snelson invented an intriguing structure: a few sticks suspended rigidly in mid air without touching each other. It seemed like a magic trick. He showed this to the entrepreneur, builder, visionary, and self-styled mathematician, R. Buckminster Fuller, who called it a *tensegrity* because of its "tensional integrity." Fuller talked about tensegrities and wrote about them extensively. Snelson went on to build a great variety of fascinating tensegrity sculptures all over the world, including the 60-foot work of art at the Hirschhorn Museum in Washington, DC. shown in Figure 21.1.

Why did these tensegrities hold up? What were the geometric principles? They were often under-braced, and they seemed to need a lot of tension for their stability. So Fuller's name, tensegrity, is quite appropriate.

In this chapter I will show how to describe the stability of most of the tensegrities that Snelson and others have built, and how to predict their stability. I begin with a set of principles that can be used to understand many of the Snelson-like tensegrities. This relies on the properties of the stress matrix, a symmetric matrix that defines a kind of potential energy that we would like to minimize at a given configuration. This leads to a lot of interesting examples one can build, assured that the structure will not fall down.

Then I will show how the rank of a stress matrix can predict how generic configurations of bar tensegrities (usually called bar frameworks) predict a strong sort of rigidity that I call global rigidity, and mention some exciting new results where the stress matrix plays a central role. There are several quite interesting applications of the theory of tensegrities. Of course, there is a natural application to structural engineering, where the pin-jointed bar-and-joint model is appropriate for an endless collection of structures. In computational geometry, there was the carpenter's rule conjecture, inspired by a problem in robot arm manipulation. This proposes that a non-intersecting polygonal chain in the plane can be straightened, keeping the edge lengths fixed, without creating any self-intersections. The key idea in that problem uses basic tools in the theory of tensegrity structures and stresses. Granular materials of hard spherical disks can be reasonably modeled as tensegrities, where all the members are struts. Again the theory of tensegrities can be applied to predict behavior and provide the mathematical basis for computer simulations as well as to predict the distribution of internal stresses.

Terminology and Notation

To define a tensegrity we first define a tensegrity graph G with vertices $1, 2, \ldots, n$ in d-dimensional space E^d. Edges are denoted as unordered pairs of different vertices $\{i, j\}$, where

R. Connelly
Cornell University, 433 Malott Hall,
Ithaca, NY 14853, USA
e-mail: connelly@math.cornell.edu; http://www.math.cornell.edu/~connelly/

M. Senechal (ed.), *Shaping Space*, DOI 10.1007/978-0-387-92714-5_21,

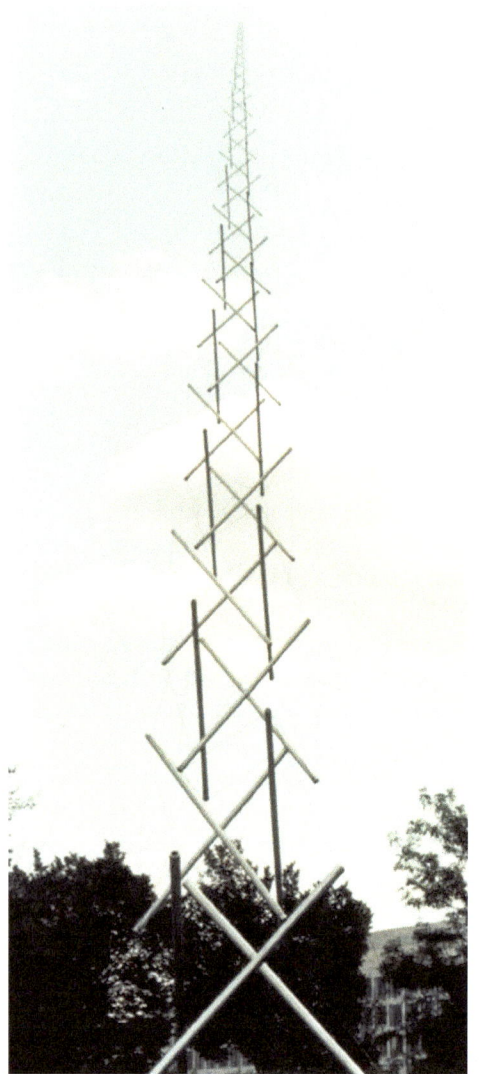

Figure 21.1. Kenneth Snelson, Needle Tower, 1968, aluminum and stainless steel, $60 \times 20 \times 20$ ft. — $18.2 \times 6 \times 6$ m.

$i \neq j$. Each edge of G is declared a cable, strut, bar or not connected. The cables, struts, and bars are *members* of the tensegrity, and are denoted by dashed line segments, bold line segments, or thin line segments respectively.

The next ingredient is a configuration $p = (p_1, \ldots, p_n)$ of n points or nodes in the Euclidean plane or space, where each p_i corresponds to a vertex i of G. Nodes connected by cables are allowed to get closer

together (or stay the same distance apart), the nodes connected by struts are allowed to get farther apart (or stay the same distance apart), but bars must always stay the same distance apart.

The graph together with the corresponding configuration is a tensegrity $G(p)$. This notation is useful since we sometimes want to consider the abstract graph G without referring to a particular configuration, or the configuration p without referring to any particular graph. The notation is meant to suggest, for example, that there are physical cables connecting pairs of nodes that have cables connecting the corresponding vertices of G.

A continuous motion of the nodes, starting at the given configuration of a tensegrity, is called a *flex* of the tensegrity. A *trivial flex* is one that moves the configuration as a whole without changing any distances between nodes, for example rotations and translations, in space. If the tensegrity has only trivial flexes, then it is said to be *rigid* in E^d. Otherwise it is *flexible*. Note that, theoretically, members can cross one another (i.e., intersect). For the purposes of our mathematical model, the tensegrity is a purely geometric object, but many of the rigid tensegrities shown here can be built with rubber (or plastic) bands for cables, and dowel rods with a slot at their ends for struts or bars. Figure 21.2 shows some examples of rigid and flexible tensegrities in the plane and space.

The rigid tensegrity in space in Figure 21.2 is one of Snelson's original objects. It is quite simple but suspends three sticks, the struts, rigidly without any pair of them touching. Indeed, Snelson does not like to call an object made of cables and struts a tensegrity unless all the struts are completely disjoint, even at their nodes. A tensegrity with all its struts disjoint and with no bars (i.e., all the other members are cables) will be called a *pure* tensegrity.

Now let's discuss techniques for computing the rigidity of tensegrities. As a by-product, the definition and analysis of global rigidity will emerge naturally. The *stress* associated to

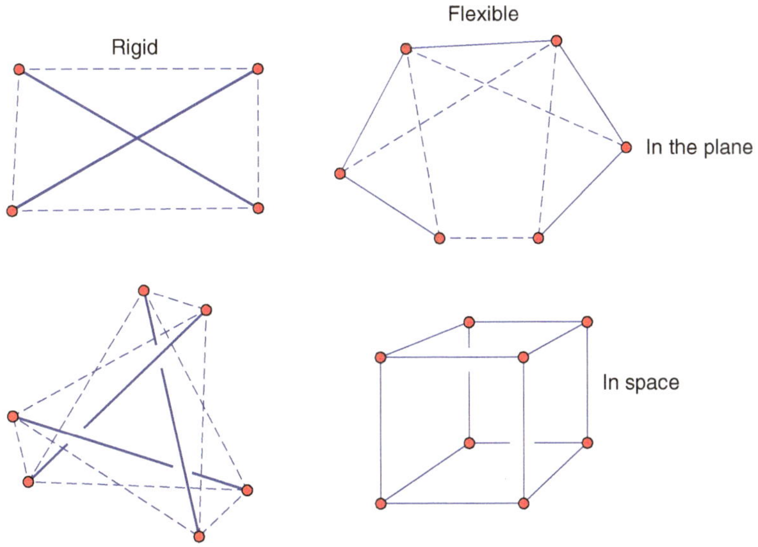

Figure 21.2. Nodes are denoted by *small round* points, cables by *dashed line* segments, struts by *solid line* segments, and bars by *thin line* segments.

a tensegrity is the vector $\omega = (\ldots, \omega_{ij}, \ldots)$, where $\omega_{ij} = \omega_{ji}$ is a scalar associated to the member $\{i, j\}$ of G. A stress $\omega = (\ldots, \omega_{ij}, \ldots)$ is *proper* if $\omega_{ij} \geq 0$ for a cable $\{i, j\}$ and $\omega_{ij} \leq 0$ for a strut $\{i, j\}$. There is no condition when $\{i, j\}$ is a bar. We set $\omega_{ij} = 0$ if the nodes $\{i, j\}$ are not connected. We say a proper stress ω is *strict* if $\omega_{ij} \neq 0$ when $\{i, j\}$ is a cable or strut.

Let $\omega = (\ldots, \omega_{ij}, \ldots)$ be a proper stress for a tensegrity graph G. For any configuration p of nodes in E^d, define the *stress-energy* associated to ω as

$$E_\omega(p) = \sum_{i<j} \omega_{ij}(p_i - p_j)^2, \qquad (21.1)$$

where the product of vectors is the ordinary dot product, and the square of a vector is the square of its Euclidean length.

Now think of the configuration p as fixed. We want to compare other configurations q to p. Let us say the tensegrity $G(p)$ *dominates* the tensegrity $G(q)$, and write $G(q) \leq G(p)$ for two configurations q and p if

$$|p_i - p_j| \geq |q_i - q_j| \qquad \text{for } \{i,j\} \text{ a cable,}$$
$$|p_i - p_j| \leq |q_i - q_j| \qquad \text{for } \{i,j\} \text{ a strut and}$$
$$|p_i - p_j| = |q_i - q_j| \qquad \text{for } \{i,j\} \text{ a bar.}$$

$$(21.2)$$

These are the *tensegrity constraints* for the configuration p with respect to the tensegrity graph G. So if $G(p)$ dominates $G(q)$ and ω is a proper stress for G, then $E_\omega(p) \geq E_\omega(q)$, because the terms of (21.2) that correspond to cables have a positive stress and can only decrease since the cable lengths can only decrease, while the terms corresponding to struts have a negative coefficient and the strut lengths can only increase.

Local and Global Rigidity

A tensegrity $G(p)$ is locally rigid if the only continuous flexes of $G(p)$ that satisfy the tensegrity constraints of Equation (21.2) are congruences. There is a good body of work devoted to the detection and understanding of local rigidity.

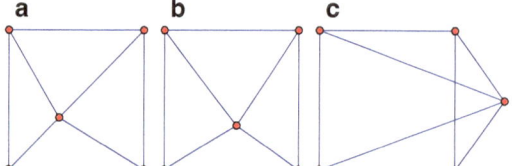

Figure 21.3. Three examples of planar rigid bar frameworks; see the text for details.

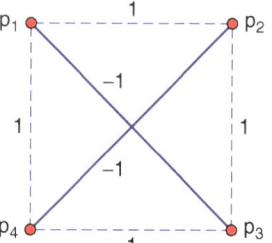

Figure 21.4. A square tensegrity with its diagonals, where a proper equilibrium stress is indicated.

However, most of the structures made by Snelson and other artists are *globally* rigid. This means that for any other configuration q of the same labeled nodes in E^d, $G(q) \leq G(p)$ implies that q is congruent to p. Even more strongly, regard $E^d \subset E^D$, for $d \leq D$. If, even though $G(p)$ is in E^d, it is true that $G(p)$ is globally rigid in E^D, for all $D \geq d$, then we say $G(p)$ is *universally globally rigid*. For example, both rigid tensegrities in Figure 21.2, are universally globally rigid. The example in Figure 21.3a is rigid in the plane, but not globally rigid in the plane, since it can reflect the upper left node around a diagonal. Figure 21.3b is globally rigid in the plane but not universally globally rigid, since it is flexible in three-space. Figure 21.3c is universally globally rigid. These are all bar frameworks.

What can we say about more complicated tensegrities? The energy function E_ω described above helps. The idea is to look for situations in which the configuration p is a minimum for the functional E_ω. The first step is to determine when p is a critical point for E_ω, i.e., when all directional derivatives given by $p' = (p'_1, \ldots, p'_n)$ starting at p are 0. This means that the following equilibrium vector equation must hold for each node i:

$$\sum_j \omega_{ij}(p_j - p_i) = 0. \qquad (21.3)$$

In this case, ω is an equilibrium stress for p, called just a stress when the equilibrium is clear from the context. To get an understanding of how this works, consider the example of a square in the plane as in Figure 21.4. It is easy to see that the vector equilibrium equation (21.3) holds for the three vectors at each node.

If a configuration p were the unique minimum, up to rigid congruences, for E_ω, we would have a global rigidity result immediately, but unfortunately this is almost never the case. We must deal with affine transformations.

Affine Transformations

An *affine transformation* or *affine map* of E^d is determined by a $d \times d$ matrix A and a vector $b \in E^d$. If $p = (p_1, \ldots, p_n)$ is any configuration in E^d, an affine image is given by $q = (q_1, \ldots, q_n)$, where $q_i = A p_i + b$. If the configuration p is in equilibrium with respect to the stress ω, then so is any affine transformation q of p.

This brings us to the question: for a tensegrity $G(p)$ in E^d, when is there an affine transformation that preserves the member constraints (21.2)? It is clear that the matrix A is the only relevant part, and it turns out that we also only need to consider the members that are bars. A preserves the length of bar $\{i, j\}$ if and only if the following holds:

$$(p_i - p_j)^2 = (q_i - q_j)^2$$
$$= (A p_i - A p_j)^2$$
$$= [A(p_i - p_j)]^T A(p_i - p_j)$$
$$= (p_i - p_j)^T A^T A(p_i - p_j),$$

or equivalently,

$$(p_i - p_j)^T (A^T A - I^d)(p_i - p_j) = 0 \quad (21.4)$$

where $()^T$ is the transpose operation, I^d is the $d \times d$ identity matrix, and vectors are regarded as column vectors in this calculation. If Equation (21.4) holds for all bars in G, we say that it has a *bar preserving affine image*, which is non-trivial if A is not orthogonal. Similarly, G has a *non-trivial affine flex* if there is a continuous family of d-by-d matrices A_t, where $A_0 = I^d$, for t in some interval containing 0 such that each A_t satisfies Equation (21.4) for t in the interval.

This suggests the following definition. If $v = \{v_1, \ldots, v_k\}$ is a collection of vectors in E^d, we say that they lie on a *quadric at infinity* if there is a non-zero symmetric d-by-d matrix Q such that for all $v_i \in v$

$$v_i^T Q v_i = 0. \tag{21.5}$$

Notice that since the definition of an orthogonal matrix A is that $A^T A - I^d = \mathbf{0}$, the affine transformation defines a quadric at infinity if and only if the affine transformation is not a congruence.

Call the *bar directions of a bar tensegrity* the set $\{p_i - p_j\}$, for $\{i, j\}$ a bar of G. With this terminology, Equation (21.5) says that if A preserves bar length, then the member directions of a bar tensegrity lie on a quadric at infinity. We can prove:

Proposition 21.1. *If $G(p)$ is a bar framework in E^d, such that the nodes do not lie in a $(d-1)$-dimensional hyperplane, then it has a non-trivial bar preserving affine image if and only if it has a non-trivial bar-preserving affine flex if and only if the bar directions lie on a quadric at infinity.*

Proof. The "only if" direction is shown above. Conversely suppose that the member directions of a bar tensegrity $G(p)$ lie on a quadric at infinity in e^d given by a non-zero symmetric matrix Q. By the spectral theorem for symmetric matrices, we know that there is an orthogonal d-by-d matrix $X = (X^T)^{-1}$ such that:

$$X^T Q X = \begin{pmatrix} \lambda_1 & 0 & 0 & \cdots & 0 \\ 0 & \lambda_2 & 0 & \cdots & 0 \\ 0 & 0 & \lambda_3 & \cdots & 0 \\ \vdots & \vdots & \vdots & \ddots & \vdots \\ 0 & 0 & 0 & \cdots & \lambda_d \end{pmatrix}.$$

Let λ_- be the smallest λ_i, and let λ_+ be the largest λ_i. Note $\infty \leq 1/\lambda_- < 1/\lambda_+ \leq \infty$, λ_- is non-positive, and λ_+ is non-negative when Q defines a non-empty quadric and when $1/\lambda_- \leq t \leq 1/\lambda_+$, $1 - t\lambda_i \geq 0$ for all $i = 1, \ldots, d$. Working Equation (21.4) backwards for $1/\lambda_- \leq t \leq 1/\lambda_+$ we define:

$$A_t = X^T \begin{pmatrix} \sqrt{1-t\lambda_1} & 0 & 0 & \cdots & 0 \\ 0 & \sqrt{1-t\lambda_2} & 0 & \cdots & 0 \\ 0 & 0 & \sqrt{1-t\lambda_3} & \cdots & 0 \\ \vdots & \vdots & \vdots & \ddots & \vdots \\ 0 & 0 & 0 & \cdots & \sqrt{1-t\lambda_d} \end{pmatrix} X. \tag{21.6}$$

Substituting A_t from Equation (21.6) into Equation (21.4), we see that it provides a non-trivial affine flex of $G(p)$. If the configuration is contained in a lower dimensional hyperplane, we should really restrict to that hyperplane since there are non-orthogonal affine transformations that are rigid when restricted to the configuration itself.

When do bar tensegrities have bar directions that lie on a quadric at infinity? In E^2, the quadric at infinity consists of two distinct directions. So a parallelogram or a grid of parallelograms has a non-trivial affine flex. In E^3 it is more interesting. The quadric at infinity is a conic in the projective plane, and such a conic is determined by 5 points. An interesting example is the bar tensegrity in Figure 21.5. The surface is obtained by taking the line $(x, 1, x)$ and rotating it about the z-axis. This creates a ruling of the surface by disjoint lines. Similarly $(x, 1, -x)$ creates another ruling. Each line in one ruling intersects each line in the other ruling or they are parallel. A bar tensegrity is obtained by placing nodes where a line on one ruling intersects a line on the other ruling, and bars such that they join every pair of nodes that lie on same line on either ruling.

Alexander Barvinok asked when a framework in a space of D dimensions can also be realized in a subspace of d dimensions, $d < D$. He proved,

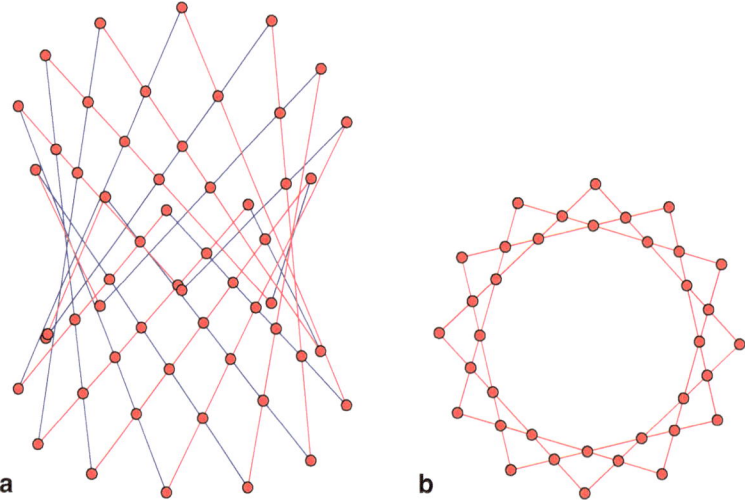

Figure 21.5. Figure (**a**) is the ruled hyperboloid given by $x^2 + y^2 - z^2 = 1$. Figure (**b**) is the flattened version after an affine flex.

Theorem 21.2. *If $G(p)$ is a bar framework in E^D with less than $d(d + 1)/2$ bars, then it has a realization in E^d with the same bar lengths.*

Proof. The space of d-by-d symmetric matrices is of dimension $d + (d^2 - d)/2 = d(d + 1)/2$. So if the vector directions of the members of a tensegrity are less than $d(d + 1)/2$, then it is possible to find a non-zero d-by-d symmetric matrix that satisfies Equation (21.5), and then flex it into a lower dimensional subspace by using Equation (22.4) until one of the diagonal entries becomes 0.

If every realization $G(p)$ of a bar graph G, where the p's are configurations in E^D, can be realized in E^d with the same bar lengths, then we say that G is *d-realizable*. Note that this is a property of the graph G: in order to qualify for being d-realizable, one has to be able to push a realization in E^D down to a realization E^d for ALL realizations in E^D. For example, the 1-realizable graphs are forests, graphs with no cycles. In particular, a triangle is not 1-realizable.

This is inspired by a problem in nuclear magnetic resonance (NMR) spectroscopy. The atoms of a protein are tagged and some of the pairwise distances are known. The problem is to identify a configuration in E^3 that satisfies those distance constraints. Finding such a configuration in E^D, for some large D, is computationally feasible, and if G is 3-realizable, one can expect to find another configuration in E^3 that satisfies the distance constraints.

A graph H is a *minor* of a graph G if it can be obtained from G by edge contractions or deletions. If a minor of a graph G is not d-realizable, then G itself is not d-realizable. It is easy to show that a graph is 1-realizable if and only if it does not have a triangle as a minor. In other words, the triangle is the one and only *forbidden minor* for 1-realizability. It is not too hard to show that the graph K_4, the tetrahedron, is the only forbidden minor for 2-realizability. Maria Belk and I showed:

Theorem 21.3. *A complete list of forbidden minors for 3-realizability is the set of two graphs, K_5 and edge graph of the regular octahedron.*

There is a reasonable algorithm to detect 3-realizablity for an abstract graph and, when the edge lengths are given, to find a realization in E^3. Tensegrity techniques are used in a significant way.

The Stress Matrix and the Fundamental Theorem

The stress-energy function E_ω defined by (21.1) is a quadratic form, and it is an easy matter to compute the matrix associated to it

$$\Omega = \sum_{i<j} \omega_{ij} \Omega(i,j)$$

where the (i,j) entry is $-\omega_{ij}$ for $i \neq j$, and the diagonal entries are such that the row and column sums are 0. (Recall that any stress ω_{ij} not designated in the vector form $\omega = (\cdots, \omega_{ij}, \cdots)$ is assumed to be 0.) With this terminology we can regard a configuration $p = (p_1, \ldots, p_n)$ in E^d as a column vector. Then

$$E_\omega(p) = \sum_{i<j} \omega_{ij}(p_i - p_j)^2$$

$$= \sum_{i<j} \omega_{ij}(x_i - x_j)^2$$

$$+ \sum_{i<j} \omega_{ij}(y_i - y_j)^2 + \ldots$$

$$= (x_1 \ x_2 \ \cdots \ x_n) \, \Omega \begin{pmatrix} x_1 \\ x_2 \\ \vdots \\ x_n \end{pmatrix}$$

$$+ (y_1 \ y_2 \ \cdots \ y_n) \, \Omega \begin{pmatrix} y_1 \\ y_2 \\ \vdots \\ y_n \end{pmatrix} + \ldots,$$

where each $p_i = (x_i, y_i, \ldots)$, for $i = 1, \ldots, n$. So we see that E_ω is essentially given by the matrix Ω repeated d times. The tensor product of matrices (or sometimes the Kronecker product) gives the matrix of E_ω as $\Omega \otimes I^d$, and

$$E_\omega(p) = (p)^T \Omega \otimes I^d \, p.$$

To rewrite the equilibrium condition (21.3) in terms of matrices, define the *configuration matrix* P for the configuration p as

$$P = \begin{pmatrix} p_1 & p_2 & \cdots & p_n \\ 1 & 1 & \cdots & 1 \end{pmatrix}.$$

P is a $(d+1)$-by-n matrix, and the equilibrium condition (21.3) is equivalent to

$$P \, \Omega = 0.$$

Each coordinate of P as a row vector multiplied on the right by Ω represents the equilibrium condition in that coordinate. The last row of ones of P represent the condition that the column sums (and therefore the row sums) of Ω are $\mathbf{0}$. It is also easy to see that the linear rank of P is the same as the dimension of the affine span of $p_1, \ldots, p_n$ in E^d.

Suppose that we add rows to P until all the rows span the co-kernel of Ω. The corresponding configuration p will be called a *universal configuration* for ω (or equivalently Ω).

Proposition 21.4. *If p is a universal configuration for ω, any other configuration q which is in equilibrium with respect to ω is an affine image of p.*

Proof. Let Q be the configuration matrix for q. Since the rows of P are a basis for the co-kernel of Ω, and the rows of Q are, by definition, in the co-kernel of Ω, there is a $(d+1)$-by-$(d+1)$ matrix A such that $AP = Q$. Since P and Q share the last row of ones, we know that A takes the form

$$A = \begin{pmatrix} A_0 & b \\ 0 & 1 \end{pmatrix},$$

where A_0 is a d-by-d matrix, b is a 1-by-d matrix (a vector in E^d), and the last row is all 0's except for the 1 in the lower right hand entry. Then we see that for each $i = 1, \ldots, n$, $q_i = A_0 p_i + b$, as desired. $\square$

The stress matrix plays a central role in this theory. Note that when the configuration $p = (p_1, \ldots, p_n)$ in E^d is universal (i.e., its affine span is all of E^d), for the stress ω, the dimension of the co-kernel (which is the dimension of the kernel) of Ω is d, and the rank of Ω is $n - d - 1$. But even when the configuration

p is not universal for ω, it is the projection of a universal configuration, and so the rank $\Omega \leq n - d - 1$.

Now we come to one of the basic tools for showing that specific tensegrities are globally rigid and more. If ω is a proper equilibrium stress for the tensegrity $G(p)$, then the difference $p_i - p_j$, where $\omega_{ij} \neq 0$, is called a *stressed direction* and the member $\{i, j\}$ is called a *stressed member*. Note that if $G(q) \leq G(p)$, $\omega_{ij} \neq 0$, and $|p_i - p_j| \neq |q_i - q_j|$, then $E_\omega(q) < E_\omega(p)$. So if p is a configuration for the minimum of E_ω, the stressed members are effectively bars.

Theorem 21.5. *Let $G(p)$ be a tensegrity, where the affine span of $p = (p_1, \ldots, p_n)$ is all of E^d, with a proper equilibrium stress ω and stress matrix Ω. Suppose further that*

1. Ω *is positive semi-definite.*
2. *The configuration p is universal with respect to the stress ω. (In other words, the rank of Ω is $n - d - 1$.)*
3. *The stressed directions of $G(p)$ do not lie on a quadric at infinity.*

Then $G(p)$ is universally globally rigid in all dimensions.

Proof. Suppose that q is a configuration such that $G(q) \leq G(p)$. Then $E_\omega(q) \leq E_\omega(p)$. By Condition 1, $E_\omega(q) = E_\omega(p) = 0$, and ω is an equilibrium stress for the configuration q as well as p. By Condition 2 and Proposition 21.4, q is an affine image of p. By Condition 3 and Proposition 21.1, q is congruent to p.

Notice that in view of Proposition 21.1, Condition 3 can be replaced by the condition that $G(p)$ has no affine flexes in E^d. For example, if it is rigid in E^d, that would be enough.

With this in mind, we say that a tensegrity is *super stable* if it has a proper equilibrium stress ω such that Conditions (1), (2) and (3) hold. If just Conditions 1 and 3 hold and ω is strict (all members stressed), then we say $G(p)$ is *unyielding*. An unyielding tensegrity, essentially, has all its members replaced by bars.

Examples

The Square Tensegrity

The stress matrix for the square of Figure 21.4 is

$$\Omega = \begin{pmatrix} +1 & -1 & +1 & -1 \\ -1 & +1 & -1 & +1 \\ +1 & -1 & +1 & -1 \\ -1 & +1 & -1 & +1 \end{pmatrix}. \qquad (21.7)$$

So Ω has rank $1 = 4 - 2 - 1 = n - d - 1$, and since its trace is 4, its single eigenvalue is 4, and it is positive semi-definite. This makes it unyielding, and since the underlying graph is the complete graph, it is universally globally rigid. It is also super stable. There are several ways to generalize this example.

Polygon Tensegrities

I showed that a tensegrity, obtained from a planar convex polygon by putting a node at each vertex, a cable along each edge, and struts connecting other nodes such that the resulting tensegrity has some proper equilibrium stress, is always super stable. Figure 21.6 shows some examples.

Radon Tensegrities

Radon's Theorem says that if $p = (p_1, \ldots, p_{d+2})$ are $d + 2$ points in E^d, no $d + 1$ in a hyperplane, then they can be separated into two simplices σ^i and σ^{d-i} of dimension i and $d - i$ such that their intersection is a common point, which is a relative interior point of each simplex. They can also be used to define a super stable tensegrity as well. Write $\sum_{k=1}^{d+2} \lambda_k p_k = \mathbf{0}$, where $\sum_{k=1}^{d+2} \lambda_k = 0$, while $\lambda_k > 0$ for $k = 1, \ldots, i + 1$, and $\lambda_k < 0$ for $k = i + 2, \ldots, d + 2$. Then the stress matrix is

$$\Omega = \begin{pmatrix} \lambda_1 \\ \lambda_2 \\ \vdots \\ \lambda_{d+2} \end{pmatrix} (\lambda_1 \; \lambda_2 \; \cdots \; \lambda_{d+2}), \qquad (21.8)$$

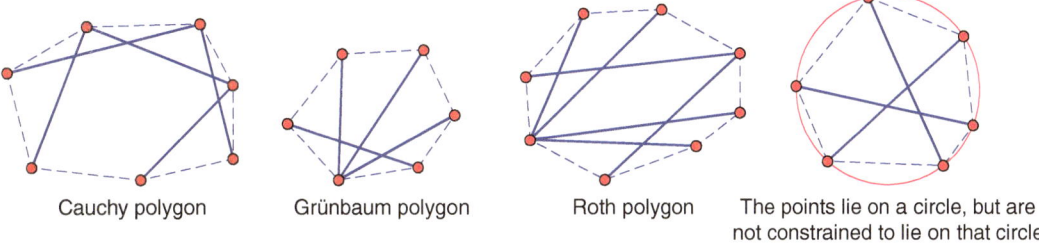

Figure 21.6. Some planar superstable polygons.

since for the configuration matrix P,

$$P \begin{pmatrix} \lambda_1 \\ \lambda_2 \\ \vdots \\ \lambda_{d+2} \end{pmatrix} = \mathbf{0}. \qquad (21.9)$$

So the stress $\omega_{ij} = -\lambda_i \lambda_j$. The edges of σ^i and σ^{d-i} are struts, while all the other members are cables. Since the rank is $d + 2 - d - 1 = 1$, and Ω is positive semi-definite, the tensegrity is super stable. Figure 21.7 shows the two examples in the plane and in three-space.

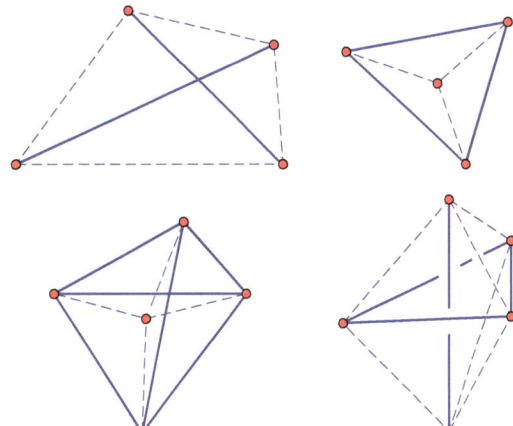

Figure 21.7. Some superstable Radon tensegrities.

Centrally Symmetric Polyhedra

L. Lovasz showed that if one places nodes at the vertices of a centrally symmetric convex polytope, cables along its edges, and struts between its antipodal points, the resulting tensegrity has a strict proper equilibrium stress, and any such stress will have a stress matrix such that Conditions 1 and 2 hold, while condition 3 is easy to check. Thus such a tensegrity is super stable and universally globally rigid. Figure 21.8 shows such an example for the cube, which is easy to check independently.

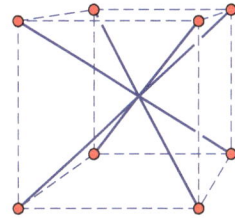

Figure 21.8. A cube with cables along its edges and struts connecting antipodal nodes, which is super stable.

Prismatic Tensegrities

Consider a tensegrity in E^3 formed by two regular polygons $(p_1, \ldots, p_n)$ and $(p_{n+1}, \ldots, p_{2n})$ in distinct parallel planes, each symmetric about the same axis. Cables are placed along the edges of each polygon. Each node of each polygon is connected by a cable to a corresponding node in the other polygon, maintaining the rotational symmetry. Similarly, each node of each polygon in connected to a corresponding node in the other polygon by a strut, maintaining the rotational symmetry. The ends of the cable and strut are k steps apart where $1 \le k \le n - 1$. This describes a *prismatic tensegrity* $P(k, n)$. Each $P(n, k)$ is super stable when the angle of the twist from a node in the top polygon to the projection of the node at the other end of the strut is $\pi(1/2 + i/n)$. Figure 21.9 shows $P(6, 1)$.

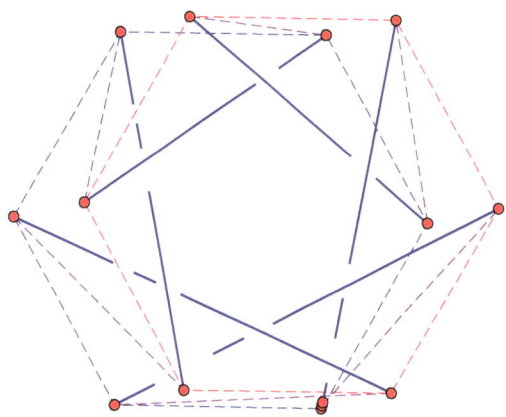

Figure 21.9. The prismic tensegrity $P(6, 1)$.

The Snelson tensegrity in the introduction is $P(3, 1)$.

Highly Symmetric Tensegrities

Many of the tensegrities created by artists such as Snelson have the super stable property discussed here. They need the stress for their stability. Their tensional integrity is part of their stability. Symmetry seems to a natural part of art, so I thought it would be interesting to see what symmetric tensegrities were super stable. It turns out that the symmetry simplifies the calculation of the rank and definiteness of the stress matrix. In addition, the theory of the representations of finite groups is a natural tool that can be used to decompose the stress matrix. With Allen Back and later Robert Terrell, we created a website where one can view and rotate the pictures of these tensegrities.

The tensegrity graph G is chosen so that there is an underlying finite group Γ acting on the tensegrity such that the action of Γ takes cables to cables and struts to struts, and the following conditions hold:

1. The group Γ acts transitively and freely on the nodes. In other words, for each pair p_i, p_j of nodes, there is a unique element $g \in \Gamma$ such that $gp_i = p_j$.
2. There is one transitivity class of struts. In other words, if $\{p_i, p_j\}$ and $\{p_k, p_l\}$ are struts, then there is $g \in \Gamma$ such that $\{gp_i, gp_j\} = \{p_k, p_l\}$ as sets.

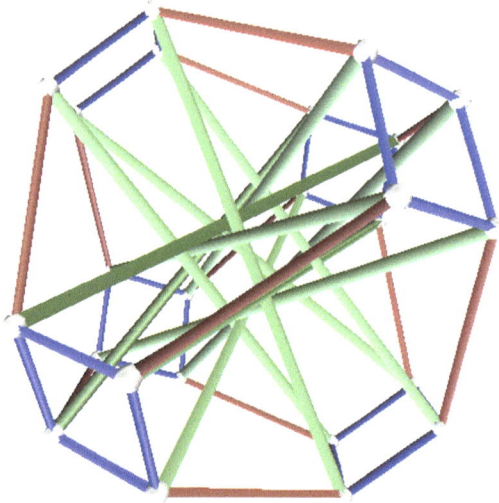

Figure 21.10. A super stable tensegrity from the catalog at http://www.math.cornell.edu. In the catalog, the struts are colored *green*, one cable transitivity class is colored *red*, and the other *blue*. In this example, the cables lie on the convex hull of the nodes, and struts are inside.

3. There are exactly two transitivity classes of cables. In other words, all the cables are partitioned into two sets, where Γ permutes the elements of each set transitively, but no group element takes a cable from one partition to the other.

The user must choose the abstract group, the group elements that correspond to the cables, the group element that corresponds to the struts, and the ratio of the stresses on the two classes of cables. Then the tensegrity is rendered. Figure 21.10 shows a typical picture from the catalog at http://www.math.cornell.edu.

Compound Tensegrities

The sum of positive semi-definite matrices is positive semi-definite. So we can glue two super stable tensegrities along some common nodes, and maintain Condition 1. Condition 3 is no problem. The rank Condition 2 may be violated, but each of the individual tensegrities will remain globally rigid, even if some of the stresses vanish on overlapping members.

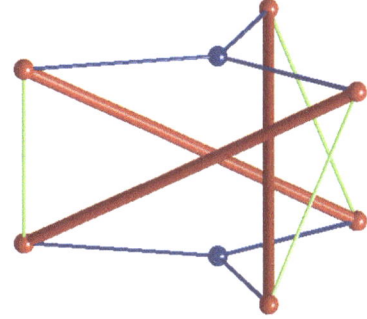

Figure 21.11. Two superstable tensegrities are added to get a third.

One example of this process is the delta-Y transformation. If one super stable tensegrity has a triangle of cables in it, one can add a tensegrity of the form in the upper right of Figure 21.7 so the stresses on the overlap of the three struts exactly cancel with the three cable stresses in the other tensegrity. So the three triangle cables replace the three other cables joined to a new node inside the triangle. In this case the resulting tensegrity is still super stable since the radon tensegrity is planar and using Condition 3. Figure 21.11 shows how this might work for the top triangle of the Snelson tensegrity of Figure 21.2. Figure 21.12 shows this replacement on both triangles.

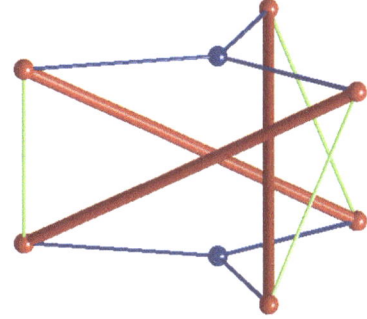

Figure 21.12. The $\Delta - Y$ transformation applied to a Snelson Tensegrity.

If the replacement as in Figure 21.12 is done for a polygon of with four or more vertices, the resulting tensegrity may not be super stable or even rigid, but if the polygons have an odd number of vertices and the struts are placed as far away from the vertical cables as possible, then the resulting tensegrity is super stable. In other words, if the star construction is done on $P(2k + 1, k)$ as in Figure 21.13 the resulting tensegrity is super stable.

It is also possible to put two (or more) super stable tensegrities together on a common polygon to create a tensegrity with a stress matrix that satisfies Condition 1 while the universal configuration is 4-dimensional instead of 3-dimensional. But each of the original pieces is universally globally rigid. The 4-dimensional realization has an affine flex around the 2-dimensional polygon used to glue the two pieces together. So the tensegrity has two non-congruent configurations in E^3 as one piece rotates about the other in E^4.

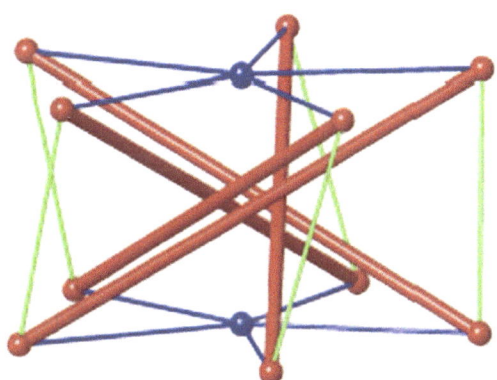

Figure 21.13. A flexible tensegrity.

Meanwhile struts and cable stresses can be arranged to cancel, and thus those members are not needed in the compound tensegrity. Figure 21.14 shows this with two Snelson tensegrities combined along a planar hexagonal tensegrity.

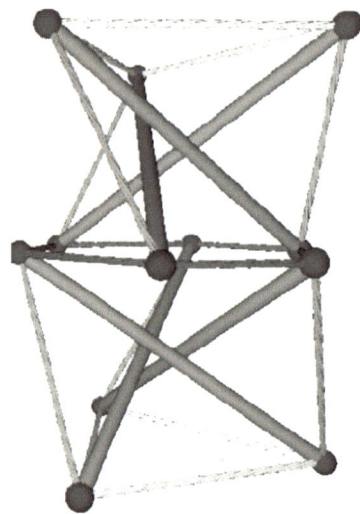

Figure 21.14. A compound rigid but not superstable tensegrity.

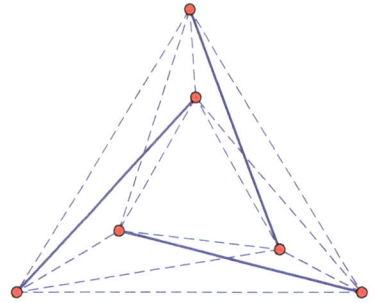

Figure 21.15. A flexible tensegrity in the plane occurs when the struts don't intersect.

This is something like the start of the Snelson tower of Figure 21.1, but the hexagonal polygon in the middle is planar, which seems a bit surprising. This tensegrity is unyielding and rigid, but not super stable. But possibly to create more stability Snelson includes more cables from one unit to the other, and this destroys the planarity of the hexagon.

There are many different ways to combine super stable units, possibly erasing some of the members in the basic units to get similar rigid tensegrities.

Pure and Flexible Examples

Recall that a pure tensegrity is one that has only cables and struts and the struts are all disjoint. We have seen several examples in E^3 of pure tensegrities, the simplest being Snelson's original as in Figure 21.11 on the left. But what about the plane? One might be tempted to think that the tensegrity of Figure 21.15 is rigid, but it isn't.

Indeed, there are no pure rigid tensegrities in the plane. This follows from the proof of the carpenter's rule property. This theorem says that any chain of non-overlaping edges in the plane can be continuously expanded (flexed) until it is straight. This result also allows for disjoint edges and, at least for a short time, the expansion can be run backwards to be a contraction, keeping the struts at a fixed length.

There's more — much more — to say about this rapidly-evolving subject; see the notes at the end of the book.

22

Ten Problems in Geometry

Moritz W. Schmitt and Günter M. Ziegler

Geometry is a field of knowledge, but it is at the same time an active field of research—our understanding of space, about shapes, about geometric structures develops in a lively dialogue, where problems arise, new questions are asked every day. Some of the problems are settled nearly immediately, some of them need years of careful study by many authors, still others remain as challenges for decades. In this chapter, we describe ten problems waiting to be solved.

1.   Unfolding Polytopes

Albrecht Dürer's famous geometry masterpiece "Underweysung der Messung mit dem Zirckel und Richtscheyt" was published in Nuremberg in 1525. The fourth part of this book contains many drawings of nets of 3-dimensional polytopes and, implicitly, the following conjecture:

> Every 3-dimensional convex polytope can be cut open along a spanning tree of its graph and then unfolded into the plane without creating overlaps.

M.W. Schmitt
Institute of Mathematics, Freie Universität Berlin, Arnimallee 2, D-14195 Berlin, Germany
e-mail: mws@math.fu-berlin.de;
http://userpage.fu-berlin.de/mws/

G.M. Ziegler
Institute of Mathematics, Freie Universität Berlin, Arnimallee 2, D-14195 Berlin, Germany
email: ziegler@math.fu-berlin.de;
http://page.mi.fu-berlin.de/gmziegler/

This conjecture was posed explicitly by the British mathematician Geoffrey C. Shephard in 1975. It has captured many geometers' attention since then—and led to many interesting results and insights in this area, many of them described in Chapter 6 in this book.

One important insight is that the spanning tree must be chosen with care. Given any polytope, it is easy to find some spanning tree in its graph. After cutting the boundary along the edges of this tree, there is a unique way to unfold it into a planar figure. However, the problem is that overlaps could occur, and indeed they do occur. Figure 22.1a shows a prism, once cut open along a good spanning tree, once cut open along a bad spanning tree. Moreover, perhaps surprisingly, Makoto Namiki observed that even an unfolding of a tetrahedron can result in an overlap: see Figure 22.1b.

The conjecture has been verified for certain somewhat narrow classes of polytopes. For example, it holds for so-called *prismoids*. These are built by taking the convex hull of two polygons that lie in parallel planes, have the same number of sides and the same angles and are positioned in such a way that their corresponding edges are parallel. Another class of polytopes for which the conjecture has been established are the *domes*. A dome has a base face and all its other faces share an edge with this base.

One approach to the problem is algorithmic. For a proof that all polytopes can be unfolded, we need a good strategy to choose a suitable spanning tree. One could look, for example, for

a

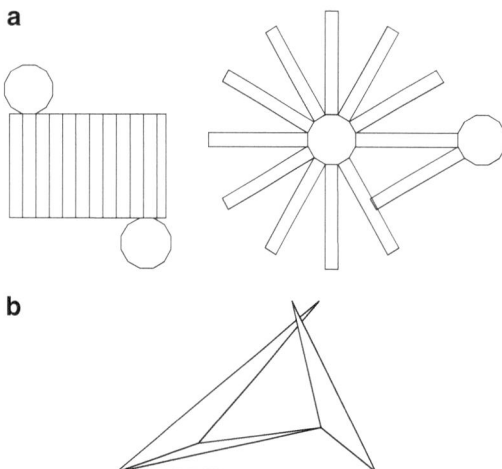

b

Figure 22.1. (a) Two unfoldings of a prism. (b) A bad unfolding of a tetrahedron.

a shortest or a longest spanning tree, which minimizes or maximizes the sum of the lengths of the edges, respectively. Or one can place the polytope in space such that no edge is horizontal and the highest vertex is unique, and then from any other vertex choose the steepest edge or the "rightmost" edge that points upwards. Such rules are motivated by and derived from various "pivot rules" of linear programming, which describe local strategies to move from any given vertex of a polyhedron along edges to the highest vertex. Extensive experiments with such rules were performed by Wolfram Schlickenrieder for his 1997 diploma thesis. None of the strategies tested by him worked for all examples, but for all examples *some* of his strategies worked.

Further interesting studies motivated by the unfolding conjecture concern relaxations of the problem. For example, there are unfolding techniques that do not cut only along edges, but may cut into faces, such as the *source unfolding* and the *star unfolding* discovered by Alexandrov. Here we only sketch the latter technique: For a star unfolding one picks one point on the boundary of the polytope such that it has a unique shortest path to every vertex. The union of these paths form a tree that connects all the vertices. If one cuts the polytope boundary open along this tree, then this is an unfolding that provably has no overlaps.

2. Almost Disjoint Triangles

How complicated can polyhedral structures be in 3-dimensional space? For example, we are interested in triangulated surfaces on n vertices in $\mathbb{R}^3$, such as the boundary of a tetrahedron, which has $n = 4$ vertices and $n = 6$ edges, of an octahedron with $n = 6$ vertices and $e = 12$ edges, or of an icosahedron with $n = 12$ vertices and $e = 30$ edges.

But what is the maximal number of edges for a triangulated surface in $\mathbb{R}^3$ on n vertices? Certainly it cannot have more than $\binom{n}{2}$ edges. This bound is not tight for all n, since for a triangulated surface the number of edges is divisible by 3. Indeed, every triangle is bounded by three edges, while each edge is contained in two triangles, the number e of edges and the number f of triangles satisfy the equation $3f = 2e$ and thus f is even and e is divisible by three. Another constraint comes from the fact that the surfaces we look at are embedded in $\mathbb{R}^3$. They have an "inside" and an "outside", so they are *orientable*, which implies that the *Euler characteristic* $n - e + f$ is even (it equals $2 - 2g$, where g is known as the *genus* of the surface). Nevertheless, this leads to only slight improvements of the upper bounds. If n is congruent to $0, 3, 4$, or 7 modulo 12, then it seems entirely possible that a surface with $e = \binom{n}{2}$ exists; its face numbers would be given by $(n, e, f) = (n, \binom{n}{2}, \frac{2}{3}\binom{n}{2})$.

Is there such a "neighborly" triangulated surface for all these parameters? For small values of n it seems so: For $n = 4$ we have the tetrahedron, and for $n = 7$ there is a triangulated surface, known as the *Császár torus*, which consists of 14 triangles and $\binom{7}{2} = 21$ edges. At the next n where we get integer parameters, $n = 12$ and $e = 66$ and $g = 6$, it is known that combinatorial schemes for suitable triangulated surfaces exist; indeed, this was established for all $n \equiv 0, 3, 4$, or $7 \pmod{12}$ as part of the so-called *Map Color Theorem* by Ringel et al. (1974). However, *none* of the 59 combinatorial schemes for such a surface can be realized as a triangulated surface in $\mathbb{R}^3$. This was established by Jürgen Bokowski, Antonio Guedes de Oliveira, and finally Lars Schewe quite recently.

But let us get beyond the small parameters. What can we expect when n gets large? Will the maximal number of edges in a triangulated surface with n vertices grow quadratically with n, or much slower? All we know at the moment is that the maximal e grows at least as fast as $n \log n$; this can be seen from surfaces that were constructed by Peter McMullen, Christoph Schulz, and Jörg Wills in 1983.

However, Gil Kalai has proposed studying a closely-related problem that is even easier to state, and may be similarly fundamental:

Given n points in 3-dimensional space, how many triangles could they span that are disjoint, except that they are allowed to share vertices?

So for Kalai's problem the triangles are not allowed to share an edge, and they are not allowed to intersect in any other way (see Figure 22.2). Let us call this *almost disjoint triangles*.

Without loss of generality we may assume that the n points that we use as vertices lie in a general position, no three of them on a line and no four of them in a plane. Clearly the maximal number $T(n)$ of vertex disjoint triangles on n points is not larger than $\frac{1}{3}\binom{n}{2}$. But is

$$T(n) \leq \frac{1}{3}\binom{n}{2}$$

a tight upper bound for infinitely many values of n? Does $T(n)$ grow quadratically when n gets large? All we know is that there is a lower bound that grows like $n^{3/2}$: Gyula Károlyi and

Jozsef Solymosi in 2002 presented a very simple and elegant method to position $n = m^2 + \binom{m}{2}$ points in $\mathbb{R}^3$ that span $m\binom{m}{2}$ almost-disjoint triangles.

3. Representing Polytopes with Small Coordinates

A famous theorem by Steinitz from 1922 characterizes the graphs of 3-dimensional convex polytopes. It states:

Theorem 22.1. *A finite graph G is the edge graph of a polytope P if and only if G is planar and 3-connected.*

Obtaining a polytope from a given such graph is a construction problem. Here one is especially interested in nice realizations. Of course, the meaning of "nice" depends on the context. One possibility is to ask for a polytope that has all its edges tangent to a sphere. Such a realization exists and it is essentially unique. This can be derived from the Koebe–Andreev–Thurston circle packing theorem. However, the edge tangent realizations are not combinatorial, and in general they have irrational vertex coordinates. One can also ask for rational realizations, such that all vertex coordinates are rational, or equivalently (after multiplication with a common denominator) for integral realizations. The existence of such realizations can be derived from Steinitz' original proofs. Just how large would the integers have to be?

Figure 22.3 shows a dodecahedron realized with very small integer coordinates due to Francisco Santos.

The big open problem is:

Can every 3-dimensional convex polytope with n vertices be realized with its vertex coordinates in the integer grid $\{0, 1, \ldots, f(n)\}^3$, where f is a polynomial?

All we know at the moment are exponential upper bounds on $f(n)$. The first such bounds were derived by Shmuel Onn and Bernd Sturmfels in 1994; they were subsequently improved to

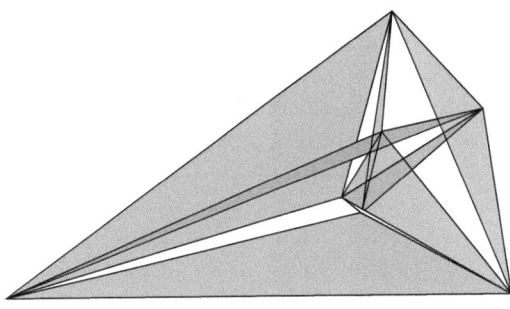

Figure 22.2. Almost disjoint triangles spanned by 7 points.

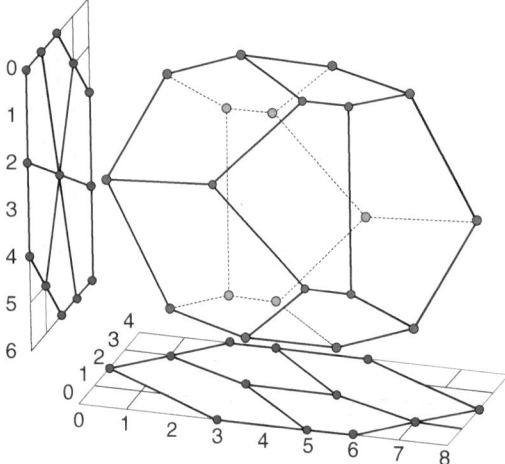

Figure 22.3. Small coordinates for the dodecahedron.

$f(n) < 148^n$. But indeed we know of no lower bounds that would exclude that all combinatorial types can be realized with $f(n) < n^2$. A recent result by Erik Demaine and André Schulz from 2010 is that for the very special case of stacked polytopes (that is, obtained from a tetrahedron by repeatedly stacking a flat pyramid onto a facet), realizations with polynomially-bounded integer coordinates exist. But do these exist for all graphs of 3-polytopes?

4. Polyhedra that Tile Space

An innocent-sounding question is

> Which convex polytopes can be used to tile 3-dimensional space?

Unfortunately, answering this question seems to be quite difficult. Indeed, not even the 2-dimensional version of this problem has been solved completely, although this has been claimed and believed several times, starting with a paper by Karl Reinhardt from 1918. Nevertheless, for tilings of the plane it is not hard to see that any convex polygon that admits a tiling of the plane—that is, such that the plane can be completely covered by congruent copies of this polygon, without gaps and without overlapping interiors—can have at most six sides. The reason

for this is topological and can be connected to Euler's polyhedron formula. Clearly the regular hexagon can be used to tile the plane (any bee knows that), but many other types of convex hexagons admit such a tiling as well.

One dimension higher, we all know the tiling of space by congruent cubes, which have 8 faces. However, it is also not hard to see that translates of the so-called permutahedron with its 14 faces and 24 vertices tile space face-to-face. So this begs the question:

> What is the maximal number of faces for a convex polytope that allows for a tiling of 3-space by congruent copies?

In 1980, the crystallographer Peter Engel produced four types of polytopes with 38 faces that tile space, and up to now this record apparently has not been topped. On the other hand, no finite upper bound is known, and the answer may as well be that there is no finite upper bound. The problem seems to be that the only effective method to produce such tilings is to look at dot-patterns (discrete point sets) in $\mathbb{R}^3$ that have a transitive symmetry group, that is, such that for any two points in the pattern there is a symmetry of space that moves one point to any other one. For such a point configuration $S \subset \mathbb{R}^3$ all the *Voronoi domains*

$$V_S := \{x \in \mathbb{R}^3 : \|x - s\| \le \|x - s'\| \text{ for all } s' \in S\}$$

for $s \in S$ are congruent. The Voronoi domain of s collects all points in space for which no other point in S is closer. Figure 22.4 shows an excerpt of a symmetric dot pattern with its Voronoi cells.

The Voronoi construction applied to symmetric dot patterns is very effective in producing tilings by congruent polytopes. Indeed, Engel's four examples with 38 faces were produced this way. However, it is also known that for tilings of this type the number of faces is bounded.

So symmetry helps to construct tilings. We should, however, not rely on this too much. In his famous list of 23 problems from the 1900 International Congress of Mathematicians in Paris, David Hilbert had asked as part of his

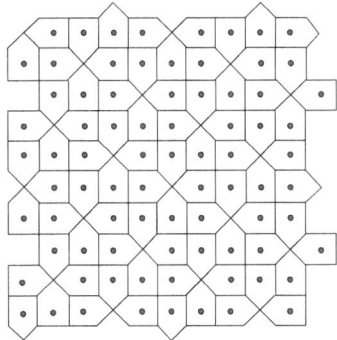

Figure 22.4. A tiling by congruent pentagons generated by the Voronoi construction applied to a symmetric dot pattern.

18th problem whether there could be a convex polytope that tiles 3-dimensional space, but such that there is no tiling that would have a symmetry group that moves tiles to tiles. The answer has long been known to be *yes*: such tilings exist. For example, various types of quasicrystals (Nobel Prize in Chemistry 2011!) demonstrate this. This shows that even though symmetry helps a lot in constructing tilings, it should not be used as our only resource.

5.   Fatness

Whereas *regular* convex polytopes (the Platonic solids) have been studied since antiquity, *general* convex polytopes came into the focus of attention much later. To Descartes and Euler we owe the "Euler polyhedron formula." In modern notation, where we write f_i for the number of i-dimensional faces of a convex polytope, it states that every 3-dimensional convex polytope satisfies

$$f_0 - f_1 + f_2 = 2.$$

In 1906 Ernst Steinitz characterized the set $\mathcal{F}_3$ of all possible triples (f_0, f_1, f_2) for convex polytopes:

$$\mathcal{F}_3 = \{(f_0, f_1, f_2) \in \mathbb{R}^3 : f_0 - f_1 + f_2 = 2,$$
$$f_2 \leq 2f_0 - 4, f_0 \leq 2f_2 - 4\}.$$

More than one hundred years later, no similarly complete description is available for the possible sequences of face numbers, or f-*vectors*, of d-dimensional polytopes for any $d > 3$. Indeed, we know that the f-vectors of d-dimensional convex polytopes satisfy essentially only one linear *equation*, the so-called *Euler–Poincaré equation*:

$$f_0 - f_1 + f_2 - \cdots + (-1)^{d-1} f_{d-1} = 1 - (-1)^d.$$

However, we do not know all the linear *inequalities*. In particular, we would be interested in linear inequalities that hold with equality for the d-dimensional simplex, which has $f_i = \binom{d+1}{i+1}$, as these special inequalities describe the "cone of f-vectors."

To make this concrete, let us concentrate on the case $d = 4$. Here everything boils down to the question whether the parameter called *fatness*,

$$\Phi := \frac{f_1 + f_2 - 20}{f_0 + f_3 - 10},$$

can be arbitrarily large for 4-polytopes. Can it be, say, larger than 10? Or is it true that

$$f_1 + f_2 - 20 \leq 10(f_0 + f_3 - 10)$$

for all convex 4-dimensional polytopes? This parameter is called "fatness" because it measures how "fat" an f-vector (f_0, f_1, f_2, f_3) is in the middle, i.e., how big the sum of the entries f_1 and f_2 is in comparison to the sum of f_0 and f_3.

Indeed, it is not hard to show that the fatness parameter Φ ranges between 2.5 and 3 for simple and simplicial polytopes. However, it is $\Phi = 4.52$ for a fascinating 4-dimensional regular polytope known as the "24-cell," which has $f_0 = 24$ vertices and $f_3 = 24$ facets, which are regular octahedra; its complete f-vector is $(f_0, f_1, f_2, f_3) = (24, 96, 96, 24)$. An even higher value of $\Phi = 5.021$ is achieved for the "dipyramidal 720-cell" constructed in 1994 by Gabor Gévay, which has f-vector $(720, 3600, 3600, 720)$. Finally, a class of polytopes named "projected deformed products of polygons," constructed by the second author in 2004, get arbitrarily close to $\Phi = 9$. That's where we stand at the time of writing. But is there a finite upper bound at all?

This may read like a problem of 4-dimensional geometry and thus outside our range of visualization, but it isn't really, since the boundary of a 4-dimensional polytope is of dimension 3. Thus one can relate the question to problems about polytopal tilings in 3-space. Here is one such problem:

> Are there normal face-to-face tilings of 3-space by convex polytopes in which
> (1) all tiles have many vertices, and
> (2) each vertex is in many tiles?

For example, in the usual tiling of space by unit cubes all tiles have 8 vertices and each vertex is in 8 tiles. For a *normal* tiling we require that there be a lower bound for the inradius and an upper bound for the circumradius of the tiles. This is satisfied, for example, if there are only finitely many types of tiles. It is not too hard to show that either of the two conditions (1) and (2) can be satisfied. But can they be satisfied by the same tiling, at the same time? If no, then fatness Φ for 4-polytopes is bounded.

6.   The Hirsch Conjecture

One of the biggest mysteries in convex geometry is about the graphs of convex polytopes and their diameters. The *graph* of a polytope is a combinatorial model which captures vertex-edge incidences. Such a graph has the vertices of the polytope as nodes and two nodes are adjacent in the graph if they are connected by an edge as vertices in the polytope. Figure 22.5 shows the octahedron and its graph.

The *diameter* of a graph is the greatest distance of two vertices in the graph, where the *distance* of two vertices is the length of the shortest path connecting them. In 1957 Warren Hirsch raised the question:

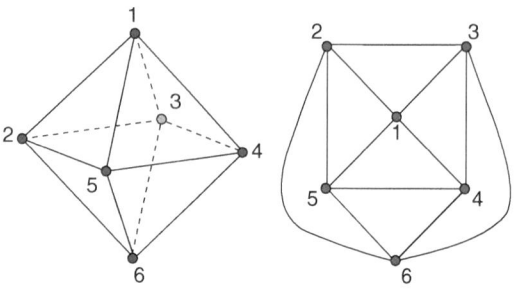

Figure 22.5. A polytope and its graph.

What is the maximal diameter of the graph of a d-polytope with n facets?

He conjectured that

$$\Delta(d,n) \leq n - d,$$

where $\Delta(d,n)$ denotes the above maximal diameter. Even though decades of research went into a solution of this problem, for over 50 years little progress was made. Finally, in May 2010, Francisco Santos announced a counterexample. By an explicit construction he could demonstrate that $\Delta(43, 86) > 43$.

While this certainly was a breakthrough, it is merely a first step in answering the above question. Santos' construction does not even rule out an upper bound on the diameter that is linear in $n - d$. Many researchers in discrete geometry believe that the real question is whether there is a polynomial bound in n and d. The best upper bound for general d-polytopes was derived by Gil Kalai and Daniel Kleitman in 1992. By using a strictly combinatorial approach they were able to prove that

$$\Delta(d,n) \ \leq \ n^{2+\log_2 d},$$

but of course this is still very far from a polynomial bound. Furthermore, by a result of Larman it follows that if one fixes the dimension d, then there is a bound that is linear in n. In general it is sufficient to prove an upper bound for simple polytopes: The facets of a non-simple polytope can be perturbed such that one gets a simple polytope. This new polytope has a graph whose diameter is at least as large as for the original graph.

Besides its importance for polyhedral geometry, the question also relates closely to linear programming. The maximal graph diameter $\Delta(d, n)$ is a lower bound for the number of steps that the simplex algorithm would need on a problem with n constraints in d variables for *any* pivot rule that would select the edges. Thus researchers from Operations Research and Mathematical Optimization are interested in the Hirsch Conjecture as well.

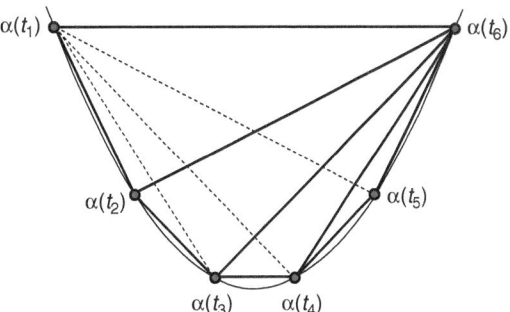

Figure 22.6. Construction of a cyclic 3-polytope with 6 vertices.

7.   Unimodality

In 1970 McMullen settled the long-standing open question, "What is the maximal number of k-faces of a d-polytope on n vertices?" He confirmed that neighborly simplicial polytopes are extremal with regard to their f-vectors: A *neighborly* polytope is a polytope such that every subset of the vertices of cardinality at most $\lfloor d/2 \rfloor$ is the vertex set of a face. One well-known class of such polytopes is the cyclic polytopes. These can be defined as the convex hull of finitely many points on the *moment curve*

$$\alpha : \mathbb{R} \longrightarrow \mathbb{R}^d, \quad t \longmapsto (t, t^2, \ldots, t^d).$$

That is, one chooses n different reals, $t_1 < \cdots < t_n$, and calls

$$C_d(n) = \mathrm{conv}(\alpha(t_1), \ldots, \alpha(t_n))$$

the d-dimensional *cyclic polytope* on n vertices. (A simple analysis, using the Vandermonde determinant, shows that cyclic polytopes are simplicial, and that the combinatorial type does not depend on the particular parameters t_i chosen.) Figure 22.6 shows a realization of $C_3(6)$.

Despite McMullen's Upper Bound Theorem, there are other questions about the face numbers of polytopes, also closely connected to the cyclic polytopes, which are not well understood, yet — the most tantalizing ones connected to unimodality. A sequence of numbers is called *unimodal* if it first increases and then decreases with no "dips" in-between. It was proved only

recently, by László Major, that the f-vectors of all cyclic polytopes are unimodal. (This was a long-standing open problem despite the fact that we have explicit formulas for the f-vectors of cyclic polytopes.) However, in 1981 Björner was able to construct polytopes with non-unimodal f-vectors by cleverly modifying a cyclic polytope of dimension at least 20.

This was quite a surprise as already in the late 1950s Theodore Motzkin had conjectured that *all* f-vectors of convex polytopes are unimodal. (Later, in 1972, Dominic Welsh came up with the same conjecture again.) Unfortunately, it is not quite that easy. At a conference in Graz, Austria, in 1964 Ludwig Danzer presented a first construction for very high-dimensional polytopes with non-unimodal f-vector. Later Jürgen Eckhoff came up with another construction that is ingeniously easy, and gets us down to dimension 8. Indeed, here is a sketch for his construction. For this we again start with a cyclic polytope, $C_8(25)$, which is a simplicial polytope on 25 vertices and 7125 facets, with f-vector

$$(25, 300, 2300, 12650, 33750, 44500,$$

$$28500, 7125).$$

Its dual polytope $C_8(25)^*$ is a simple polytope that has 7125 vertices and 25 facets. Now one can cut off one of the simple vertices of the dual polytope, which yields a simplex facet. One can then "glue" these two polytopes along one of the facets of $C_8(25)$ and the simplex facet of the modified $C_8(25)^*$. The resulting polytope, called a "connected sum" and denoted by $C_8(25)\#C_8(25)^*$, has the f-vector

$$(7149, 28800, 46800, 46400, 46400, 46800,$$

$$28800, 7149)$$

with a small but noticable dip in the middle — it is not unimodal!

So far no one succeeded in constructing a polytope with non-unimodal f-vector of dimension less then 8. It is only known that all f-vectors of polytopes up to dimension 5 are unimodal. Furthermore, all known examples of polytopes with non-unimodal f-vector have lots of vertices; the current record is 1320 vertices. All this begs the questions:

Are there "small" polytopes that have non-unimodal f-vectors?
That is, are there polytopes of dimension smaller than 8?
And are there such polytopes with much fewer than, say, 1000 vertices?

One can also speculate how large the "dips" in f-vectors polytopes can be. Are they always tiny? Indeed, Imre Bárány asked the following intriguing question, which tries to exclude dramatically deep dips:

Is it true that the smallest face number of a polytope is always given by the number of vertices, or the number of vertices (or both)? That is, do we always have

$$f_i(P) \geq \min\{f_0(P), f_{d-1}(P)\}$$

for the face numbers of a d-dimensional polytope P?

We don't know. And indeed currently no-one seems to be able to even prove that

$$f_i(P) \geq \frac{1}{1000} \min\{f_0(P), f_{d-1}(P)\}$$

holds in general, for all d-polytopes P and $0 < i < d - 1$. The unimodality questions, and in particular Bárány's problem, demonstrate strikingly how little we know about the face numbers of polytopes.

8. Decompositions of the Cube

Consider the d-dimensional cube $I_d = [0, 1]^d$ and define the following parameters:

- Let $C(I_d)$ be the minimal number of d-dimensional simplices needed for a cover of I_d. A *cover* of I_d is a collection of simplices such that the union of all simplices is I_d. The interiors of simplices are allowed to intersect.
- If all vertices of a cover are also vertices of I_d, we speak of a *vertex cover*. The minimal cardinality of a vertex cover will be denoted by $C^v(I_d)$.
- The minimal number of dissections of I_d will be abbreviated by $D(I_d)$. A *dissection* is a decomposition of I_d into d-dimensional simplices whose interiors are pairwise disjoint but that do not necessarily intersect in a common face. So simplices are allowed to touch but the interiors must not intersect.
- $D^v(I_d)$ is the same as $D(I_d)$, except that we again require the vertices of the simplices to be vertices of I_d—such a dissection is called a *vertex dissection*.
- We define $T(I_d)$ to be the size of the minimal triangulation of I_d, where *triangulation* means decomposition of I_d into pairwise disjoint d-simplices which intersect in a common face or not at all.
- Finally, $T^v(I_d)$ is defined analogously to $D^v(I_d)$.

Three obvious questions are:

1. Given the dimension, what are the values for the parameters above?
2. Can we give good estimates of the parameters for large d?
3. And what is their relationship among each other?

For the rest of this description we will write C instead of $C(I_d)$, etc. With regard to the last question, we can easily sum up what is currently known:

$$C \leq C^v, D \leq T, D^v \leq T^v.$$

The only non-trivial relation is $C^v \leq T$, but this follows from a result by Bliss and Su.

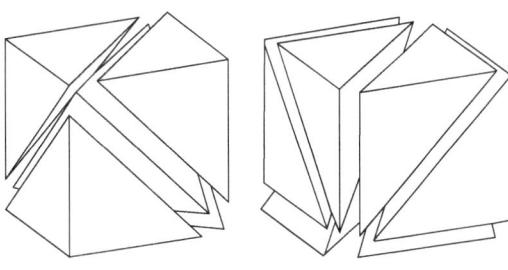

Figure 22.7. Two triangulations of a 3-cube.

The status of the first two questions cannot be summarized so concisely. Best studied seems to be the parameter T^v. The case $d = 2$ is straightforward but already $d = 3$ allows vertex triangulations of different cardinality, as demonstrated by Figure 22.7.

To get an upper bound on T^v one considers the so-called *standard triangulation*. It is of size $d!$ and one constructs it by linking a simplex to each permutation $\pi \in S_d$ by using the following description

$$\Delta_\pi = \{(x_1, \ldots, x_d) \in \mathbb{R}^d :$$

$$0 \le x_{\pi(1)} \le \cdots \le x_{\pi(d)} \le 1\}.$$

This triangulation is maximal among those that only use vertices of the cube, but minimal only for $d = 2$. For lower bounds one directly looks at the more general case of C^v. To get asymptotic estimates for example, the following idea is applied: If $V(d)$ denotes the maximal determinant of a 0/1-matrix, then $V(d)/d!$ is an upper bound of the volume of the largest simplex in I_d and we get

$$C^v \ge \frac{d!}{V(d)}.$$

Determining $V(d)$ is not easy but one can use the Hadamard inequality to bound it. By using hyperbolic volume instead of Euclidean volume, Smith obtained in 2000 the bound

$$C^v, D, T, D^v, T^v \ge \frac{6^{d/2}d!}{2(d+1)^{(d+1)/2}}.$$

Glazyrin recently improved this bound for T^v:

$$T^v \ge \frac{d!}{(\sqrt{d}/2)^d}.$$

The following table sums up lower bounds which are results of several research articles. Bold entries denote optimal bounds.

Dimension	D	C^v, T	D^v	T^v
3	**5**	**5**	**5**	**5**
4	15	**16**	**16**	**16**
5	48	60	61	**67**
6	174	252	270	**308**
7	681	1,143	1,175	**1,493**
8	2,863	5,104	5,522	5,522
9	12,811	22,616	26,593	26,593
10	60,574	98,183	131,269	131,269
11	300,956	520,865	665,272	665,272
12	1,564,340	$2.9276 \cdot 10^6$		

9.   The Ball-and-Cube Problem

Consider the d-dimensional ball

$$B_d = \{x \in \mathbb{R}^d : \|x\| \le 1\}$$

and let $P_d \subseteq \mathbb{R}^d$ be a convex polytope of dimension d with $2d$ facets that contains B_d. One example of such a polytope is the d-dimensional unit cube

$$C_d = \{(x_1, \ldots, x_d) \in \mathbb{R}^d :$$

$$|x_i| \le 1 \text{ for } i = 1, \ldots, d\}.$$

Furthermore, for such a polytope P_d let $\sigma(P_d)$ be the maximal distance between some point of P_d and the origin $0 \in \mathbb{R}^d$. Then a conjecture of Chuanming Zong (who also suggested that we include this problem), states that

$$\sigma(P_d) \ge \sqrt{d},$$

where equality is supposed to hold if and only if P_d is congruent to C_d. It is not difficult to verify this conjecture for $d = 2$: Assume that it is not correct, i.e., there exists a quadrilateral P_2 that contains the unit disc but has $\sigma(P_2) < \sqrt{2}$. Using the center of the disc as a vertex, one can dissect it into four triangles whose vertices are

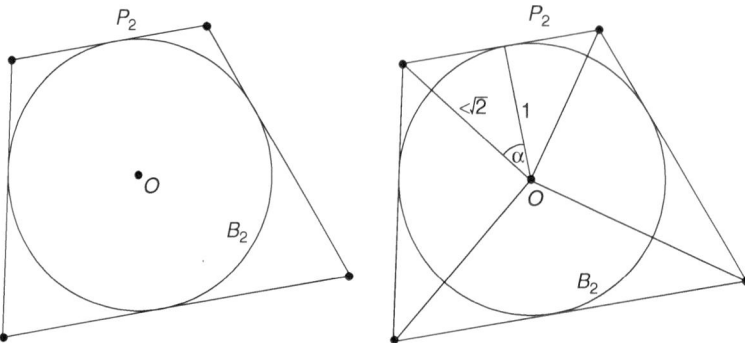

Figure 22.8. Solution of the ball-and-cube problem for dimension 2.

the corners of the quadrilateral and the origin. Our assumption in particular means that the edge length of an origin-corner edge is strictly less than $\sqrt{2}$. (Compare Figure 22.8)

Since the function $\arccos\colon [-1, 1] \to [0, \pi]$ is strictly monotonically decreasing, we have for the angle α in the drawing

$$\cos\alpha > \frac{1}{\sqrt{2}} \implies \alpha < \arccos\frac{1}{\sqrt{2}} = \frac{\pi}{4}.$$

In total we have an angle sum of strictly less than $8 \cdot \frac{\pi}{4} = 2\pi$ for a whole tour around the origin—clearly a contradiction. Besides this easy case not much is known. László Fejes Toth was able to prove an equivalent conjecture for $d = 3$ and Dalla et al. verified the statement for $d = 4$. All higher-dimensional cases are still open at the time of writing (2011).

10. The 3^d Conjecture

The last century there has been amazing progress in the understanding of face numbers of convex polytopes. As discussed in Problem 5 (Fatness), the case $d = 3$ was solved by Steinitz in 1906: The possible f-vectors are

$$\{(f_0, f_1, f_2) \in \mathbb{Z}^3 : f_0 - f_1 + f_2 = 2,$$
$$f_2 \le 2f_0 - 4, \; f_0 \le 2f_2 - 4\}.$$

The first condition is Euler's equation, the first inequality is satisfied by equality for polytopes

where all faces are triangles, while the second inequality characterizes polytopes where all vertices have degree 3 as the extreme case. In the case $d = 4$ one basic problem that remains is the fatness problem discussed above. A complete answer for d-dimensional simple or simplicial polytopes is available via the so-called g-Theorem proved by Billera–Lee and Stanley in 1980.

In contrast to this, it is amazing how little we know about *centrally-symmetric* convex polytopes, that is, polytopes that are left unchanged by a reflection in the origin in $\mathbb{R}^d$. Let's first look at the 3-dimensional case again. Here the possible f-vectors can be described as

$$\{(f_0, f_1, f_2) \in (2\mathbb{Z})^3 : f_0 - f_1 + f_2 = 2,$$
$$f_2 \le 2f_0 - 4, \; f_0 \le 2f_2 - 4, \; f_0 + f_2 \ge 14\}.$$

All the face numbers of a centrally-symmetric polytope are even, thus we have $(f_0, f_1, f_2) \in (2\mathbb{Z})^3$. We recognize Euler's equation and the two inequalities from above. And then there is an additional relation, which for $d = 3$ is easy to prove: A centrally-symmetric 3-polytope has at least 6 vertices, and if it has only 6 vertices, then it must be an octahedron which has 8 facets. If, however, the centrally-symmetric 3-polytope has at least 8 vertices, then it also has at least 6 facets with the only extreme case of an affine cube. In summary, this yields the third inequality, which by Euler's equation we can rewrite as

$$f_0 + f_1 + f_2 + f_3 \ge 27,$$

with equality if and only if the polytope is either a cube or an octahedron. In 1989 Gil Kalai asked whether a similar statement was true in all dimensions:

> Does every d-dimensional centrally-symmetric polytope have at least 3^d non-empty faces?

Kalai's question fits into a series of three basic conjectures:

The 3^d conjecture   (Kalai 1989)
> Every centrally-symmetric d-dimensional polytope satisfies $f_0 + f_1 + \cdots + f_d \geq 3^d$.

The flag conjecture   (Kalai 2008)
> Every centrally-symmetric d-dimensional polytope satisfies $f_{0,1,2,\ldots,d-1} \geq 2^d d!$.

The Mahler conjecture   (Mahler 1939)
> Every centrally-symmetric convex body K satisfies $\mathrm{Vol}(K) \cdot \mathrm{Vol}(K^*) \geq 4^d/d!$, where K^* is the polar of K.

These three conjectures are remarkable since they seem basic; they have been around for quite a while, but we know so little about them. The 3^d conjecture was proved for $d \leq 4$ by Sanyal et al. in 2009, but is open beyond this. The flag conjecture is not even known for $d = 4$. Yet worse, the Mahler conjecture has been an object of quite some scrutiny, but it seems open even for $d = 3$.

The three conjectures belong together since we believe we know the answer—the same answer for all of them. Indeed, the class of *Hanner polytopes* introduced by Olof Hanner in 1956 is obtained by starting with a single centrally-symmetric interval such as $[-1, 1] \subset \mathbb{R}$ and then taking products and polars—or equivalent, taking products and direct sums of polytopes. It is easy to compute that all d-dimensional Hanner polytopes have exactly 3^d non-empty faces, they have exactly $2^d d!$ complete flags of faces, and they have exactly Mahler Volume $\mathrm{Vol}(P) \cdot \mathrm{Vol}(P^*) = 4^d/d!$. But are they the only centrally-symmetric polytopes with these properties? And can't there be any other polytopes with even smaller values? This is not known.

Notes and References

Notes and References for Chapter 2

Page 14 *limit yourself to using only one or two different shapes* ... These types of starting points were suggested by Adrian Pine. For many more suggestions for polyhedra-making activities, see Pinel's 24-page booklet *Mathematical Activity Tiles Handbook*, Association of Teachers of Mathematics, Unit 7 Prime Industrial Park, Shaftesbury Street, Derby DEE23 8YB, United Kingdom website; www.atm.org.uk/.

Page 17 *"What-If-Not Strategy"* ... S. I. Brown and M. I. Walter, *The Art of Problem Posing*, Hillsdale, N.J.; Lawrence Erlbaum, 1983.

Page 18 *easy to visualize.* ... Marion Walter, "On Constructing Deltahedra," *Wiskobas Bulletin, Jaargang 5/6*, I.O.W.O., Utrecht, Aug. 1976.

Page 19 *may be ordered* ... from the Association of Teachers of Mathematics (address above).

Page 28 *hexagonal kaleidocycle* ... D. Schattschneider and W. Walker, *M. C. Escher Kaleidocycles*, Pomegranate Communications, Petaluma, CA; Taschen, Cologne, Germany.

Page 33 *we suggest you try it* ... If you get stuck, see Jim Morey, YouTube: theAmatour's channel, http://www.youtube.com/user/theAmatour, instructional videos of balloon polyhedra since November 6, 2007. See also http://moria.wesleyancollege.edu/faculty/morey/.

Page 34 *a practical construction* ... see Morey above, and also Jeremy Shafer, "Icosahedron balloon ball," *Bay Area Rapid Folders Newsletter*, Summer/Fall 2007, and Rolf Eckhardt, Asif Karim, and Marcus Rehbein, "Balloon molecules," http://www.balloonmolecules.com/, teaching molecular models from balloons since 2000. First described in *Nachrichten aus der Chemie* 48:1541–1542, December 2000. The German website, http://www.ballonmolekuele.de/, has more information including actual constructions.

Page 37 *family of computationally intractable "NP-complete" problems* ... Michael R. Garey and David S. Johnson, *Computers and Intractability: A Guide to the Theory of NP-Completeness*, W. H. Freeman & Co., 1979.

Page 37 *the graph's bloon number* ... Erik D. Demaine, Martin L. Demaine, and Vi Hart. "Computational balloon twisting: The theory of balloon polyhedra," in *Proceedings of the 20th Canadian Conference on Computational Geometry*, Montréal, Canada, August 2008.

Page 37 *tetrahedron requires only two balloons* ... see Eckhardt et al., above.

Page 37 *a snub dodecahedron from thirty balloons* ... see Mishel Sabbah, "Polyhedrons: Snub dodecahedron and snub truncated icosahedron," Balloon Animal Forum posting,

March 7, 2008, http://www.balloon-animals. com/forum/index.php?topic=931.0. See also the YouTube video http://www.youtube.com/watch? v=Hf4tfoZC_L4.

Page 37 *Preserves all or most of the symmetry of the polyhedron* ... see Demaine et al., above.

Page 37 *family of difficult "NP-complete" problems* ... see Demaine et al., above.

Page 37 *regular polylinks* ... Alan Holden, *Orderly Tangles: Cloverleafs, Gordian Knots, and Regular Polylinks*, Columbia University Press, 1983.

Page 39 *the six-pentagon tangle and the four-triangle tangle* ... see George W. Hart. "Orderly tangles revisited," in *Mathematical Wizardry for a Gardner*, pages 187–210, 2009, and http://www.georgehart.com/orderly-tangles-revisited/tangles.htm.

Page 37 *computer software to find the right thicknesses of the pieces* ... see Hart, above.

Page 39 *polypolyhedra* ... Robert J. Lang. "Polypolyhedra in origami," in *Origami*[3]: *Proceedings of the 3rd International Meeting of Origami Science, Math, and Education*, pages 153–168, 2002, and http://www.langorigami. com/science/polypolyhedra/polypolyhedra.php4.

Page 39 *balloon twisting* ... Carl D. Worth. "Balloon twisting," http://www.cworth. org/balloon_twisting/, June 5, 2007, and Demaine et al., above.

Page 39 *video instructions* ... see, in particular, Morey (above) and Eckhardt (above).

Notes and References for Chapter 3

Page 42 *a very nice book on the history of these things* ... B. L. van der Waerden *Scientific Awakening, rev. ed* (Leiden: Noordhoff International Publishing; New York, Oxford University Press, 1974).

Page 42 *Federico gives the details of this story.* ... P. J. Federico, *Descartes on Polyhedra*, (New York, Springer-Verlag, 1982).

Page 44 *very nicely described in The Sleepwalkers by Arthur Koestler* ... A. Koestler *The Watershed* [from Koestler's larger book, *The Sleepwalkers*] (Garden City, NY, Anchor Books, 1960).

Page 45 *In his little book Symmetry,* ... H. Weyl, *Symmetry* Princeton, NJ, Princeton University Press, 1952.

Page 45 *Felix Klein did a hundred years ago* ... F. Klein, *Lectures on the Icosahedron*, English trans., 2nd ed., New York, Dover Publications, 1956.

Page 51 *may be any number except one.* ... Grünbaum and T. S. Motzkin "The Number of Hexagons and the Simplicity of Geodesics on Certain Polyhedra," *Canadian Journal of Mathematics* 15, (1963): 744–51.

Notes and References for Chapter 4

Page 53 *two papyri* ... Gillings, R. J., *Mathematics in the Time of the Pharaohs*. Cambridge, Mass.: MIT Press, 1972.

Page 54 *volume of the frustrum of a pyramid* ... Gillings, R. J., "The Volume of a Truncated Pyramid in Ancient Egyptian Papyri," *The Mathematics Teacher 57* (1964):552–55.

Page 55 *singling them out for study* ... William C. Waterhouse, "The Discovery of the Regular Solids," *Archive for History of Exact Sciences 9* (1972):212–21.

Page 55 *By the time we get to Euclid* ... Heath, T. L. *The Thirteen Books of Euclid's Elements.* Cambridge, England: Cambridge University Press, 1925; New York: Dover Publications, 1956.

Page 56 *"semiregular" solids discovered by Archimedes* ... Pappus. *La Collection Mathématique.* P. Ver Eecke, trans. Paris: De Brouwer/Blanchard, 1933. See also Como, S., *Pappus of Alexandria and the Mathematics of Late Antiquity*, New York: Cambridge U. Press, 2000.

Page 56 *They may date from the Roman period* ... Charrière, "Nouvelles Hypothèses sur les dodécaèdres Gallo-Romains." *Revue Archéologique de l'Est et du Centre-Est 16* (1965):143–59. See also Artmann, Benno, "Antike Darstellungen des Ikosaeders [Antique representations of the icosahedron]," Mitt. Dtsch. Math.-Ver. 13 (2005): 46–50.

Page 56 *that have been cited by scholars* ... Conze, *Westdeutsche Zeitschrift für Geschichte und Kunst 11* (1892):204–10; Thevenot, E., "La mystique des nombres chez les Gallo-Romains, dodécaèdres boulets et taureauxtricomes," *Revue Archéolgique de l'Est et du Centre-Est 6* (1955):291–95.

Page 56 *they were used as surveying instruments* ... Friedrich Kurzweil, "Das Pentagondodekaeder des Museum Carnuntinun und seine Zweckbestimmung," *Carnuntum Jahrbuch 1956* (1957):23–29.

Page 56 *the best guess is that they were candle holders* ... F. H. Thompson, "Dodecahedrons Again," *The Antiquaries Journal 50,* pt. I (1970):93–96.

Page 56 *A large collection of such polyhedral objects* ... F. Lindemann, "Zur Geschichte der Polyeder und der Zahlzeichen," *Sitzungsberichte der Mathematisch-Physikalische Klasse der Koeniglich Akademie der Wissenschaften [Munich] 26* (1890):625–783 and 9 plates.

Page 57 *situation during the Renaissance is extremely complicated* ... Schreiber, Peter and Gisela Fischer, Maria Luise Sternath, "New light on the rediscovery of the Archimedean solids during the Renaissance." *Archive for History of Exact Sciences* 62 2008:457–467.

Page 57 *Pacioli actually made polyhedron models* ... Pacioli, L. *Divina Proportione,* Milan: 1509; Milan: Fontes Ambrosioni, 1956.

Page 57 *related to the emergent study of perspective* ... Descarques, P. *Perspective,* New York: Van Nostrand Reinhold, 1982.

Page 57 *Professor Coxeter discusses* ... Coxeter, H. S. M., "Kepler and Mathematics," Chapter 11.3 in Arthur Beer and Peter Beer, eds., *Kepler: Four Hundred Years, Vistas in Astronomy,* vol. 18 (1974), pp. 661–70.

Page 57 *Dürer made nets for the dodecahedron* ... Dürer, Albrecht, *Unterweysung der Messung mit dem Zyrkel und Rychtscheyd,* Nürnberg: 1525.

Page 58 *two "new" "regular" (star) polyhedra* ... Field J. V., "Kepler's Star Polyhedra," *Vistas in Astronomy 23* (1979):109–41.

Page 58 *Kepler's work is constantly cited* ... Kepler, Johannes, *Harmonices Mundi,* Linz, 1619; *Opera Omnia,* vol. 5, pp. 75–334, Frankfort: Heyden und Zimmer, 1864; M. Caspar, ed., *Johannes Kepler Gesammelte Werke,* vol. 6, Munich: Beck, 1938.

Page 58 *the papers of Lebesgue* ... Lebesgue, H., "Remarques sur les deux premières demonstrations du théorème d'Euler, rélatif aux polyèdres," *Bulletin de la Société Mathématique de France 52* (1924):315–36.

Page 58 *the same statement in Pólya* ... George Pólya, *Mathematical Discovery* (New York: John Wiley and Sons, 1981), combined ed., vol. 2, p. 154.

Page 58 *nice paper by Peter Hilton and Jean Pedersen* ... P. Hilton and J. Pedersen, "Descartes, Euler, Poincaré, Pólya and Polyhedra," *L'Enseignment Mathématique 27* (1981):327–43.

Page 58 *an extensive summary of this debate* ... P. J. Federico, *Descartes on Polyhedra.* New York: Springer-Verlag, 1982.

Page 58 *evident as they were, I had not perceived them* ... Jacques Hadamard, *An Essay on the Psychology of Invention in the Mathematical Field* (Princeton, N.J.: Princeton University Press, 1945), 51.

Page 59 *elongated square gyrobicupola (Johnson solid J37)* ... Norman W. Johnson, "Convex Solids with Regular Faces", *Canadian Journal of Mathematics,* 18, 1966, pages 169–200.

Page 59 *we can continue to honor his memory* ... Grünbaum, Branko, "An Enduring Error," *Elemente der Mathematik* 64 (2009):89–101.

Page 59 *Euler, although he found his formula, was not successful in proving it* ... Euler, Leonhard, "Elementa Doctrinae Solidorum," *Novi Commentarii Academiae Scientiarum Petropolitanae 4* (1752–53):

109–40 (*Opera Mathematica,* vol. 26, pp. 71–93); Euler, Leonhard, "Demonstratio Nonnullarum Insignium Proprietatus Quibus Solida Hedris Planis Inclusa Sunt Praedita," *Novi Commentarii Academiae Scientiarum Petropolitanae 4* (1752–53):140–60 (*Opera Mathematica,* vol. 26, pp. 94–108).

Page 59 *The first proof was provided by Legendre* ... Legendre, A., *Éléments de géometrie.* 1st ed. Paris: Firmin Didot, 1794; or to him: there are claims that Meyer Hirsch gave a correct proof (Hirsch, M. *Sammlung Geometrischer Aufgaben,* part 2, Berlin, 1807) prior to Legendre (Malkevitch, Joseph, "The First Proof of Euler's Theorem," *Mitteilungen, Mathematisches Seminar, Universität Giessen 165* (1986):77–82.) I believe that this is an error that came about due to a misreading (Steinitz, Ernst, and H. Rademacher, *Vorlesungen über die Theorie des Polyeder,* Berlin: Springer-Verlag, 1934) of something in Max Brückner's book, Brückner, M, *Vielecke und Vielfläche,* Leipzig: Teubner, 1900.

Page 59 *providing different proofs* ... Lakatos, I., *Proofs and Refutations.* New York: Cambridge University Press, 1976; see also Richeson, D. S., *Euler's Gem,* Princeton: Princeton U. Press, 2008.

Page 60 *Poinsot's work on the star polyhedra* ... Poinsot, L., "Mémoire sur les polygones et les polyèdres," *Journal de l'École Polytechnique 10* (1810):16–48; Poinsot, L, "Note sur la théorie des polyèdres," *Comptes Rendus 46* (1858):65–79.

Page 60 *Other contributors* ... Cauchy, A. L, "Sur les polygones et les polyédres," *J. de l'École Polytechnique 9* (1813):87–98; J. Bertrand (1858), and Cayley, A., "On Poinsot's Four New Regular Solids," *The London, Edinburgh, and Dublin Philosophical Magazine and Journal of Science,* ser. 4, *17* (1859): 123–28.

Page 60 *Cauchy made major contributions* ... Cauchy, A. L, "Recherches sur les polyèdres," *Journal de l'École Imp. Polytechnique 9* (1813):68–86; *Oeuvres Complètes,* ser. 2, vol. 1, pp. 1–25, Paris: 1905. See also Cauchy's paper cited above.

Page 58 *duals of the Archimedean polyhedra* ... Catalan, M. E., "Mémoire sur la théorie des polyèdres," *Journal de l'École Imp. Polytechnique 41* (1865):1–71.

Page 60 *Max Brückner published* ... see Brückner, above.

Page 60 *a paper in 1905 of D. M. Y. Sommerville* ... Sommerville, D. M. Y., "Semiregular Networks of the Plane in Absolute Geometry," *Transactions of the Royal Society* [Edinburgh] *41* (1905):725–47 and plates I–XII.

Page 61 *challenges to mathematics in the future* ... Hilbert, David, Mathematical problems, Lecture delivered at the Intern. Congress of Mathematicians at Paris in 1900, *Bull. Amer. Math. Soc.* (N.S) 37(2000), no. 4, 407–436, reprinted from *Bull. Amer. Math. Soc.* 8 (1902), 437–479.

Page 61 *cutting a polyhedron into pieces and assembling them into another polyhedron* ... Boltianskii, V., *Hilbert's Third Problem,* New York: John Wiley and Sons, 1975.

Page 61 *these pieces can be assembled* ... Boltyanskii, V., *Equivalent and Equidecomposable Figs.,* Boston: D.C. Heath, 1963.

Page 61 *extended Dehn's work* ... see Hadwiger, Hugo and Paul Glur, "Zerlegungsgleichheit ebener Polygone," *Elemente der Mathematik* 6 1951:97106.

Page 61 *the theory of equidecomposability* ... Rajwade, A. R., *Convex Polyhedra with Regularity Conditions and Hilbert's Third Problem,* New Delhi: Hindustan Book Agency, 2001.

Page 61 *can not be decomposed, using existing vertices, into tetrahedra* ... Schoenhardt, E., "Uber die Zerlegung von Dreieckspolyedern in Tetraeder," *Mathematische Annalen,* 98 1928:309–312; Toussaint, Godfried T. and Clark Verbrugge, Cao An Wang, Binhai Zhu, "Tetrahedralization of Simple and Non-Simple Polyhedra," *Proceedings of the Fifth Canadian Conference on Computational Geometry,* 1993:24–29.

Page 61 *Lennes polyhedra* ... Lennes, N.J., "On the simple finite polygon and polyhedron," *Amer. J. Math.* 33 (1911): 37–62.

Page 61 *the most important early twentieth-century contributor to the theory of polyhedra* ... Steinitz, Ernst, "Polyeder und Raumeinteilungen," in W. F. Meyerand and H. Mohrmann, eds., *Encyklopädie der mathematischen Wissenschaften,* vol. 3. Leipzig: B. G. Teubner, 1914–31.

Page 61 *Lennes polyhedra ...* see Rademacher, above.

Page 62 *if one starts with a* 1 × 1 *square* ... see Alexander, Dyson, and O'Rourke, "The convex polyhedra foldable from a square," *Proc. 2002 Japan Conference on Discrete Computational Geometry. Volume 2866, Lecture Notes in Computer Science*, pp. 38–50, Berlin: Springer, (2003). Good sources of relatively recent work on a wide variety of aspects of polyhedra (convex and non-convex) from a combinatorial point of view (e.g rigidity, Bellow's conjecture, nets (folding to polyhedra), and unfolding algorithms, Steinitz's Theorem (including the circle packing approach to proving this seminal theorem, as well as the dramatic extension due to the late Oded Schramm)) are Geometric Folding Algorithms: Linkages, Origami, and Polyhedra by Eric Demaine and Joseph O'Rourke (Demaine, E. and J. O'Rourke, *Geometric Folding Algorithms*, New York: Cambridge, U. Press. 2007) and Pak, I. *Lectures on Discrete and Polyhedral Geometry* (to appear).

Page 62 *the stellated icosahedra ...* Coxeter, H. S. M., P. du Val, H. T. Flather, and J. F. Petrie, *The Fifty-nine Icosahedra,*. New York: Springer-Verlag, 1982 (reprint of 1938 edition.)

Page 62 *famous work on uniform polyhedra* ... Coxeter, H. S. M., M. S. Longuet-Higgins, and J. C. P. Miller, *Uniform Polyhedra, Philosophical Transactions of the Royal Society* [London], sec. A, *246* (1953/56):401–50; Skilling, J., "The Complete Set of Uniform Polyhedra," *Philosophical Transactions of the Royal Society* [London], ser. A, *278*, (1975):111–35.

Page 62 *many important results in several branches of mathematics ...* Coxeter, H. S. M. *Regular Polytopes.* 3rd ed. New York: Dover Publications, 1973; Coxeter, H. S. M., *Regular Complex Polytopes.* London: Cambridge University Press, 1974.

Page 62 *O. Rausenberger shows ...* Rausenberger, O., "Konvex pseudoregulare Polyeder." *Zeitschr. fur math. u. maturwiss.* Utr-

erricht 1915:135–142. Probably independently discovered by Hans Freudenthal and B. L. van der Waerden, "Over een bewering van Euclides," *Simon Stevin* 25 (1946/47) 115–121.

Page 62 *Johnson, Grünbaum, V. A. Zalgaller et al. prove Johnson's conjecture.* ... Grünbaum, Branko, and N. W. Johnson. "The Faces of a Regular-faced Polyhedron," *Journal of the London Mathematical Society 40* (1965):577–86; Johnson, N. W., "Convex Polyhedra with Regular Faces." *Canadian Journal of Mathematics 18* (1966):169–200; Zalgaller, V. A., *Convex Polyhedra with Regular Faces,* New York: Consultants Bureau, 1969. See also Cromwell, P., *Polyhedra,* New York: Cambridge U. Press, 2008.

Page 63 *his beautiful book ...* Grünbaum, Branko. *Convex Polytopes.* New York: John Wiley and Sons, 1967. Second Edition, New York: Springer, 2003.

Page 63 *he published an article in 1977* ... Grünbaum, Branko, "Regular Polyhedra–Old and New." *Aequationes Mathematicae 16* (1977):1–20.

Page 63 *list of regular polyhedra that Grünbaum gave is complete ...* Dress, Andreas W. M., "A Combinatorial Theory of Grünbaum's New Regular Polyhedra. Part I. Grünbaum's New Regular Polyhedra and Their Automorphism Groups," *Aequationes Mathematicae 23* (1981):252–65; Dress, Andreas W. M., "A Combinatorial Theory of Grünbaum's New Regular Polyhedra, Part II. Complete Enumeration," *Aequationes Math.* 29 (1985) 222–243.

Notes and References for Chapter 5

Page 67 *Dividing Equation 5.1 by* $2E$ *and then substituting ...* The division of both sides of Equation 5.1 by the summation is based on the assumption that r_{av} and n_{av} remain finite even in the limit of infinite number of vertices and edges. The detailed justification of this assumption is beyond the scope of this paper, but it is limited to surfaces in which the density of vertices and edges (averaged over a region whose area is large compared to that of a face) is reasonably uniform throughout the structure. Obviously this assumption is valid for periodically repeating structures.

Page 68 *the number of hexagons can be any positive integer except 1* ... B. Grünbaum and T.S. Motzkin, "The Number of Hexagons and the Simplicity of Geodesics on Certain Polyhedra," *Canadian Journal of Mathematics 15* (1963): 744–51.

Page 68 *Equation 5.6 still yields an enumerable set of solutions* ... Arthur L. Loeb, *Space Structures: Their Harmony and Counterpoint*, Reading, Mass. Addison-Wesley, Advanced Book Program, 1976.

Notes and References for Chapter 6

Page 77 *masterwork on geometry* ... Albrecht Dürer. *The painter's manual: A manual of measurement of lines, areas, and solids by means of compass and ruler assembled by Albrecht Dürer for the use of all lovers of art with appropriate illustrations arranged to be printed in the year MDXXV*, New York: Abaris Books, 1977, 1525, English translation by Walter L. Strauss of 'Unterweysung der Messung mit dem Zirkel un Richtscheyt in Linien Ebnen uhnd Gantzen Corporen'.

Page 79 *sharp mathematical question* ... Geoffrey C. Shephard, "Convex polytopes with convex nets," *Math. Proc. Camb. Phil. Soc.*, 78:389–403, 1975.

Page 79 *researchers independently discovered* ... Alexey S. Tarasov, "Polyhedra with no natural unfoldings," *Russian Math. Surv.*, 54(3):656–657, 1999; Marshall Bern, Erik D. Demaine, David Eppstein, Eric Kuo, Andrea Mantler, and Jack Snoeyink, "Ununfoldable polyhedra with convex faces," *Comput. Geom. Theory Appl.*, 24(2):51–62, 2003; Branko Grünbaum, "No-net polyhedra," *Geombinatorics*, 12:111–114, 2002.

Page 83 *number is exponential in the square root of the number of faces* ... Erik D. Demaine and Joseph O'Rourke, *Geometric Folding Algorithms: Linkages, Origami, Polyhedra*, Cambridge University Press, July 2007, http://www.gfalop.org.

Page 83 *every prismoid has a net* ... Joseph O'Rourke, Unfolding prismoids without overlap, Unpublished manuscript, May 2001.

Page 84 *three different proofs* ... Demaine and O'Rourke, above; Alex Benton and Joseph O'Rourke, Unfolding polyhedra via cut-tree truncation, In *Proc. 19th Canad. Conf. Comput. Geom.*, pages 77–80, 2007; Val Pincu, "On the fewest nets problem for convex polyhedra," In *Proc. 19th Canad. Conf. Comput. Geom.*, pages 21–24, 2007.

Page 85 *we describe here* ... (see Demaine and O'Rourke, above, for another).

Page 85 *This concept was introduced by Alexandrov* ... Aleksandr Danilovich Alexandrov, *Vupyklue Mnogogranniki. Gosydarstvennoe Izdatelstvo Tehno-Teoreticheskoi Literaturu*, 1950. In Russian; Aleksandr D. Alexandrov, *Convex Polyhedra*, Springer-Verlag, Berlin, 2005, Monographs in Mathematics. Translation of the 1950 Russian edition by N. S. Dairbekov, S. S. Kutateladze, and A. B. Sossinsky.

Page 85 *Alexandrov unfolding* ... Ezra Miller and Igor Pak, "Metric combinatorics of convex polyhedra: Cut loci and nonoverlapping unfoldings, *Discrete Comput. Geom.*," 39:339–388, 2008.

Page 85 *proved to be non-overlapping* ... Boris Aronov and Joseph O'Rourke, "Nonoverlap of the star unfolding," *Discrete Comput. Geom.*, 8:219–250, 1992.

Page 86 *when Q shrinks down to x* ... Jin-ichi Itoh, Joseph O'Rourke, and Costin Vîlcu, "Star unfolding convex polyhedra via quasigeodesic loops," *Discrete Comput. Geom.*, 44, 2010, 35–54.

Page 86 *This enticing problem remains unsolved today* ... See Demaine and O'Rourke, above.

Page 86 *unfolds any orthogonal polyhedron* ... Mirela Damian, Robin Flatland, and Joseph O'Rourke, "Epsilon-unfolding orthogonal polyhedra," *Graphs and Combinatorics*, 23[Suppl]:179–194, 2007. Akiyama-Chvátal Festschrift.

Page 86 *Unsolved problems abound in this topic* ... see Demaine and O'Rourke, Chapter 25, above; and Joseph O'Rourke, "Folding polygons to convex polyhedra," In Timothy V. Craine and Rheta Rubenstein, editors, *Understanding Geometry for a Changing World: National Council of Teachers of Mathematics, 71st Yearbook*, pages 77–87. National Council of Teachers of Mathematics, 2009.

Notes and References for Chapter 8

Page 123 *variations on a theme is really the crux of creativity* ... Douglas R. Hofstadter, "Metamagical Themas: Variations on a Theme As the Essence of Imagination," *Scientific American 247*, no. 4 (October 1982): 20–29.

Page 123 *You see things, and you say 'why?'* ... George Bernard Shaw, *Back to Methuselath*, London: Constable (1921).

Notes and References for Chapter 9

Page 126 *forms that are simultaneously mathematical and organic* ... George W. Hart, website, http://www.georgehart.com

Page 126 *black and white version became official* ... http://www.soccerballworld.com/History.htm accessed, June 2009.

Page 128 *particle detectors spherically arranged with the structure of GP(2,1)* ... N.G. Nicolis, "Polyhedral Designs of Detection Systems for Nuclear Physics Studies", *Meeting Alhambra*, Proceedings of *Bridges 2003*, J. Barrallo et al. eds., University of Granada, Granada, Spain, pp. 219–228.

Page 129 *spherical allotropes of carbon* ... Aldersey-Williams, Hugh, *The Most Beautiful Molecule: The Discovery of the Buckyball*. Wiley, 1995; P. W. Fowler, D. E. Manolopoulos, *An Atlas of Fullerenes*, Oxford University Press, 1995.

Page 129 *truncated icosahedron was already a well-known mathematical structure* ... see Aldersey-Williams, above.

Page 129 *If we want to have a structure, we have to have triangles* ... R. Buckminster Fuller, *Synergetics: Explorations in the Geometry of Thinking*, MacMillan, 1975, Section 610.12

Page 132 *the basic geometric ideas* ... see Clinton, J.D., 1971, *Advanced Structural Geometry Studies Part I - Polyhedral Subdivision Concepts for Structural Applications*, Washington, D.C.: NASA Tech. Report CR-1734., 1971; H.S.M. Coxeter, "Virus Macromolecules and Geodesic Domes," in *A Spectrum of Mathematics*, J.C. Butcher (editor), Aukland, 1971; J. Francois Gabriel, *Beyond the Cube: The Architecture of Space Frames and Polyhedra*, Wiley, 1997; Hugh Kenner, *Geodesic Math and How to Use It*, University of California Press, 1976; Magnus J. Wenninger, *Spherical Models*, Cambridge, 1979, (Dover reprint 1999).

Page 133 *In the optimal solution* ... Michael Goldberg, "The Isoperimetric Problem for Polyhedra," *Tohoku Mathematics Journal*, 40, 1934/5, pp. 226–236.

Page 133 *Details of one algorithm for the general (a,b) case* ... George W. Hart, "Reticulated Geodesic Constructions" *Computers and Graphics*, Vol 24, Dec, 2000, pp. 907–910.

Page 133 *the center of gravity of these tangency points is the origin* ... Gunter M. Ziegler, *Lectures on Polytopes*, Springer-Verlag, 1995, p. 118.

Page 133 *a simple fixed-point iteration* ... George W. Hart, "Calculating Canonical Polyhedra," *Mathematica in Education and Research*, Vol 6 No. 3, Summer 1997, pp. 5–10.

Page 134 *any desired number of different forms* ... Michael Goldberg, "A Class of Multi-Symmetric Polyhedra," *Tohoku Mathematics Journal*, 43, 1937, pp. 104–108.

Page 134 *Paper models ...* See Magnus J. Wenninger, "Artistic Tessellation Patterns on the Spherical Surface," *International Journal of Space Structures*, Vol. 5, 1990, pp. 247–253.

Page 135 *models far too tedious to cut with scissors ...* George W. Hart, "Modular Kirigami," *Proceedings of Bridges Donostia*, San Sebastian, Spain, 2007, pp. 1–8.

Page 135 *an equal-area projection ...* Snyder, John P. (1992), "An equal-area map projection for polyhedral globes," *Cartographica* 29: 10–21.

Page 137 *models of two-layer spheres ...* George W. Hart, Two-layer dome instructions, http://www.georgehart.com/MathCamp-2008/dome-two-layer.html

Notes and References for Chapter 10

Page 139 *nonperiodic arrangements exist in the solid state ...* Concerning crystals with 5-fold symmetry, see D. Shechtman, I. Gratias, and J. W. Cahn ["Metallic Phase with Long-range Orientational Order and No Translational Symmetry," *Physical Review Letters 53* (1984): 1951–53]. Their paper has changed our views of the solid state. For an overview of the rapidly-developing mathematical theory—including a role for polyhedra—see M. Senechal, *Quasicrystals and Geometry*, Cambridge University Press, corrected paperback edition, 1996; reprinted in 2009.

Page 142 *the methods are used for the description of crystal structures ...* see, for example, A. L. Loeb, "A Systematic Survey of Cubic Crystal Structures," *Journal of Solid State Chemistry I* (1970):237–67.

Page 142 *a different kind of basic unit ...* G. Lejeune Dirichlet, "Über die Reduction der positiven quadratischen Formen mit drei unbestimmten ganzen Zahlen," *Journal für die reine und angewandte Mathematik 40* (1850):209–27.

Page 145 *all cubic crystal structures ...* For details of the classification see Chung Chieh, "The Archimedean Truncated Octahedron, and Packing of Geometric Units in Cubic

Crystal Structures," *Acta Crystallographica*, sec. A, 35 (1979):946–52; "The Archimedean Truncated Octahedron II. Crystal Structures with Geometric Units of Symmetry $\bar{4}$3m," *Acta Crystallographica*, sec. A, 36(1980):819–26. "The Archimedean Truncated Octahedron III. Crystal Structures with Geometric Units of Symmetry m3m," *Acta Crystallographica*, sec. A, *38* (1982):346–49. Chung Chieh, Hans Burzlaff, and Helmuth Zimmerman, "Comments on the Relationship between 'The Archimedean Truncated Octahedron, and Packing of Geometric Units in Cubic Crystal Structures' and 'On the Choice of Origins in the Description of Space Groups'," *Acta Crystallographica*, sec. A, 38(1982):746–47.

Page 145 *crystal structure of a γ brass ...* M. H. Booth, J. K. Brandon, R.Y. Brizard, C. Chieh, and W. B. Pearson, "γ Brasses with F Cells," *Acta Crystallographica*, sec. B, 33 (1977):30–30 and references therein.

Page 146 *Dirichlet domains of tetragonal, rhombohedral, and hexagonal lattices ...* Chung Chieh, "Geometric Units in Hexagonal and Rhombohedral Space Groups," *Acta Crystallographica*, sec. A, 40(1984):567–71.

Page 148 *geometric plan of the tetragonal space groups ...* Chung Chieh, "Geometric Units in Tetragonal Crystal Structures," *Acta Crystallographica*, sec. A, 39 (1983):415–21.

Page 148 *a series of organometallic compounds ...* Chung Chieh, "Crystal Chemistry of Tetraphenyl Derivatives of Group IVB Elements," *Journal of the Chemical Society* [London] *Dalton Transactions* (1972):1207–1208.

Notes and References for Chapter 11

Page 153 *Coxeter has said ...* H.S.M Coxeter, "Preface to the First Edition," in *Regular Polytopes* 3rd ed., New York: Dover Publications, 1973.

Page 153 *termed polyhedral chemistry. ...* A more extensive discussion of the shapes and symmetries of molecules and crystals can be found in M. Hargittai and I. Hargittai, *Symmetry through the Eyes of a Chemist*, Third Edition, Springer, hardcover, 2009; softcover, 2010.

Page 153 *Muetterties movingly described his attraction . . .* Earl L. Muetterties, ed., *Boron Hydride Chemistry*, New York: Academic Press, 1975.

Page 154 *Two independent studies . . .* V.P Spiridonov and G. I. Mamaeva, "Issledovanie molekuli bordigrida tsirkoniya metodom gazovoi elektronografii," *Zhurnal Strukturnoi Khimii, 10* (1969), 133–35. Vernon Plato and Kenneth Hedberg, "An Electron-Diffraction Investigation of Zirconium Tetraborohydride, $Zr(BH_4)_4$," *Inorganic Chemistry 10* (1970), 590–94. In one interpretation (Spiridonov and Mamaeva, "Issledovanie molekuli bordigrida tsirkoniya metodom gazovoi elektronografii") there are four $Zr - B$ bonds; according to the other . . . there is no $Zr - B$ bond (Plato and Hedberg, "An Electron-Diffraction Investigation of Zirconium Tetraborohydride, $Zr(BH_4)_4$.")

Page 154 *sites are taken by carbon atoms . . .* see Muetterties, above.

Page 154 *removing one or more polyhedral sites . . .* Robert E. Williams, "Carboranes and Boranes; Polyhedra and Polyhedral Fragments," *Inorganic Chemistry 10* (1971), 210–14; Ralph W. Rudolph, "Boranes and Heteroboranes: A Paradigm for the Electron Requirements of Clusters?" *Accounts of Chemical Research 9* (1976), 446–52.

Page 155 *energetically most advantageous . . .* A. Greenberg and J.F. Liebman, *Strained Organic Molecules*, New York: Academic Press, 1978.

Page 155 *an attractive and challenging playground . . .* See the previous reference and Lloyd N. Ferguson, "Alicyclic Chemistry: The Playground for Organic Chemists," *Journal of Chemical Education 46*, (1969), 404–12.

Page 155 *amazingly stable . . .* Günther Maier, Stephan Pfriem, Ulrich Shäfer, and Rudolph Matush, "Tetra-*tert*-butyltetrahedrane," *Angewandte Chemie, International Edition in English 17* (1978), 520–21.

Page 155 *known for some time . . .* Philip E. Eaton and Thomas J. Cole, "Cubane," *Journal of the American Chemical Society 86* (1964), 3157–58.

Page 155 *prepared more recently . . .* Robert J. Ternansky, Douglas W. Balogh, and Leo A. Paquette, "Dodecahedrane," *Journal of the American Chemical Society 104* (1982), 4503–04.

Page 155 *was predicted . . .* H.P. Schultz, "Topological Organic Chemistry. Polyhedranes and Prismanes," *Journal of Organic Chemistry 30* (1965), 1361–64

Page 157 *Triprismane . . .* Thomas J. Katz and Nancy Acton, "Synthesis of Prismane," *Journal of the American Chemical Society 95* (1973), 2738–39; Nicholas J. Turro, V. Ramamurthy, and Thomas J. Katz, "Energy Storage and Release. Direct and Sensitized Photoreactions of Dewar Benzene and Prismane," *Nouveau Journal de Chimie1* (1977), 363–65.

Page 157 *and pentaprismane . . .* Philip E. Eaton, Yat Sun Or, and Stephen J. Branca, "Pentaprismane," *Journal of the American Chemical Society 103* (1981), 2134–36.

Page 157 *only a few are stable . . .* Dan Farcasiu, Erik Wiskott, Eiji Osawa, Wilfried Thielecke, Edward M. Engler, Joel Slutsky, Paul v. R. Schleyer, and Gerald J. Kent, "Ethanoadamantane. The Most Stable $C_{12}H_{18}$ Isomer," *Journal of the American Chemical Society 96*(1974), 4669–71.

Page 157 *The name iceane was proposed by Fieser . . .* Louis F. Fieser, "Extensions in the Use of Plastic Tetrahedral Models," *Journal of Chemical Education 42* (1965), 408–12.

Page 158 *The challenge was met. . . .* Chris A. Cupas and Leonard Hodakowski, "Iceane," *Journal of the American Chemical Society 96* (1974), 4668–69.

Page 158 *Diamond has even been called . . .* Chris Cupas, P. von R. Schleyer, and David J. Trecker, "Congressane," *Journal of the American Chemical Society 87* (1965) 917–18. Tamara M. Gund, Eiji Osawa, Van Zandt Williams, and P. v R. Schleyer, "Diamantane. I. Preparation of Diamantane. Physical and Spectral Properties." *Journal of Organic Chemistry 39* (1974) 2979–87. The high symmetry of adamantane is emphasized when its structure is described by four imaginary cubes: Here it is assumed that the two different kinds of C – H bonds in adamantane have equal length. See I. Hargittai and K. Hedberg, in S.J Cyvin, ed., *Molecular Structures and Vibrations* (Amsterdam: Elsevier, 1972).

Page 159 *Joined tetrahedra are even more obvious* ... A. F. Wells, *Structural Inorganic Chemistry* 5th Edition (New York: Oxford University Press, 1984).

Page 159 *similar consequences in molecular shape and symmetry* ... M. Hargittai and I. Hargittai, *The Molecular Geometries of Coordination Compounds in the Vapour Phase* (Budapest: Akadémiai Kiadó; Amsterdam: Elsevier, 1977).

Page 159 *The* $H_3N \cdot AlCl_3$ *donor–acceptor complex* ... Magdolna Hargittai, István Hargittai, Victor P. Spiridonov, Michel Pelissier, and Jean F. Labarre, "Electron Diffraction Study and CNDO/2 Calculations on the Complex of Aluminum Trichloride with Ammonia," *Journal of Molecular Structure 24* (1975), 27–39.

Page 160 *the model with two halogen bridges* ... E. Vajda, I. Hargittai, and J. Tremmel, "Electron Diffraction Investigation of the Vapour Phase Molecular Structure of Potassium Tetrafluoro Aluminate," *Inorganic Chimica Acta 25* (1977), L143–L145.

Page 160 *the barrier to rotation and free-energy difference* ... A. Haaland and J.E. Nilsson, "The Determination of Barriers to Internal Rotation by Means of Electron Diffraction. Ferrocene and Ruthenocene," *Acta Chemica Scandinavica 22* (1968), 2653–70.

Page 160 *structures with beautiful and highly symmetric polyhedral shapes* ... F. A. Cotton and R. A. Walton, *Multiple Bonds between Metal Atoms*, New York: Wiley-Interscience, 1982.

Page 160 *One example is* ... F. A. Cotton and C. B. Harris, "The Crystal and Molecular Structure of Dipotassium Octachlorodirhenate(III) Dihydrate, $K_2[Re_2Cl_8] \cdot 2H_2O$," *Inorganic Chemistry 4* (1965), 330–33; V.G. Kuznetsov and P. A. Koz'min, "A Study of the Structure of (PyH)HReCl_4," *Zhurnal Strukturnoi Khimii 4* (1963), 55–62.

Page 160 *the paddlelike structure of dimolybdenum tetra-acetate* ... M. H. Kelley

and M. Fink, "The Molecular Structure of Dimolybdenum Tetra-acetate," *Journal of Chemical Physics 76* (1982), 1407–16.

Page 160 *hydrocarbons called paddlanes* ... E. H. Hahn, H. Bohm, and D. Ginsburg, "The Synthesis of Paddlanes: Compounds in Which Quaternary Bridgehead Carbons Are Joined by Four Chains," *Tetrahedron Letters* (1973), 507–10.

Page 160 *in which interactions between bridgehead carbons have been interpreted* ... Kenneth B. Wiberg and Frederick H. Walker, "[1.1.1] Propellane," *Journal of the American Chemical Society 104* (1982), 5239–40; James E. Jackson and Leland C. Allen, "The C_1–C_3 Bond in [1.1.1] Propellane," *Journal of the American Chemical Society 106* (1984), 591–99.

Page 160 *distances are equal within experimental error* ... Vernon Plato, William D. Hartford, and Kenneth Hedberg, "Electron-Diffraction Investigation of the Molecular Structure of Trifluoramine Oxide, F_3NO," *Journal of Chemical Physics 53* (1970), 3488–94.

Page 160 *nonbonded distances* ... István Hargittai, "On the Size of the Tetra-fluoro 1,3-dithietane Molecule," *Journal of Molecular Structure 54* (1979), 287–88.

Page 161 *remarkably constant at 2.48Å* ... István Hargittai, "Group Electronegativities: as Empirically Estimated from Geometrical and Vibrational Data on Sulphones," *Zeitschrift für Naturforschung*, part Λ 34 (1979), 755–60.

Page 161 *depending on the X and Y ligands* ... I. Hargittai, *The Structure of Volatile Sulphur Compounds* (Budapest: Akadémiai Kiadó; Dordrecht: Reidel, 1985).

Page 162 *the SO bond distances differ by 0.15Å* ... Robert L. Kuczkowski, R. D. Suenram, and Frank J. Lovas "Microwave Spectrum, Structure and Dipole Moment of Sulfuric Acid," *Journal of the American Chemical Society 103* (1981), 2561–66.

Page 162 *such molecules are nearly regular tetrahedral* ... K. P. Petrov, V. V. Ugarov, and N. G. Rambidi, "Elektronograficheskoe issledovanie stroeniya molekuli Tl_2SO_4," *Zhurnal Strukturnoi Khimii 21* (1980), 159–61.

Page 162 *simple and successful model ...* R. J. Gillespie, *Molecular Geometry*, New York: Van Nostrand Reinhold, 1972.

Page 163 *nicely simulated by balloons ...* H. R. Jones and R. B. Bentley, "Electron-Pair Repulsions, A Mechanical Analogy," *Proceedings of the Chemical Society* [London]. (1961), 438–440; *... nut clusters on walnut trees* see Gavril Niac and Cornel Florea, "Walnut Models of Simple Molecules," *Journal of Chemical Education 57* (1980), 4334–35.

Page 165 *attention to angles of lone pairs ...* I. Hargittai, "Trigonal-Bipyramidal Molecular Structures and the VSEPR Model," *Inorganic Chemistry 21* (1982), 4334–35.

Page 165 *results from quantum chemical calculations ...* Ann Schmiedekamp, D. W. J. Cruickshank, Steen Skaarup, Péter Pulay, István Hargittai, and James E. Boggs, "Investigation of the Basis of the Valence Shell Electron Pair Repulsion Model by Ab Initio Calculation of Geometry Variations in a Series of Tetrahedral and Related Molecules," *Journal of the American Chemical Society 101* (1979), 2002–10; P. Scharfenberg, L. Harsányi, and I. Hargittai, unpublished calculations, 1984.

Page 165 *equivalent to a rotation ...* R. S. Berry, "The New Experimental Challenges to Theorists," in R. G. Woolley, ed., *Quantum Dynamics of Molecules*, New York: Plenum Press, 1980.

Page 165 *lifetime of a configuration and the time scale of the investigating technique ...* E. L Muetterties, "Stereochemically Nonrigid Structures," *Inorganic Chemistry 4* (1965), 769–71.

Page 165 *experimentally determined structures occur ...* Zoltan Varga, Maria Kolonits, Magdolna Hargittai, "Gas-Phase Structures of Iron Trihalides: A Computational Study of all Iron Trihalides and an Electron Diffraction Study of Iron Trichloride." *Inorganic Chemistry 43* (2010):1039–45; Zoltan Varga, Maria Kolonits, Magdolna Hargittai, "Iron dihalides: structures and thermodynamic properties from computation and an electron diffraction study of iron diiodide" *Structural Chemistry*, 22 (2011) 327–36; Magdolna Hargittai, "Metal Halide Molecular

Structures" *Chemical Reviews* 100 (2000): 2233–301.

Page 165 *different kinds of carbon positions ...* W. v. E. Doering and W. R. Roth, "A Rapidly Reversible Degenerate Cope Rearrangement. Bicyclo[5.1.0]octa-2, 5-diene," *Tetrahedron 19* (1963), 715–737; G. Schroder, "Preparation and Properties of Tricyclo[3, 3, 2, $0^{4,6}$]deca-2, 7, 9-triene (Bullvalene)," *Angewandte Chemie, International Edition in English 2* (1963), 481–82; Martin Saunders, "Measurement of the Rate of Rearrangement of Bullvalene," *Tetrahedron Letters* (1963), 1699–1702.

Page 165 *hydrocarbon whose trivial name was chosen ...* J. S. McKennis, Lazaro Brener, J. S. Ward, and R. Pettit, "The Degenerate Cope Rearrangement in Hypostrophene, a Novel $C_{10}H_{10}$ Hydrocarbon," *Journal of the American Chemical Society 93* (1971), 4957–58.

Page 166 *discovered by Berry ...* R. Stephen Berry, "Correlation of Rates of Intramolecular Tunneling Processes, with Application to Some Group V Compounds," *Journal of Chemical Physics 32* (1960), 933–38.

Page 167 *similar pathway was established ...* George M. Whitesides and H. Lee Mitchell, "Pseudorotation in $(CH_3)_2NPF_4$," *Journal of the American Chemical Society 91* (1969), 5384–86.

Page 167 *permutation of nuclei in five-atom polyhedral boranes ...* Earl L. Muetterties and Walter H. Knoth, *Polyhedral Boranes*, New York: Marcel Dekker, 1968. In one mechanism (W. N. Lipscomb, "Framework Rearrangement in Boranes and Carboranes," *Science 153* (1966), 373–78); illustrated by interconversion (R. K. Bohn and M. D Bohn, "The Molecular Structure of 1,2-,1,7-, and 1,12-Dicarba-*closo*-dodecaborane(12), $B_{10}C_2H_{12}$," *Inorganic Chemistry 10* (1971), 350–55.). A similar model has been proposed (Brian F. G. Johnson and Robert E. Benfield, "Structures of Binary Carbonyls and Related Compounds. Part 1. A New Approach to Fluxional Behaviour," *Journal of the Chemical Society* [London] *Dalton Transactions* (1978), 1554–68.)

Page 168 *even in the solid state ...* Brian E. Hanson, Mark J. Sullivan, and Robert J. Davis,

"Direct Evidence for Bridge-Terminal Carbonyl Exchange in Solid Dicobalt Octacarbonyl by Variable Temperature Magic Angle Spinning ^{13}C NMR Spectroscopy," *Journal of the American Chemical Society 106* (1984), 251–53.

Page 168 *polyhedron whose vertices are occupied by carbonyl oxygens* ... Robert E. Benfield and Brian F. G. Johnson, "The Structures and Fluxional Behaviour of the Binary Carbonyls; A New Approach. Part 2. Cluster Carbonyls $M_m(CO)_n$ (n = 12, 13, 14, 15, or 16)," *Journal of the Chemical Society* [London] *Dalton Transactions* (1980), 1743–67.

Page 168 *Kepler's planetary system* ... Johannes Kepler, *Mysterium Cosmographicum*, 1595.

Page 168 *observed in mass spectra* ... E. A. Rohlfing, D. M. Cox, and A. Kaldor, "Production and Characterization of Supersonic Carbon Cluster Beams," *Journal of Chemical Physics 81* (1984), 3322–3330; the structure was assigned to it (H. W. Kroto, J. R. Heath, S. C. O'Brien, R. F. Curl, and R. E. Smalley, "C_{60}:Buckminsterfullerene," *Nature 318* (1985), 162–163); produced for the first time (W. Krätschmer, L. D. Lamb, K. Fostiropoulos, and D. R. Huffman, "Solid C_{60}: a new form of carbon," *Nature 347* (1990), 354–358).

Notes and References for Chapter 12

Page 172 *Lucretius (99–50 B.C.E.) argued that motion is impossible* ... Titus Lucretius Carus, *De Rerum Natura:* book I, lines 330–575; book II, lines 85-308; see Cyril Bailey, trans., *Lucretius on the Nature of Things* (Oxford: Oxford University Press, 1910), pp. 38–45, 68–75.

Page 173 *William Rankine (1820–1872) proposed a theory of molecular vortices* ... William John Macquorn Rankine, "On the Hypothesis of Molecular Vortices, or Centrifugal Theory of Elasticity, and Its Connexion with the Theory of Heat," *Philosophical Magazine*, ser. 4, *10* (1855):411.

Page 173 *Hermann von Helmholtz (1821–1894) derived mathematical expressions* ... H. Helmholtz, "On Integrals of the Hydrodynamical Equations, Which Express Vortex-Motion," *Philosophical Magazine*, ser. 4, *33* suppl. (1867):485; "from Crelle's *Journal*, LV (1858), kindly communicated by Professor Tait."

Page 173 *Lord Kelvin* ... Sir William Thomson, "On Vortex Atoms," *Philosophical Magazine*, ser. 4, *34* (1867):15.

Page 173 *a lecture demonstration of smoke rings by Tait* ... Cargill Gilston Knott, *Life and Scientific Work of Peter Guthrie Tait*, Cambridge, England: Cambridge University Press, 1911, p. 68.

Page 173 *Tait described an apparatus to produce smoke rings* ... P. G. Tait, *Lectures on Some Recent Advances in Physical Science*, London, 1876, p. 291.

Page 173 *investigations on the analytic geometry of knots* ... Peter Guthrie Tait, *Scientific Papers*, 2 vols., Cambridge University Press, 1898, I, pp. 270–347; papers originally published 1867–85.

Page 173 *his search after the true interpretation of the phenomena* ... J. C. Maxwell, "On Physical Lines of Force. Part II. —The Theory of Molecular Vortices applied to Electric Currents." *Philosophical Magazine*, ser. 4, *21* (1861): 281–91.

Page 174 *demonstrating that there is no aether* ... Loyd S. Swenson, Jr., "The Michelson-Morley-Miller Experiments before and after 1905," *The Journal for the History of Astronomy 1* (1970):56–78.

Page 174 *introduced into the mainstream of European chemistry* ... Joseph Le Bel, *Bulletin de la Société Chimique* [Paris], *22* (November 1874), 337–47; J. H. van 't Hoff, *The Arrangement of Atoms in Space*, 2nd ed., London: Longmans, Green, 1898.

Page 175 *Alfred Werner (1866–1919) used octahedra to model metal complexes* ... A. Werner, "Beitrag zur Konstitution anorganischer Verbindungen," *Zeitschrift für anorganische Chemie 3* (1893):267–330.

Page 175 *useful in teaching general chemistry* ... G. N. Lewis, *Valence and The Structure* *of Atoms and Molecules*, New York: The Chemical Catalog Company, 1923, p. 29.

Page 175 *developed the cubic model in a more formal manner* … G. N. Lewis, "The Atom and the Molecule," *Journal of the American Chemical Society 38*, (1916): 762–85.

Page 175 *60-carbon cluster molecule* … H. W. Kroto, J. R. Heath, S. C. O'Brien, R. F. Curl, and R. E. Smalley, "C_{60}: Buckminsterfullerene," *Nature 318* (1985): 162–63.

Page 176 *Francis Crick and James Watson suggested* … F. H. C. Crick and J. D. Watson, "Structure of Small Viruses," *Nature 177* (1956):473–75.

Page 176 *the design is embodied in the specific bonding properties of the parts* … D. L. D. Caspar and A. Klug. "Physical Principles in the Construction of Regular Viruses." *Cold Spring Harbor Symposia on Quantitative Biology 27* (1962):1–24.

Page 177 *a radical departure from the idea of quasi-equivalence* … I. Rayment, T. S. Baker, D. L. D. Caspar, and W. T. Murakami, "Polyoma Virus Capsid Structure at 22.5 Å Resolution," *Nature 295* (1982):110–15.

Page 178 *collection of plasticene balls* … See J. D. Bernal, "A Geometrical Approach to the Structure of Liquids," *Nature 183* (1959):141–47; "The Structure of Liquids," *Scientific American 203* (1960):124–34. See also J. L. Finney, "Random Packings and the Structure of Simple Liquids. I. The Geometry of Random Close Packing," *Proceedings of the Royal Society* [London], *sec. A, 319* (1970):479–93.

Page 178 *Stephen Hales (1677–1761) used this strategy* … Stephen Hales, *Vegetable Staticks,* London: W. & J. Innys, 1727. Reprint (Canton, Mass.: Watson, Neale, 1969).

Page 178 *Extending experimental methods used by Plateau* … Joseph A. F. Plateau, *Statique expérimentale et théorique des liquides soumis aux seules forces moléculaires* (Paris: Gand, 1873).

Page 179 *William Barlow (1845–1934) observed* … W. Barlow, "Probable Nature of the Internal Symmetry of Crystals," *Nature 29* (1883–84):186–88, 205–07, 404.

Page 179 *close-packing spheres with oriented binding sites* … Sir William Thomson, "Molecular Constitution of Matter," *Proceedings of the Royal Society of Edinburgh 16* (1888–89):693–724.

Page 179 *Pauling developed in detail the model of atoms-as-spheres* … Linus Pauling, "Interatomic Distances and Their Relation to the Structure of Molecules and Crystals," ch. 5 in *The Nature of the Chemical Bond* (Ithaca, N.Y.: Cornell University Press, 1939).

Page 179 *polyhera as models for plant cells* … Ralph O. Erickson, "Polyhedral Cell Shapes," in Grzegorz Rozenberg and Arto Salomaa, eds, *The Book of L* (Berlin: Spring-Verlag, 1986), pp. 111–124.

Page 179 *Erickson described experiments* … James W. Marvin. "Cell Shape Studies in the Pith of *Eupatorium purpureum*," *American Journal of Botany 26* (1939):487–504. A survey of other similiar work is given in Edwin B. Matzke and Regina M. Duffy, "The Three-Dimensional Shape of Interphrase Cells within the Apical Meristem of *Anacharis densa*," *American Journal of Botany 42* (1955): 937–45.

Page 179 *Previous investigators had taken Kelvin's tetrakaidecahedra* … Sir William Thomson. "On the Division of Space with Minimum Partitional Area." *The London, Edinburgh, and Dublin Philosophical Magazine and Journal of Science,* ser. 5, *24* (1887): 503–14.

Page 180 *In a series of Kepler-type experiments, Matzke and Marvin* … Edwin B. Matzke, "Volume-Shape Relationships in Lead Shot and Their Bearing on Cell Shapes," *American Journal of Botany 26* (1939):288–95; James W. Marvin, "The Shape of Compressed Lead Shot and its Relation to Cell Shape," *American Journal of Botany 26* (1939):280–88.

Page 180 *Matzke also used another model* … E. B. Matzke, "The Three-dimensional Shape of Bubbles in Foam—An Analysis of the Rôle of Surface Forces in Three-dimensional Cell Shape Determination," *American Journal of Botany 33.*

Page 184 *Dynamic computational geometry* ... Godfried T. Toussaint. "On Translating a Set of Polyhedra." McGill University School of Computer Science Technical Report No. SOCS-84.6, 1984. See also Godfried T. Toussaint, "Movable Separability of Sets," in G. T. Toussaint, ed., *Computational Geometry*, (Amsterdam: North-Holland, 1985), pp. 335–75.

Page 185 *translated without collisions* ... Godfried T. Toussaint, "Shortest Path Solves Translation Separability of Polygons," McGill University School of Computer Science Technical Report No. SOCS-85.27, 1985.

Page 185 *The answer is yes.* ... Binay K. Bhattacharya and Godfried T. Toussaint, "A Linear Algorithm for Determining Translation Separability of Two Simple Polygons," McGill University School of Computer Science Technical Report No. SOCS-86.1, 1986.

Page 185 *for any set of circles of arbitrary sizes* ... G. T. Toussaint, "On Translating a Set of Spheres," Technical Report SOCS-84.4, School of Computer Science, McGill University, March 1984.

Page 185 *This problem can be generalized* ... G. T. Toussaint, "The Complexity of Movement," *IEEE International Symposium on Information Theory*, St. Jovite, Canada, September 1983; J.-R. Sack and G. T. Toussaint, "Movability of Objects," *IEEE International Symposium on Information Theory*, St. Jovite, Canada, September 1983. G. T. Toussaint and J.-R. Sack, "Some New Results on Moving Polygons in the Plane," *Proceedings of the Robotic Intelligence and Productivity Conference*, Detroit: Wayne State University (1983) pp. 158–63.

Page 186 *no translation ordering exists for some directions.* ... L. J. Guibas and Y. F. Yao, "On Translating a Set of Rectangles." *Proceedings of the 12th Annual ACM Symposium on Theory of Computing*, Association for Computing Machinery Symposium on Theory of Computing, *Conference Proceedings, 12* (1980), pp. 154–160.

Page 186 *translated in any direction without disturbing the others.* ... Robert Dawson, "On Removing a Ball Without Disturbing the Others," *Mathematics Magazine 57*, no. 1 (1984): 27–30.

Page 186 *no one can be moved without disturbing the others.* ... K. A. Post, "Six Interlocking Cylinders with Respect to All Directions," unpublished paper, University of Eindhoven, The Netherlands, December 1983.

Page 187 *can be separated with a single translation* ... Toussaint, "On Translating a Set of Spheres."

Page 187 *star-shaped polyhedra in 3-space* ... Toussaint, "On Translating a Set of Spheres."

Page 187 *separated with a single translation* ... Godfried T. Toussaint and Hossam A. El Gindy, "Separation of Two Monotone Polygons in Linear Time," *Robotica 2* (1984):215–20.

Page 187 *known in Japanese carpentry as the ari-kake joint* ... Kiyosi Seike, *The Art of Japanese Joinery*, New York: John Weatherhill, 1977.

Page 189 *examined (by numerical calculations)* ... H. M. Princen and P. Levinson, "The Surface Area of Kelvin's Minimal Tetrakaidecahedron: The Ideal Foam Cell," *Journal of Colloid and Interface Science*, vol. 120, no. 1, pp. 172–175 (1978).

Page 189 *alternative unit cell* ... D. Weaire and R. Phelan, "A Counter-Example to Kelvin's Conjecture on Minimal Surfaces," *Philosophical Magazine Letters*, vol. 69, no. 2, pp. 107–110 (1994).

Notes and References for Chapter 13

Page 193 *Proofs and Refutations* ... Imre Lakatos, *Proofs and Refutations*, New York: Cambridge University Press, 1976.

Page 194 *"Branko Grünbaum's definition of "polyhedron"* ... B. Grünbaum, "Regular Polyhedra, Old and New," *Aequationes Mathematicae 16* (1977):1–20.

Page 197 *the theory is still evolving.* ... B. Grünbaum and G. C. Shephard, *Tilings and Patterns*, San Francisco: W. H. Freeman, 1986.

Page 198 *The numbers of combinatorial types ...* P. Engel, "On the Enumeration of Polyhedra," *Discrete Mathematics 41* (1982):215–18.

Page 198 *surprisingly low bounds for the number of combinatorially distinct polytopes ...* J. E. Goodman and R. Pollack, "There Are Asymptotically Far Fewer Polytopes Than We Thought," *Bulletin (New Series) American Mathematical Society 14*, no. 1 (January 1986):127–29.

Notes and References for Chapter 14

Page 202 *SolidWorks ...* Solidworks: 3d mechanical design and 3d cad software, http://www.solidworks.com/

Page 202 *Cauchy's famous theorem (1813) ...* A. L. Cauchy, "Sur le polygones et les polyèdres, seconde mémoire," *Journal de l'Ecole Polytechnique*, XVIe Cahier (Tome IX):87–90, 1813, and *Oevres complètes*, IIme Série, Vol. I, Paris, 1905, 26–38. For modern presentations, see P. R. Cromwell, *Polyhedra*, Cambridge University Press, 1997, and M. Aigner and G. M. Ziegler, *Proofs From the Book*, Springer-Verlag, Berlin, 1998.

Page 202 *theorems and implications due to Dehn, Weyl and A.D. Alexandrov ...* Dehn, Weyl and Alexandrov obsrved that Cauchy's proof can be adapted to yield an *infinitesimal rigidity* theorem; see M. Dehn, "Über die Starrheit konvexer Polyeder," *Mathematische Annalen*, 77:466–473, 1916; H. Weyl, "Über die Starrheit der Eiflächen und konvexer Polyeder," In *Gesammelte Abhandlungen*, Springer Verlag, Berlin, 1917; and A. D. Alexandrov, *Convex Polyhedra*, Springer Monographs in Mathematics. Springer Verlag, Berlin Heidelberg, 2005, English Translation of Russian edition, Gosudarstv. Izdat.Tekhn.-Teor. Lit., Moscow-Leningrad, 1950.

Page 202 *characterized by Steinitz's theorem ...* E. Steinitz, *Polyeder und Raumeinteilungen*, volume 3 (Geometrie) of *Enzyklopädie der Mathematischen Wissenschaften*, 1–139, 1922.

Page 203 *3-connected planar graphs. ...* We refer to them as *skeleta of convex polyhedra*.

Page 203 *a later theorem due to Gluck ...* H. Gluck, Almost all simply connected closed surfaces are rigid, *Lecture Notes in Mathematics*, 438:225–239, 1975.

Page 204 *the Tay graph ...* T.-S. Tay, "Rigidity of multigraphs I: linking rigid bodies in n-space," *Journal of Combinatorial Theory, Series B*, 26:95–112, 1984; T.-S. Tay, "Linking $(n-2)$-dimensional panels in n-space II: $(n-2,2)$-frameworks and body and hinge structures," *Graphs and Combinatorics*, 5:245–273, 1989.

Page 204 *Tay and Whiteley proved ...* T.-S. Tay and W. Whiteley, "Recent advances in the generic rigidity of structures," *Structural Topology*, 9:31–38, 1984.

Page 204 *Their guess was proved true ...* N. Katoh and S. Tanigawa, "A proof of the molecular conjecture," in *Proc. 25th Symp. on Computational Geometry (SoCG'09)*, pages 296–305, 2009, http://arxiv.org/abs/0902.0236.

Page 205 *the structure theorem for (k,k)-sparse graphs ...* A. Lee and I. Streinu, "Pebble game algorithms and sparse graphs," *Discrete Mathematics*, 308(8):1425–1437, April 2008.

Page 206 *the resulting graph G' is sparse ...* For further structural properties of sparse graphs, especially the (6, 6) case, see A. Lee and I. Streinu, cited above.

Page 206 *It relies on the pebble game algorithm ...* A. Lee and I. Streinu, cited above.

Page 208 *a single-vertex origami ...* I. Streinu and W. Whiteley, "Single-vertex origami and spherical expansive motions," in J. Akiyama and M. Kano, editors, *Proc. Japan Conf. Discrete and Computational Geometry (JCDCG 2004)*, volume 3742 of *Lecture Notes in Computer Science*, pages 161–173, Tokai University, Tokyo, 2005. Springer Verlag.

Notes and References for Chapter 15

Page 212 *the eighteenth-century works of Euler and Meister ...* see L. Euler, "Elementa doctrinae solidorum," *Novi Comm. Acad. Sci. Imp. Petropol. 4* (1752–53):109–40; see also in *Opera Mathematica,* vol. 26, pp. 71–93. P. J. Federico, *Descartes on Polyhedra,* New York: Springer-Verlag, 1982, pp. 65–69. A. L. F. Meister, "Commentatio de solidis geometricis," *Commentationes soc. reg. scient. Gottingensis,* cl. math. 7 (1785). M. Brückner, *Vielecke und Vielfläche* (Leipzig: Teubner, 1900), p. 74.

Page 212 *each face of one is the polar* ... see H. S. M. Coxeter, *Regular Polytopes,* 3rd ed., New York: Dover Publications, 1973, p. 126.

Page 213 *are duals of each other* ... see B. Jessen, "Orthogonal Icosahedron," *Nordisk Matematisk Tidskrift 15* (1967):90–96; J. Ounsted, "An Unfamiliar Dodecahedron," *Mathematics Teaching, 83* (1978):45–47. B. M. Stewart, *Adventures among the Toroids,* 2nd ed. (Okemos, Mich.: B. M. Stewart, 1980), p. 254, and B. Grünbaum and G. C. Shephard, "Polyhedra with Transitivity Properties," *Mathematical Reports of the Academy of Science* [Canada] *6,* (1984): 61–66.

Page 213 *Császár polyhedron* ... see, for example, "A Polyhedron without Diagonals," *Acta Scientiarum Mathematicarum 13* (1949):140–42. B. Grünbaum, *Convex Polytopes* (London Interscience, 1967), p. 253. M. Gardner, "On the Remarkable Császár Polyhedron and its Applications in Problem Solving," *Scientific American 232,* no. 5 (May 1975):102–107. Stewart, *Adventures among the Toroids,* p. 244.

Page 213 *Szilassi polyhedron* ... see L. Szilassi, "A Polyhedron in Which Any Two Faces Are Contiguous" [In Hungarian, with Russian summary], *A Juhasz Gyula Tanarkepzo Foiskola Tudomanyos Kozlemenyei,* 1977, Szeged. M. Gardner, "Mathematical Games, in Which a Mathematical Aesthetic Is Applied to Modern Minimal Art," *Scientific American 239,* no. 5 (November 1973):22–30. Stewart, *Adventures among the Toroids,* p. 244. P. Gritzmann, "Polyedrische Realisierungen geschlossener 2-dimensionaler Mannigfaltigkeiten im R^3," Ph.D. thesis, Universität Siegen, 1980.

Page 213 *this goal may be unattainable* ... B. Grünbaum and G. C. Shephard, "Polyhedra with Transitivity Properties."

Page 216 *this generalization of the concept of a polyhedron* ... See, for example, A. F. Möbius, "Über die Bestimmung des Inhaltes eines Polyeders" (1865), in *Gesammelte Werke,* vol. 2 (1886), pp. 473–512. E. Hess, *Einleitung in die Lehre von der Kugelteilung* (Leipzig: Teubner, 1883). Brückner, *Vielecke und Vielfläche,* p. 48. H. S. M. Coxeter, *Regular Polytopes.* H. S. M. Coxeter, M. S. Longuet-Higgins, and J. C. P. MIller, "Uniform Polyhedra," *Philosophical Transactions of the Royal Society* [London], sec. A. *246* (1953–54):401–450. J. Skilling, "The Complete Set of Uniform Polyhedra," *Philosophical Transactions of the Royal Society* [London], sec. A. *278* (1975):111–135. M. J. Wenninger, *Polyhedron Models* (London: Cambridge University Press, 1971). M. Norgate, "Non-Convex Pentahedra," *Mathematical Gazette 54* (1970): 115–24.

Notes and References for Chapter 16

Page 218 *combinatorial analogue of the tiling problem for higher dimensions* ... see L. Danzer, B. Grünbaum, and G. C. Shephard, "Does Every Type of Polyhedron Tile Three-space?" *Structural Topology* 8 (1983):3–14, and D. G Larman and C. A. Rogers, "Durham Symosium on the Relations between Infinite-dimensional and Finitely-dimensional Convexity," *Bulletin of the London Mathematical Society* 8 (1976):1–33.

Page 220 *we can prove this generalization* ... E. Schulte, "Tiling Three-space by Combinatorially Equivalent Convex Polytopes," *Proceedings of the London Mathematical Society 49,* no. 3, (1984):128–40.

Page 220 *operation, due to Steinitz* ... Branko Grünbaum, *Convex Polytopes,* New York: John Wiley and Sons, 1967.

Page 221 *all simplicial polytopes give locally finite face-to-face tilings* ... B. Grünbaum, P. Mani-Levitska, and G. C. Shephard, "Tiling Three-dimensional Space with Polyhedral Tiles of a Given Isomorphism Type," *Journal of* the *London Mathematical Society 29,* no. 2 (1984):181–91.

Page 221 *the reverse is true* ... see E. Schulte, "Tiling Three-space by Combinatorially Equivalent Convex Polytopes," *Proceedings of the London Mathematical Society 49,* no. 3, (1984):128–40; E. Schulte, "The Existence of Nontiles and Nonfacets in Three Dimensions," *Journal of Combinatorial Theory,* ser. A, 38 (1985):75–81; E. Schulte, "Nontiles and Nonfacets for the Euclidean Space, Spherical Complexes and Convex Polytopes." *Journal für die Reine und Angewandte Mathematik,* 352 (1984):161–83.

Page 221 *the fundamental region for the symmetry group of the regular tessellation of E^3 by cubes is a 3-simplex* ... H. S. M. Coxeter, *Regular Polytopes,* 3rd ed., New York: Dover Publications, 1973

Page 222 *examples of space-filling toroids* ... D. Wheeler and D. Sklar, "A Space-filling Torus," *The Two-Year College Mathematics Journal 12,* no. 4 (1981):246–48.

Page 222 *Polyhedra with this property have recently been studied* ... P. McMullen, C. Schulz, and J. M. Wills, "Polyhedral Manifolds in E^3 with Unusually Large Genus," *Israel Journal of Mathematics 46* (1983):127–44.

Page 222 *monotypic tilings of M by topological polytopes or tiles of another homeomorphism type* ... E. Schulte, "Regular Incidence-polytopes with Euclidean or Toroidal Faces and Vertex-figures," *Journal of Combinatorial Theory, ser. A, 40* (1985): 305–30.

Notes and References for Chapter 17

Page 223 *infinitely many equivelar manifolds* ... B. Grünbaum and G. Shephard, "Polyhedra with Transitivity Properties," *Mathematical Reports of the Academy of Science* [Canada] 6 (1984):61–66; P. McMullen, C. Schulz, and J. M. Wills, "Polyhedral 2-Manifolds in E^3 with Unusually Large Genus," *Israel Journal of Mathematics 46* (1983):124–44.

Page 224 *there exist no face-transitive polyhedra with $g > 0$* ... see Grünbaum and Shephard, above.

Page 224 *first vertex-transitive polyhedron with $g > 0$* ... U. Brehm and W. Kühnel, "Smooth Approximation of Polyhedral Surfaces Regarding Curvatures," *Geometriae Dedicata* 12 (1982):438.

Page 224 *two more polyhedra with vertex-transitivity under symmetries* ... see Günbaum and Shephard, above.

Page 225 *an impression of four Platonohedra* ... J. M. Wills, "Semi-Platonic Manifolds" in P. Gruber and J. M. Wills, eds., *Convexity and Its Applications* (Basel: Birkhäuser, 1983); J. M. Wills,"On Polyhedra with Transitivity Properties," *Discrete and Computational Geometry 1* (1986):195–99.

Page 225 *a figure of $\{3, 8; 5\}$ from which a three-dimensional construction is possible* ... see Grünbaum and Shephard, above.

Page 225 *a precise construction of $\{3, 8; 3\}$.* ... E. Schulte and J. M. Wills, "Geometric Realizations for Dyck's Regular Map on a Surface of Genus 3," *Discrete and Computational Geometry 1* (1986):141–53.

Page 226 *the last four can be realized in E^4* ... H. S. M. Coxeter, "Regular Skew Polyhedra in Three and Four Dimensions, and Their Topological Analogues," *Proceedings of the London Mathematical Society,* ser. 2, 43 (1937):33–62.

Page 226 *The geometric construction traces back to Alicia Boole-Stott* ... A. Boole-Stott, *Geometrical Reduction of Semiregular from Regular Polytopes and Space Fillings,* Amsterdam: Ver. d. K. Atkademie van Wetenschappen, 1910; P. McMullen, C. Schulz, and J. M. Wills, "Equivelar Polyhedral Manifolds in E^3," *Israel Journal of Mathematics 41* (1982):331–46. They are projections of Coxeter's regular skew polyhedra (see Schulte and Wills, above).

Page 226 *polyhedral realization of Felix Klein's famous quartic* ... E. Schulte and J. M. Wills, "A Polyhedral Realization of Felix Klein's Map $\{3, 7\}_8$ on a Riemann Surface of Genus 3," *Journal of the London Mathematical Society 32* (1985):539–47.

Page 228 *Do equivelar manifolds exist with* $p \geq 5$ *and* $q \geq 5?$ ... see McMullen, Schultz, and Wills, 1982, above; see also P. McMullen, C. Schulz, and J. M. Wills, "Two Remarks on Equivelar Manifolds," *Israel Journal of Mathematics* 52 (1985):28–32.

Page 228 *for* $q = 4$ *no such manifold exists* ... see McMullen, Schultz, and Wills, 1985, above.

Page 229 *a nice and interesting proof* ... Günter Ziegler and Michael Joswig "Polyhedral surfaces of high genus," Oberwolfach seminars, vol. 38 (2009), *Discrete Differential Geometry* (Bobenko, Sullivan, Schrder, Ziegler).

Page 229 *This polyhedron was found by David McCooey in 2009.* ... David McCooey, A non-selfintersecting polyhedral realization of the all-heptagon Klein map, *Symmetry*, vol. 20, no. 1- (2009), 247–268.

Notes and References for Chapter 18

Page 231 *to determine when they were rigid* ... The technical term is *infinitesimally rigid*. A bar-and-joint framework is infinitesimally rigid if every set of velocity vectors which preserves the lengths of the bars represents a Euclidean motion of the whole space.

Page 231 *the reciprocal figure* ... J. C. Maxwell, "On Reciprocal Diagrams and Diagrams of Forces," *Philosophical Magazine*, 4, 27(1864):250–61, J.C. Maxwell, "On Reciprocal Diagrams, Frames and Diagrams of Forces." *Transactions of the Royal Society of Edinburgh* 26 (1869–72):1–40.

Page 231 *the field of graphical statics* ... L. Cremona, *Graphical Statics*, English trans., London: Oxford University Press, 1890.

Page 231 *grows from these geometric roots* ... H. Crapo and W. Whiteley, "Statics of Frameworks and Motions of Panel Structures: A projective Geometric Introduction," *Structural Topology* 6, (1982):43–82, H. Crapo and W. Whiteley, "Plane Stresses and Projected Polyhedra I, the basic pattern" *Structural*

Topology 20 (1993), 55–68. http://www-iri.upc.es/people/ros/StructuralTopology/, W. Whiteley, "Motions and Stresses of Projected Polyhedra," *Structural Topology 7* (1982):13–38.

Page 232 *Several workers independently observed* ... D. Huffman, "A Duality Concept for the analysis of Polyhedral Scenes," in E. W. Elcock and D. Michie (eds.), *Machine Intelligence 8* [Ellis Horwood, England] (1977):475–92. A. K. Mackworth, "Interpreting Pictures of Polyhedral Scenes," *Artifical Intelligence 4* (1973):121–37.

Page 232 *necessary and sufficient condition for correct pictures* ... H. Crapo and W. Whiteley, "Plane Stresses and Projected Polyhedra I, the basic pattern" *Structural Topology 20* (1993), 55–68. http://www-iri.upc.es/people/ros/StructuralTopology/.

Page 232 *projected from the point of tangency of one face* ... K. Q. Brown, "Voronoi Diagrams from Convex Hulls," *Information Processing Letters 9* (1979):223–38.

Page 232 *the diagram of centers forms a classical reciprocal figure* ... P. Ash and E. Bolker, "Recognizing Dirichlet Tessellations," *Geometriae Dedicata 19* (1985):175–206.

Page 232 *last link in the proof* ... P. Ash and E. Bolker, "Generalized Dirichlet Tessellations," *Geometriae Dedicata 20* (1986):209–43.

Page 233 *an explicit construction of a polyhedron* ... H. Edelsbrunner and R. Seidel, "Voronoi Diagrams and Arrangements," *Discrete and Computational Geometry I* (1986):25–44.

Page 235 *Delauney triangulation* ... In a Dirichlet tessellation the centers of the cells which share a vertex are equidistant from that vertex. If the centers are chosen at random, then (with probability 1) no four lie on a circle. So the resulting Dirichlet tessellation has only 3-valent vertices.

Page 236 *a proof can be found* ... P. Ash and E. Bolker, "Recognizing Dirichlet Tessellations," *Geometriae Dedicata 19* (1985):175–206.

Page 237 *Ash and Bolker proved* ... P. Ash and E. Bolker, "Generalized Dirichlet Tessellations," *Geometriae Dedicata 20* (1986): 209–43.

Page 237 *model a simple biological phenomenon* ... H. Edelsbrunner and R. Seidel,

"Voronoi Diagrams and Arrangements," *Discrete and Computational Geometry I* (1986):25–44.

Page 237 *this need not always be true...* Other examples and a more complete bibliography are given in P. Ash and E. Bolker, "Recognizing Dirichlet Tessellations," *Geometriae Dedicata 19* (1985):175–206.

Page 238 *they are rigid in the plane...* Other examples and a more complete bibliography are given in P. Ash and E. Bolker, "Recognizing Dirichlet Tessellations," *Geometriae Dedicata 19* (1985):175–206.

Page 238 *the appearance of a spider web signals that it is shaky...* R. Connelly, "Rigidity and Energy," *Inventiones Mathematicae 66* (1982):11–33.

Page 238 *cannot be made denser...* H. Crapo and W. Whiteley, "Statics of Frameworks and Motions of Panel Strucutres: A projective Geometric Introduction," *Structural Topology 6,* (1982):43–82, W. Whiteley, "Motions and Stresses of Projected Polyhedra," *Structural Topology 7* (1982):13–38.

Page 238 *we have constructed a convex reciprocal...* R. Connelly, "Rigid Circle and Sphere Packings I. Finite Packings," *Structural Topology,* 14 (1988) 43–60. R. Connelly, "Rigid Circle and Sphere Packings II. Infinite Packings with Finite Motion," *Structural Topology,* 16 (1990) 57–76.

Page 238 *Thus we have proved...* J. C. Maxwell, "On Reciprocal Diagrams and Diagrams of Forces," *Philosophical Magazine,* 4, 27(1864):250–61, H. Crapo and W. Whiteley, "Plane Stresses and Projected Polyhedra I, the basic pattern" *Structural Topology 20* (1993), 55–68. http://www-iri.upc.es/people/ros/StructuralTopology/.

Page 240 *a projective polarity about the Maxwell paraboloid...* H. Crapo and W. Whiteley, "Plane Stresses and Projected Polyhedra I, the basic pattern" *Structural Topology 20* (1993), 55–68. http://www-iri.upc. es/people/ros/StructuralTopology/.

Page 240 *the positions of the remaining planes can be deduced...* J. C. Maxwell, "On Reciprocal Diagrams and Diagrams of Forces," *Philosophical Magazine,* 4, 27(1864):250–61, H.

Crapo and W. Whiteley, "Plane Stresses and Projected Polyhedra I, the basic pattern" *Structural Topology 20* (1993), 55–68. http://www-iri.upc. es/people/ros/StructuralTopology/.

Page 241 *Choose the centers to be the points...* H. Crapo and W. Whiteley, "Plane Stresses and Projected Polyhedra I, the basic pattern" *Structural Topology 20* (1993), 55–68. http://www-iri.upc.es/people/ros/ StructuralTopology/, Section 4.

Page 242 *in the cell of the sectional Dirichlet tessellation...* H. Edelsbrunner and R. Seidel, "Voronoi Diagrams and Arrangements," *Discrete and Computational Geometry I* (1986): 25–44.

Page 242 *This completes the converse...* H. Crapo and W. Whiteley, "Plane Stresses and Projected Polyhedra I, the basic pattern" *Structural Topology 20* (1993), 55–68. http://www-iri.upc. es/people/ros/StructuralTopology/.

Page 243 *furthest-point Dirichlet tessellation of centers...* P. Ash and E. Bolker, "Generalized Dirichlet Tessellations," *Geometriae Dedicata 20* (1986):209–43.

Page 243 *the picture of some convex polyhedron...* H. Edelsbrunner and R. Seidel, "Voronoi Diagrams and Arrangements," *Discrete and Comutational Geometry I* (1986):25–44.

Page 244 *This argument and its converse prove...* see B. Roth and W. Whiteley, "Tensegrity Frameworks," *Transactions of the American Mathematical Society 265* (1981):419–45, Whiteley, "Motions and Stresses of Projected Polyhedra," *Structural Topology 7* (1982):13–38, and R. Connelly, "Rigidity and Energy," *Inventiones Mathematicae 66* (1982):11–33.

Page 244 *This gives...* R. Connelly, "Rigidity and Energy," *Inventiones Mathematicae 66* (1982):11–33.

Page 245 *positions of the vertices and the directions of the infinite edges...* Brown, "Voronoi Diagrams from Convex Hulls," above.

Page 246 *any triply connected planar graph can be realized as a convex polyhedron...* Roth and Whiteley, "Tensegrity Framworks," above.

Page 246 *conditions that must be satisfied if there is to be a stress on all members...*

B. Grünbaum, *Convex Polytopes*, New York: Interscience, 1967.

Page 246 *one for each finite cell in the graph...* N. White and W. Whiteley, "The Algebraic Geometry of Stresses in Frameworks," *S.I.A.M. Journal of Algebraic and Discrete Methods 4* (1983):481–511.

Page 246 *study of White and Whiteley...* C. Davis, "The Set of Non-Linearity of a Convex Piece-wise Linear Function," *Scripta Mathematica 24* (1959):219–28.

Page 246 *the dual bowl over this reciprocal will be a zonohedral cap...* White and Whiteley, "The Algebraic Geometry of Stresses in Frameworks," above.

Page 247 *edge graphs of convex polyhedra with inspheres...* H. S. M Coxeter, "The Classification of Zonohedra by Means of Projective Diagrams," *Journal de Mathématiques Pures et Appliquées,* ser. 9, 42(1962):137–56.

Page 247 *projection of a polyhedron which cannot have an insphere...* E. Schulte, "Analogues of Steinitz's Theorem about Non-Inscribable Polytopes," *Colloquia Mathematica Societatis Janos Bolyai* 48 (1985), 503–516.

Page 247 *a geometric characterization...* Grünbaum, *Convex Polytopes*, above.

Page 248 *the one first studied in graphical statics...* P. Ash and E. Bolker, "Recognizing Dirichlet Tessellations," *Geometriae Dedicata 19* (1985):175–206.

Page 248 *both truly belong to projective geometry...* J. C. Maxwell, "On Reciprocal Diagrams and Diagrams of Forces," *Philosophical Magazine*, 4, 27(1864):250–61, Cremona, *Graphical Statics* (above).

Page 248 *Connelly discovered...* J.C. Maxwell, "On Reciprocal Diagrams, Frames and Diagrams of Forces." *Transactions of the Royal Society of Edinburgh 26* (1869-72):1–40, W. Whiteley, "The Projective Geometry of Rigid Frameworks," in C. Baker and L. Batten, eds, *Proceedings of the Winnipeg Conference on Finite Geometries*, New York: Marcel Dekker, 1985, pp. 353–70.

Page 249 *remains true...* Connelly, "Rigid Circle and Sphere Packings: I and II," above.

Page 250 *this algebraic definition allows...* H. Edelsbrunner and R. Seidel, "Voronoi Diagrams and Arrangements," *Discrete and Comutational Geometry I* (1986):25–44.

Page 250 *in the spatial Dirichlet tessellaton it sections...* H. Crapo and W. Whiteley, "Plane Stresses and Projected Polyhedra I, the basic pattern" *Structural Topology 20* (1993), 55–68. http://www-iri.upc.es/people/ros/StructuralTopology/.

Page 250 *called a pentagrid...* N. G. de Brujin, "Algebraic Theory of Penrose's Non-periodic Tilings, I, II," *Akademie van Wetenschappen* [Amsterdam], *Proceedings,* ser. A, 43 (1981):32–52, 53–66.

Page 251 *publications on Voronoi Diagrams...* see F. Aurenhammer, "Voronoi diagrams: a survey of a fundamental geometric data structure," *ACM Computing Surveys (CSUR)* Volume 23 (1991), 345–405; A. Okabe, B. Boots, K. Sugihara, *Spatial tessellations : concepts and applications of voronoi diagrams,* Wiley & Sons, 1992; S. Fortune, "Voronoi diagrams and Delaunay triangulations," in *Computing in Euclidean Geometry,* Ding-zhu Du, Frank Hwang (eds), Lecture notes series on *computing*; vol. 1 World Scientific (1992), 193–234.

Page 251 *continuing work on weighted tessellations...* D. Letscher "Vector Weighted Voronoi Diagrams and Delaunay Triangulations," Canadian Conference on Computational Geometry 2007, Ottawa, Ontario, August 20–22, 2007.

Page 251 *advances on reciprocal diagrams in the plane...* H. Crapo and W. Whiteley. "Spaces of stresses, projections, and parallel drawings for spherical polyhedra", *Beitraege zur Algebra und Geometrie / Contributions to Algebra and Geometry 35* (1994), 259–281.

Page 251 *zonohedral tilings (cubic partial cubes)...* D. Eppstein, "Cubic partial cubes from simplicial arrangements," *The Electronic Journal of Combinatorics.* 13 (2006); R79.

Page 251 *projections of higher dimensional cell complexes...* K. Rybnikov, "Stresses and liftings of cell-complexes," *Discrete Comp. Geom.* 21, No. 4, (1999), 481–517; R. M. Erdahl,

K. A. Rybnikov, and S. S. Ryshkov, "On traces of d-stresses in the skeletons of lower dimensions of homology d-manifolds," *Europ. J. Combin.* 22, No. 6, (2001), 801–820; W. Whiteley, "Some Matroids from Discrete Geometry", in *Matroid Theory,* J. Bonin, J. Oxley and B. Servatius (eds.), AMS Contemporary Mathematics, 1996, 171–313.

Notes and References for Chapter 19

Page 253 *the coordinates are given by systematic choices ...* H. S. M. Coxeter, *Regular Polytopes*, New York: Dover Publications, 1973, pp. 50–53, 156–162.

Page 254 *see Hardy and Wright's ...* G. H. Hardy and E. M. Wright, *An Introduction to the Theory of Numbers*, 4th ed., London: Oxford University Press, 1960, pp. 221–222.

Notes and References for Chapter 20

Page 257 *the way various three-dimensional faces fit together in 4-space ...* to form regular polytopes; see for example D.M.Y. Sommerville's description of the regular polytopes in 4-space, *Geometry of n Dimensions*, London: Methuen, 1929, or H.S.M. Coxeter's treatment, *Regular Polytopes*, 3rd ed., New York: Dover Publications, 1973.

Page 258 *This polytope is complicated enough...* Similar treatments of the 120-cell and the 600-cell are implicit in the work of several mathematicians, notable Coxeter whose *Regular Complex Polytopes*, Cambridge: Cambridge University Press, 1974, is the primary source for all material of this sort.

Page 262 *In 4-space the sphere of points at unit distance ...* One direct analogy with the usual coordinate system on the 2-sphere in 3-space would be to use

$$(x, y, u, v) = (\cos(\theta)\sin(\varphi)\sin(\psi), \sin(\theta)\sin(\varphi)$$
$$\sin(\psi), \cos(\varphi)\sin(\psi), \cos(\psi)). \quad (22.3)$$

which would suggest the same sort of decomposition of the 3-sphere into "parallel 2-spheres of latitude." Such a decomposition has been carried out by several authors (including D.M.Y. Sommerville), and an early computer generated film by George Olshevsky uses such an approach to display the slices of regular polyhedra in 4-space by sequences of hyperplanes perpendicular to various coordinate axes.

Page 264 *the 16 vertices of the hypercube on the unit hypersphere ...* may be given by $\frac{1}{2}[\pm 1 \pm i, \pm 1 \pm i]$. Compare p. 37 of Coxeter, *Regular Complex Polytopes*.

Page 266 *mapped to the vertices of a regular tetrahedron inscribed in the unit 2-sphere ...* The coordinates for the 24-cell obtained in this way are very similar to those which appear in Coxeter's discussion of the 24-cell in *Twisted Honeycombs* (Regional Conference Series in Mathematics, no. 4, American Mathematical Society, Providence 1970), although he does not explicitly use the Hopf mapping in any of his constructions. Professor Coxeter pointed out that these coordinates also appear in a slightly different form in the 1951 dissertation of G.S. Shephard.

Notes and References for Chapter 21

Page 267 *The key idea in that problem ...* See R. Connelly, "Expansive motions. Surveys on discrete and computational geometry," *Contemp. Math.*, (2008) 453, 213–229 Amer. Math. Soc., Providence, RI.

Page 267 *the theory of tensegrities can be applied ...* For more about these ideas, see Aleksandar Donev; Salvatore Torquato; Frank H. Stillinger and Robert Connelly, "A linear programming algorithm to test for jamming in hard-sphere packings," *J. Comput. Phys.*, 197, (2004), no. 1, 139–166.

Page 269 *stress associated to a tensegrity ...* Note that in the paper by B. Roth and W. Whiteley, "Tensegrity frameworks," *Trans. Amer. Math. Soc.*, 265, (1981), no. 2, 419–446, a proper stress is called what is strict and proper here, whereas in my definition proper stresses need not be strict. And don't confuse this notion of stress with that used in structure analysis, physics or engineering. In those fields, stress is defined as a force per cross-sectional area. There are no cross-sections in my definition; the scalar ω_{ij} is better interpreted as a force per unit length.

Page 269 *are congruences* ... this is also called *local rigidity* as in S. Gortler, A. Healy, and D. Thurston, "Characterizing generic global rigidity," arXiv:0710.0907 v1, 2007.

Page 269 *a good body of work* ... see the surveys in András Recski, "Combinatorial conditions for the rigidity of tensegrity frameworks. Horizons of combinatorics," 163–177, *Bolyai Soc. Math. Stud*, 17, Springer, Berlin, 2008; Walter Whiteley, "Infinitesimally rigid polyhedra. I. Statics of frameworks." *Trans. Amer. Math. Soc.*, 285, (1984), no. 2, 431–465; Walter Whiteley, "Infinitesimally rigid polyhedra. II. Weaving lines and tensegrity frameworks." *Geom. Dedicata*, 30 (1989), no. 3, 255–279.

Page 270 *when all directional derivatives given by* $p' = (p'_1, \ldots, p'_n)$ *starting at* p *are* 0 ... We perform the following calculation starting from (21.1) for $0 \le t \le 1$:

$$E_\omega(p + tp') = \sum_{i<j} \omega_{ij}((p_i - p_j)^2$$
$$+ 2t(p_i - p_j)(p'_i - p'_j) + t^2(p'_i - p'_j)^2).$$

Taking derivatives and evaluating at $t = 0$, we get:

$$\frac{d}{dt} E_\omega(p+tp')|_{t=0} = 2\sum_{i<j} \omega_{ij}(p_i-p_j)(p'_i-p'_j).$$

At a critical configuration p, this equation must hold for all directions p'.

Page 270 *so is any affine transformation* ... This is seen by the following calculation:

$$\sum_j \omega_{ij}(q_j - q_i) = \sum_j \omega_{ij}(Ap_j + \mathbf{b} - Ap_i - \mathbf{b})$$

$$= A \sum_j \omega_{ij}(p_j - p_i) = 0.$$

Page 271 *quadric at infinity* ... The reason for this terminology is that real projective space $\mathbb{R}p^{d-1}$ can be regarded as the set of lines through the origin in E^d, and Equation (21.5) is the definition of a quadric in $\mathbb{R}p^{d-1}$.

Page 271 *We can prove* ... Conversely suppose that the member directions of a bar tensegrity $G(p)$ lie on a quadric at infinity in E^d given by a non-zero symmetric matrix Q. By the spectral theorem for symmetric matrices, we know that there is an orthogonal d-by-d matrix $X = (X^T)^{-1}$ such that:

$$X^T Q X = \begin{pmatrix} \lambda_1 & 0 & 0 & \cdots & 0 \\ 0 & \lambda_2 & 0 & \cdots & 0 \\ 0 & 0 & \lambda_3 & \cdots & 0 \\ \vdots & \vdots & \vdots & \ddots & \vdots \\ 0 & 0 & 0 & \cdots & \lambda_d \end{pmatrix}.$$

Let λ_- be the smallest λ_i, and let λ_+ be the largest λ_i. Note $\infty \le 1/\lambda_- < 1/\lambda_+ \le \infty$, λ_- is non-positive, and λ_+ is non-negative when Q defines a non-empty quadric and when $1/\lambda_- \le t \le 1/\lambda_+$, $1 - t\lambda_i \ge 0$ for all $i = 1, \ldots, d$. Working Equation (21.4) backwards for $1/\lambda_- \le t \le 1/\lambda_+$ we define:

$$A_t = X^T \begin{pmatrix} \sqrt{1-t\lambda_1} & 0 & 0 & \cdots & 0 \\ 0 & \sqrt{1-t\lambda_2} & 0 & \cdots & 0 \\ 0 & 0 & \sqrt{1-t\lambda_3} & \cdots & 0 \\ \vdots & \vdots & \vdots & \ddots & \vdots \\ 0 & 0 & 0 & \cdots & \sqrt{1-t\lambda_d} \end{pmatrix} X.$$
$$(22.4)$$

Substituting this expression for A_t into Equation (21.4), we see that it provides a non-trivial affine flex of $G(p)$. If the configuration is contained in a lower dimensional hyperplane, we should really restrict to that hyperplane since there are non-orthogonal affine transformations that are rigid when restricted to the configuration itself.

Page 271 *lie on same line on either ruling* ... Consider the diagonal matrix Q with diagonal entries $\lambda_1 = \lambda_2 = 1, \lambda_3 = -1$. When one node of each bar is translated to a single point, they all lie on a circle at infinity given by Q. The flex given by Formula (22.4) flexes the configuration until the nodes lie on a line when

$t = 1/\lambda_+ = 1$ because two of the eigenvalues for Q vanish for that value of t, and in the other direction, when $t = 1/\lambda_- = -1$, the nodes lie in a plane. This structure is easy to build with dowel rods and rubber bands securing the joints where the rulings intersect.

The space of d-by-d symmetric matrices is of dimension $d + (d^2 - d)/2 = d(d + 1)/2$. So if the vector directions of a tensegrity are less than $d(d + 1)/2$, then it is possible to find a non-zero d-by-d symmetric matrix that satisfies Equation (21.5), and then flex it into a lower dimensional subspace.

Page 271 *Barvinok proved* ... Barvinok, A. I., "Problems of distance geometry and convex properties of quadratic maps." *Discrete Comput. Geom.*, 13, (1995), no. 2, 189–202.

Page 272 *Maria Belk and I showed* ... Maria Belk and Robert Connelly, "Realizability of graphs," *Discrete Comput. Geom.*, 37, (2007), no. 2, 125–137; Maria Belk, "Realizability of graphs in three dimensions," *Discrete Comput. Geom.*, 37, (2007), no. 2, 139–162.

Page 272 *Tensegrity techniques are used in a significant way.* ... see Belk, cited above.

Page 273 *when the configuration* $p = (p_1, \ldots, p_n)$ *in* E^d *is universal* ... It is not difficult to prove that if p is a universal configuration for ω, any other configuration q which is in equilibrium with respect to ω is an affine image of p.

Page 274 *I showed that a tensegrity* ... Robert Connelly, "Rigidity and energy", *Invent. Math.*, 66, (1982), no. 1, 11–33. These results answered some questions Grünbaum posed in "Lectures on Lost Mathematics" (1975); notes digitized and reissued at Structural Topology Revisited Conference (2006), http://hdl.handle.net/1773/15700.

Page 275 *two examples in the plane and in three-space* ... this tensegrity is described in Károly Bezdek and Robert Connelly, Two-distance preserving functions from Euclidean space. Discrete geometry and rigidity (Budapest, 1999). *Period. Math. Hungar.*, 39, (1999), no. 1–3, 185–200.

Page 275 *L. Lovasz showed* ... László Lovász, Steinitz representations of polyhedra and the Colin de Verdière number. *J. Combin. Theory Ser. B*, 82, (2001), no. 2, 223–236.

Page 275 *such a tensegrity is super stable* ... this is explained in Károly Bezdek and Robert Connelly, Stress Matrices and M Matrices, *Oberwolfach Reports* Vol. 3, No. 1 (2006), 678–680; it answers a question of K. Bezdek.

Page 275 *P(n,k) is super stable* ... R. Connelly; M. Terrell: Tenségrités symétriques globalement rigides. [Globally rigid symmetric tensegrities] Dual French-English text. *Structural Topology*, No. 21 (1995), 59–78.

Page 276 *a website where one can view* ... this is available at http://www.math.cornell.edu/~tens/.

Page 277 *resulting tensegrity is super stable* ... J. Y. Zhang: Simon D. Guest; Makoto Ohsaki; Robert Connelly, "Dihedral 'Star' Tensegrity Structures," Int. J. Solids Struct. (2009).

Page 278 *This follows from the proof of the carpenter's rule property* ... Robert Connelly, Erik D. Demaine, and Günter Rote, Straightening polygonal arcs and convexifying polygonal cycles. U.S.-Hungarian Workshops on Discrete Geometry and Convexity (Budapest, 1999/Auburn, AL, 2000). *Discrete Comput. Geom.*, 30 (2003), no. 2, 205–239.

Page 278 *There's more – much more – to say* ... Let's begin with **Generic global rigidity**. The configurations in previous sections must be constructed carefully. What about a bar framework where the configuration is more general? It turns out that the problem of determining when a bar framework is globally rigid is equivalent to a long list of problems known to be hard. (See, for example, James B. Saxe, Embeddability of weighted graphs in k-space is strongly NP-hard. Technical report, Computer Science Department, Carnegie Mellon University, 1979.) The problem of whether a cyclic chain of edges in the line has another realization with the same bar lengths is equivalent to the uniqueness of a solution of the knapsack problem. This is one of the many problems on the list of NP complete problems.

One way to avoid this difficulty is to assume that the configuration's coordinates are *generic*. This means that the coordinates of p in E^d are *algebraically independent over the rational numbers*, which means that there is no non-zero polynomial with rational coordinates satisfied by the coordinates of p. This implies, among other things, that no $d + 2$ nodes lie in a hyperplane, for example, and a lot more. I proved ("Generic global rigidity," *Discrete Comp. Geometry* **33** (2005), pp 549–563) that:

Theorem. If $p = (p_1, \ldots, p_n)$ in E^d is generic and $G(p)$ is a rigid bar tensegrity in E^d with a non-zero stress matrix Ω of rank $n - d - 1$, then $G(p)$ is globally rigid in E^d.

Notice that the hypothesis includes Conditions 2 and 3 of Theorem 23 in the text. The idea of the proof is to show that since the configuration p is generic, if $G(q)$ has the same bar lengths as $G(p)$, then they should have the same stresses. Then Proposition 21.1 applies.

Thurston proved the converse (see Thurston, cited above):

Theorem. If $p = (p_1, \ldots, p_n)$ in E^d is generic and $G(p)$ is a globally rigid bar tensegrity in E^d, then either $G(p)$ is a bar simplex or there is stress matrix Ω for $G(p)$ with rank $n - d - 1$.

The idea here, very roughly, is to show that a map from an appropriate quotient of an appropriate portion of the space of all configurations has even topological degree when mapped into the space of edge lengths.

As Dylan Thurston pointed out, using these results it is possible to find a polynomial time numerical (probablistic) algorithm that calculates whether a given graph is generically globally rigid in E^d, and that the property of being globally rigid is a generic property. In other words, if $G(p)$ is globally rigid in E^d at one generic configuration p, it is globally rigid at all generic configurations. Interestingly, he also showned that if p is generic in E^d, and $G(q)$ has the same bar lengths in $G(p)$ in E^d, then $G(p)$ can be flexed

to $G(q)$ in E^{d+1}, similar to the discussion of compound tensegrities.

A bar graph G is *generically redundantly rigid* in E^d if $G(p)$ is rigid at a generic configuration p and remains rigid after the removal of any bar. A graph is *vertex k-connected* if it takes the removal of at least k vertices to disconnect the rest of the vertices of G. The following theorem of Hendrickson (B. Hendrickson, Conditions for unique graph realizations, *SIAM J. Comput* **21** (1992), pp 65–84) provides two necessary conditions for generic global rigidity.

Theorem. If p is a generic configuration in E^d, and the bar tensegrity $G(p)$ is globally rigid in E^d, then

1. G is vertex $(d + 1)$-connected, and
2. $G(p)$ is redundantly rigid in E^d.

Condition 1 on vertex connectivity is clear since otherwise it is possible to reflect one component of G about the hyperplane determined by some d or fewer vertices. Condition 2 on redundant rigidity is natural since if, after a bar $\{p_i, p_j\}$ is removed, $G(p)$ is flexible, one watches as the distance between p_i and p_j changes during the flex, and waits until the distance comes back to its original length. If p is generic to start with, the new configuration will be not congruent to the original configuration.

Hendrickson conjectured that Conditions 1 and 2 were also sufficient for generic global rigidity, but it turns out that the complete bipartite graph $K_{5,5}$ in E^3 is a counterexample (R. Connelly, On generic global rigidity, in *Applied Geometry and Discrete Mathematics*, DIMACS Ser. Discrete Math, Theoret. Comput. Scie **4**, AMS, 1991, pp 147–155). This is easy to see as follows.

Similar to the analysis of Radon tensegrities, for each of the nodes for the two partitions of $K_{5,5}$ consider the affine linear dependency $\sum_{i=1}^{5} \lambda_i p_i = \mathbf{0}, \sum_{i=1}^{5} \lambda_i = 0$ and $\sum_{i=6}^{10} \lambda_i p_i = \mathbf{0}, \sum_{i=6}^{10} \lambda_i = 0$, where $(p_1, \ldots, p_5)$ and $(p_6, \ldots, p_{10})$ are the two partitions of $K_{5,5}$. When the configuration $p = (p_1, \ldots, p_{10})$ is generic in E^3, then, up to a scaling factor, the stress matrix for $K_{5,5}(p)$ is

$$\Omega = \begin{pmatrix} & \mathbf{0} & & \begin{pmatrix} \lambda_6 \\ \vdots \\ \lambda_{10} \end{pmatrix} (\lambda_1 \cdots \lambda_5) \\ \begin{pmatrix} \lambda_1 \\ \vdots \\ \lambda_5 \end{pmatrix} (\lambda_6 \cdots \lambda_{10}) & & \mathbf{0} \end{pmatrix}.$$

See Bolker, E. D. and Roth, B., "When is a bipartite graph a rigid framework?" *Pacific J. Math.* **90** (1980), no. 1, 27–44. But the rank of Ω is $2 < 10 - 3 - 1 = 6$, while rank 6 is needed for generic global rigidity in this case by Theorem 25.

In E^3, $K_{5,5}$ is the only counterexample to Hendrickson's conjecture that I know of. On the other hand, a graph G is generically globally rigid in E^d if and only if the cone over G is generically globally rigid in E^{d+1} (see R. Connelly and W. Whiteley, "Global Rigidity: The effect of coning," submitted). This gives more examples in dimensions greater than 3, and there are some other bipartite graphs as well in higher dimensions by an argument similar to the one here.

Meanwhile, the situation in the plane is better. Suppose G is a graph and $\{i, j\}$ is an edge of G, determined by nodes i and j. Remove this edge, add another node k and join k to i, j, and $d - 1$ distinct other nodes not i or j. This is called a *Henneberg operation* or sometimes *edge splitting*.

It is not hard to show that edge splitting preserves generic global rigidity in E^d. When the added node lies in the relative interior of the line segment of the bar that is being split, there is a natural stress for the new bar tensegrity, and the subdivided tensegrity is also universal with respect to the new stress. If the original configuration is generically rigid, a small perturbation of the new configuration to a generic one will not change the rank of the stress matrix. Thus generic global rigidity is preserved under edge splitting. A. Berg and T. Jordán and later B. Jackson (A. Berg and T. Jordán, A proof of Connelly's conjecture on 3-connected circuits of the rigidity matroid. *J. Combinatorial Theory Ser. B.*, 88, 2003: pp 77–97) and T. Jordán (B. Jackson, and T. Jordan, Connected rigidity matroids and unique

realization graphs, *J. Combinatorial Theory B* **94** 2005, pp 1–29) solved a conjecture of mine:

Theorem. If a graph G is vertex 3-connected (Condition 23) for $d = 2$ and is generically redundantly rigid in the plane (Condtion 23) for $d = 2$ then G can be obtained from the graph K_4, by a sequence of edge splits and insertions of additional bars.

Thus Hendrickson's conjecture, that Conditions 1 and 2 are sufficient as well as necessary for generic global rigidity in the plane, is correct. This also gives an efficient non-probablistic polynomial-time algorithm for determining generic global rigidity in the plane.

Notes and References for Chapter 22

Notes on Problem 1

Albrecht Dürer's *Underweysung der Messung mit dem Zirkel und Richtscheydt*, Nürnberg, 1525 (English translation with commentary by W. L. Strauss: "The Painter's Manual: Instructions for Measuring with Compass and Ruler", New York 1977), is an exciting piece of art and science. The original source for the unfolding polytopes problem is Geoffrey C. Shephard, "Convex polytopes with convex nets," *Math. Proceedings Cambridge Math. Soc.*, 1975, 78, 389–403.

The example of an overlapping unfolding of a tetrahedron is reported by Komei Fukuda in "Strange Unfoldings of Convex Polytopes," http://www.ifor.math.ethz.ch/~fukuda/unfold_home/unfold_open.html, March/June 1997.

The first figure in our presentation is taken from Wolfram Schlickenrieder's Master's Thesis, "Nets of polyhedra," TU Berlin, 1997, with kind permission of the author.

For more detailed treatments of nets and unfolding and for rich sources of related material, see Joe O'Rourke's chapter in this volume, and Erik D. Demaine and Joseph O'Rourke, *Geometric Folding Algorithms: Linkages, Origami, Polyhedra*, Cambridge University Press, 2008.

See also Igor Pak, *Lectures on Discrete and Polyhedral Geometry* (in preparation), http://www.math.ucla.edu/~pak/book.htm, where the source and star unfoldings are presented and discussed.

disjoint triangles was posed by Gil Kalai; see Gyula Károlyi and József Solymosi, "Almost disjoint triangles in 3-space," *Discrete Comput. Geometry*. 2002, 28, 577–583, for the problem and for the lower bound of $n^{3/2}$.

Notes on Problem 2

The neighborly triangulation of the torus with 7 vertices (and $\binom{7}{2} = 21$ edges) was described by Möbius in 1861, but the first polytopal realization without self-intersections was provided by Ákos Császár in 1948, in his article "A polyhedron without diagonals," *Acta Sci. Math. (Szeged)*, 1949/50, 13, 140–142; see also Frank H. Lutz, "Császár's Torus," April 2002, Electronic Geometry Model No. 2001.02.069, http://www.eg-models.de/models/ClassicalModels/2001.02.069.

Neighborly triangulations of orientable surfaces for all possible parameters were provided by Ringel et al. as part of the Map Color Theorem: see Gerhard Ringel, *Map Color Theorem*, Springer-Verlag, New York, 1974. Beyond $n = 4$ (the boundary of a tetrahedron) and $n = 7$ (the Császár torus) the next possible value is $n = 12$: But Jürgen Bokowski and Antonio Guedes de Oliveira, in "On the generation of oriented matroids," *Discrete Comput. Geom.*, 2000, 24, 197–208, and Lars Schewe in "Nonrealizable minimal vertex triangulations of surfaces: Showing nonrealizability using oriented matroids and satisfiability solvers", *Discrete Comput. Geometry*, 2010, 43, 289–302, showed that there is no realization of any of the 59 combinatorial types of a neighborly surface with 12 vertices and $\binom{12}{2} = 66$ edges (of genus 6) without self-intersections in $\mathbb{R}^3$.

The McMullen–Schulz–Wills surfaces "with unusually large genus" were constructed by Peter McMullen, Christoph Schulz and Jörg M. Wills in "Polyhedral 2-manifolds in E^3 with unusually large genus," *Israel J. Math.*, 1983, 46, 127–144; see also Günter M. Ziegler, "Polyhedral surfaces of high genus" in *Discrete Differential Geometry*, Oberwolfach Seminars, 38, Birkhäuser, Basel 2008, 191–213. The question about almost

Notes on Problem 3

Steinitz' theorem is a fundamental result: see Ernst Steinitz, "Polyeder und Raumeinteilungen" in *Encyklopädie der mathematischen Wissenschaften, Dritter Band: Geometrie, III.1.2., Heft 9, Kapitel III A B 12*,1–139, 1922, B. G. Teubner, Leipzig, and Ernst Steinitz and Hans Rademacher, *Vorlesungen über die Theorie der Polyeder*, Springer-Verlag, Berlin, For modern treatments, see Branko Grünbaum, *Convex Polytopes*, Springer-Verlag, 2003 and Günter M. Ziegler, *Lectures on Polytopes*, Second edition, Springer, 1995, revised edition, 1998; seventh updated printing 2007.

Steinitz' proofs imply that a realization with integer vertex coordinates exists for every combinatorial type. Furthermore, there are only finitely many different combinatorial types for each n, so $f(n)$ exists and is finite. The first explicit upper bounds on $f(n)$ were derived by Shmuel Onn and Bernd Sturmfels in "A quantitative Steinitz' theorem," *Beiträge zur Algebra und Geometrie*, 1994, 35, 125–129, from the rubber band realization method of Tutte (see William T. Tutte, "Convex representations of graphs," *Proceedings London Math. Soc.*, 1960, 10, 304–320).

Since Jürgen Richter-Gebert's exposition in *Realization Spaces of Polytopes*, Springer-Verlag, Berlin Heidelberg, 1996, there has been a great deal of research to improve the upper bounds; see in particular the Ph.D. thesis of Ares Ribó Mor, "Realization and Counting Problems for Planar Structures: Trees and Linkages, Polytopes and Polyominoes," FU Berlin, http://www.diss.fuberlin.de/diss/receive/FUDISS_thesis_000000002075, and Ares Ribó Mor, Günter Rote and André Schulz, "Embedding 3-polytopes on a small grid," *Proc. 23rd Annual Symposium on Computational Geometry (Gyeongju, South Korea, June 6–8,*

2007), Association for Computing Machinery, New York, 112–118.

The upper bound $f(n) < 148^n$ can be found in Kevin Buchin and André Schulz, "On the number of spanning trees a planar graph can have," (2010) arXiv:0912:0712v2

The result about stacked polytopes was achieved by Erik Demaine and A. Schulz, "Embedding stacked polytopes on a polynomial-size grid," in: Proc. 22nd ACM-SIAM Symposium on Discrete Algorithms (SODA), San Francisco, 2011, ACM Press, 1177–1187.

A lower bound of type $f(n) \geq n^{3/2}$ follows from the fact that grids of such a size are needed to realize a convex n-gon; compare Torsten Thiele, "Extremalprobleme für Punktmengen," Master's Thesis, Freie Universität Berlin, 1991; the minimal $n \times n$-grid on which a convex m-gon can be embedded has size $n = 2\pi \left(\frac{m}{12}\right)^{3/2} + O(m \log m)$.

The theorem about edge-tangent realizations of polytopes via circle packings is detailed in Günter M. Ziegler,"Convex Polytopes: Extremal constructions and f-vector shapes," in *Geometric Combinatorics*, Proc. Park City Mathematical Institute (PCMI) 2004, American Math. Society, 2007; we refer to that exposition also for further references.

Notes on Problem 4

A survey of the theory of tilings can be found in Egon Schulte, "Tilings," in *Handbook of Convex Geometry*, v. B, North-Holland, 1993, 899–932. For tilings with congruent polytopes, we refer to the survey by Branko Grünbaum and Geoffrey C. Shephard, "Tiling with congruent tiles," *Bulletin Amer. Math. Soc.*, 3, 951–973. Their book, *Tilings and Patterns*, Freeman, New York, 1987, is a rich source of information on planar tilings.

For the problem about the maximal number of faces, see also Peter Brass, William O. J. Moser and János Pach, *Research Problems in Discrete Geometry*, Springer, New York, 2005.

Peter Engel presented his tilings by congruent polytopes with up to 38 faces in "Über Wirkungsbereichsteilungen von kubischer Symmetrie," *Zeitschrift f. Kristallographie*, 1981, 154, 199–215, and *Geometric Crystallography*, D. Reidel, 1986.

Notes on Problem 5

The fascinating history of the original work by Descartes—lost, reconstructed and rediscovered several times—is discussed in Chapters 3 and 4 of this volume.

The paper by Ernst Steinitz describing the f-vectors (f_0, f_1, f_2) of 3-polytopes completely is "Über die Eulerschen Polyederrelationen," *Archiv für Mathematik und Physik*, 1906, 11, 86–88.

The fatness parameter first appears (with a slightly different definition) in Günter M. Ziegler, "Face Numbers of 4-Polytopes and 3-Spheres," in *Proceedings of the International Congress of Mathematicians (ICM 2002, Beijing)*, 625–634, Higher Education Press, Beijing; see also Günter M. Ziegler, "Convex Polytopes: Extremal constructions and f-vector shapes," cited above for Problem 3.

The 720-cell was apparently first found and presented by Gabor Gévay, "Kepler hypersolids," in *Intuitive Geometry (Szeged, 1991)*, North-Holland, 1994, 119–129. The "projected deformed products of polygons" were introduced in Günter M. Ziegler, "Projected Products of Polygons," *Electronic Research Announcements AMS*, 2004,10,122. For a complete combinatorial analysis, see Raman Sanyal and Günter M. Ziegler, "Construction and analysis of projected deformed products," *Discrete Comput. Geom.*, 2010, 43, 412–435.

Notes on Problem 6

The Hirsch conjecture appears in George Dantzig's classic book, *Linear Programming and Extensions*, Princeton University Press, 1963. For surveys see Chapter 16 of Branko Grünbaum, *Convex Polytopes*, cited above; Victor Klee and Peter Kleinschmidt, "The d-step conjecture and its relatives," *Math. Operations Research*, 1987,

12, 718–755; Günter M. Ziegler, *Lectures on Polytopes*, cited above, and most recently Edward D. Kim and Francisco Santos, "An update on the Hirsch conjecture," Jahresbericht DMV, 2010, 112, 73–98. The Kim–Santos paper in particular explains very nicely many bad examples for the Hirsch conjecture. Santos' long-awaited counter-example to the Hirsch conjecture appears in "A counterexample to the Hirsch conjecture," *Annals of Math.*, 2012, 176, 383–412.

The Kalai–Kleitman quasipolynomial upper bound (Gil, Kalai, and Daniel J., Kleitman, "A quasi-polynomial bound for the diameter of graphs of polyhedra," *Bulletin Amer. Math. Soc.*, 1992, 26, 315–416, can also be found in *Lectures on Polytopes*. David Larman published his result in "Paths on polytopes," *Proc. London Math. Soc*, 1970, 20, 161–178.

For the connection to Linear Programming we refer to *Lectures on Polytopes*; Jiří Matoušek, Micha Sharir and Emo Welzl, "A subexponential bound for linear programming," *Proc. Eighth Annual ACM Symp. Computational Geometry* (Berlin 1992), ACM Press, 1992; Gil Kalai, "Linear programming, the simplex algorithm and simple polytopes," *Math. Programming, Ser. B*, 1997, 79, 217–233; and Volker Kaibel, Rafael Mechtel, Micha Sharir, and Günter M. Ziegler, "The simplex algorithm in dimension three," *SIAM J. Computing*, 2005, 34, 475–497.

Notes on Problem 7

The upper-bound theorem was proved by Peter McMullen in "The maximum numbers of faces of a convex polytope," *Mathematika*, 1970, 17, 179–184. Very high-dimensional simplicial polytopes with non-unimodal f-vectors were apparently first constructed by Ludwig Danzer in the 1960s. The results quoted about non-unimodal d-polytopes, $d \geq 8$ and about non-unimodal simplicial d-polytopes, $d \geq 20$, are due to Anders Björner ("The unimodality conjecture for convex polytopes," *Bulletin Amer. Math. Soc.*, 1981, 4, 187–188 and "Face numbers of complexes and polytopes," *Proceedings of the International Congress of Mathematicians* (Berkeley CA, 1986), 1408–1418; Carl W. Lee, "Bounding the numbers of faces of polytope pairs and simple polyhedra," *Convexity and Graph Theory (Jerusalem, 1981)*, North-Holland, 1984, 215–232; and Jürgen Eckhoff, "Combinatorial properties of f-vectors of convex polytopes," 1985, unpublished, and "Combinatorial properties of f-vectors of convex polytopes," *Normat*, 2006, 146–159.

For a current survey concerning general (non-simplicial) polytopes see Axel Werner, "Linear constraints on face numbers of polytopes," Ph. D. Thesis, TU Berlin, 2009, http://opus.kobv. de/tuberlin/volltexte/2009/2263/. A detailed discussion of cyclic polytopes and their properties can be found in *Lectures on Polytopes*, Sec. 8.6. The example of a non-unimodal 8-polytope glued from a cyclic polytope and its dual also appears there, as Example 841. The proof of the unimodality conjecture for cyclic polytopes was achieved only recently by László Major, Log-concavity of face vectors of cyclic and ordinary polytopes, http://arxiv.org/abs/1112.1713.

Notes on Problem 8

The relation $C^v \leq T$ was proved by Adam Bliss and Francis Su in "Lower Bounds for Simplicial Covers and Triangulations of Cubes," *Discrete Comput. Geom.*, 2005, 33, 4, 669–686. The results by Smith can be found in Warren D. Smith, "A lower bound for the simplexity of the n-cube via hyperbolic volumes," *Eur. J. of Comb.*, 2000, 21, 131–138, and those proved by Glazyrin in Alexey Glazyrin, "On Simplicial Partitions of Polytopes," *Mathematical Notes*, 2009, (6) 85, 799–806. A lot more on triangulations in general can be found in the book by Jesús De Loera, Jörg Rambau and Francisco Santos, *Triangulations*, Springer-Verlag, 2010, while Chuanming Zong's book *The Cube—A window to Convex and Discrete Geometry*, Cambridge Tracts in Mathematics, Cambridge University Press, 2006 is an in-depth look at I_d.

Notes on Problem 9

Zong posed his conjecture at a geometry meeting in 1994 in Vienna and published it in "What is known about unit cubes?", *Bulletin of the Amer. Math. Soc*, 2005, 42, 181–211. László Fejes Tóth's *Lagerungen in der Ebene, auf der Kugel und im Raum*, Springer-Verlag, 1972, contains the solutions for the case $d = 3$. See Leoni Dalla, David Larman, Peter Mani-Levitska, and Chuanming Zong, "The blocking numbers of convex bodies," *Discrete Comput. Geom.*, 2000, 24, 267–277, for the proof of Zong's conjecture in dimension 4.

Notes on Problem 10

The f-vector problem for 3-polytopes was solved by Ernst Steinitz in "Über die Eulerschen Polyederrelationen". The case $d = 4$ is surveyed in Günter M. Ziegler, "Face Numbers of 4-Polytopes and 3-Spheres," cited above.

For the g-Theorem see Section 8.6 of *Lectures on Polytopes*. Kalai's 3^d conjecture is "Conjecture A" in Gil Kalai, "The number of faces of centrally-symmetric polytopes (Research Problem)," *Graphs and Combinatorics*, 1989, 5, 389–391; the conjecture is verified for $d \leq 4$ in Raman Sanyal, Axel Werner and Günter M. Ziegler, "On Kalai's conjectures about centrally symmetric polytopes," *Discrete Comput. Geometry*, 2009, 41, 183–198.

The Hanner polytopes were introduced by Olof Hanner in "Intersections of translates of convex bodies," *Math. Scand.*, 1956, 4, 67–89; The Mahler conjecture goes back to 1939—see Mahler's original paper, K. Mahler, "Ein Übertragungsprinzip für konvexe Körper," *Casopis Pest. Mat. Fys.*, 1939, 68, 184–189; a detailed current discussion of the Mahler conjecture with recent related references appears in Terry Tao's blog, "Open question: the Mahler conjecture on convex bodies," http://terrytao.wordpress.com/2007/03/08/open-problem-the-mahler-conjecture-on-convex-bodies/.

Sources and Acknowledgments

Sources

We are grateful to all the authors, artists, museums, publishers, and copyright holders whose work appears in this volume for permission to print it.

Preface

All photographs by Stan Sherer.

Chapter 1

Figures 1.1 and 1.2. Photographs by Stan Sherer.
Figures 1.4, 1.13, and 1.18. From Henry Martyn Cundy and A. P. Rollett, *Mathematical Models*, 2nd edition, Oxford University Press, 1981.
Figures 1.5 and 1.6. ©M. C. Escher Heirs c/o Cordon Art-Baarn-Holland. Figure 1.6 is from Bruno Ernst, *The Magic Mirror of M. C. Escher*, New York: Ballantine Books, 1976.
Figure 1.7. From Peter S. Stevens, *Patterns in Nature*. (Boston: Little, Brown and Company, 1974). Reprinted by permission.
Figures 1.9 and 1.20. From Paolo Portoghesi, *The Rome of Borromini: Architecture as Language*, New York: George Braziller, 1968. Photographs courtesy of Electa, Milano.
Figure 1.11. From Cedric Rogers, *Rock and Minerals*. London, Triune Books, 1973.

Figure 1.10. From Linus Pauling and Roger Hayward, *The Architecture of Molecules*, San Francisco, W. H. Freeman, 1964.
Figure 1.8. Photograph in the Sophia Smith Collection, Smith College.
Figure 1.16. Wenzel Jamnitzer, *Perspectiva Corporum Regularium*, 1568; facsimile reproduction, Akademische Druck- u. Verlagsanstalt, 1973, Graz, Austria.
Figure 1.19. Courtesy of Scienza e Tecnica 76, Mondadori.
Figure 1.21. From Bruce L. Chilton and George Olshevsky, *How to Build a Yog-Sothoth* (George Olshevsky, P. O. Box 11021, San Diego, California, 92111–0010, 1986).

Chapter 2

The photographs in Figures 2.1–2.3 and 2.15 are by Ken R. O'Connell, and the photographs in Figures 2.2 and 2.4–2.14 are by Marion Walter. The photographs in Figures 2.18, 2.19, 2.25, 2.29, 2.33, and 2.37 are by Stan Sherer.

Chapter 3

Figure 3.2. Drawing from Johannes Kepler, *Harmonices Mundi*, 1619.
Figure 3.6. Drawing from Johannes Kepler, *Mysterium Cosmographicum*, 1595.

Figures 3.7 and 3.17–3.20. Drawn by Patrick DuVal for his book *Homographies, Quaternions, and Rotations*, London: Oxford University Press, 1964.

Figures 3.25–3.28. Photographs by Stan Sherer.

Chapter 4

Figure 4.1. The Metropolitan Museum of Art, New York. 27.122.5, Fletcher Fund, 1927.

Figure 4.2. The Metropolitan Museum of Art, New York. 37.11.3, Museum purchase, 1937.

Figure 4.3. Photograph reproduced by courtesy of the Society of Antiquaries of London.

Chapter 5

Figures 5.1–5.3. ©M.C Escher Heirs c/o Cordon Art – Baarn – Holland.

Figures 5.4, 5.10, 5.21 and 5.22. From the Teaching Collection in the Carpenter Center for the Visual Arts, Harvard University. Reproduced with permission of the Curator. The egg in Figure 5.4 was painted by Beth Saidel; the model in Figure 5.10 was created by Brett Tomlinson. The models in Figures 5.21 and 5.22 were designed and constructed by Jonathan Lesserson and photographed by C. Todd Stuart.

Table 5.1 and Figures 5.9, 5.11, 5.12, and 5.14 are from Arthur L. Loeb, *Space Structures: Their Harmony and Counterpoint*, Reading, Mass. Addison-Wesley, Advanced Book Program, 1976.

Chapter 6

Figures 6.1–6.3, 6.5–6.7, 6.10b, 6.12 are reprinted, with the permission of Cambridge University Press, from Eric D. Demaine and Joseph O'Rourke, *Geometric Folding Algorithms: Linkages, Origami, Polyhedra*, Cambridge University Press, July 2007. In that volume they are, respectively, Figures 3, 21.3, 22.7, 22.9, 22.8, 22.17, 22.32, and 22.2.

Chapter 7

Figure 7.1. Photograph by Hirmer Verlag, Munich.

Figure 7.2. U.S. Air Force Photo by Eddie McCrossan.

Figures 7.3, 7.4, 7.37, and 7.38. Photographs by Irwin Hauer.

Figure 7.5. Photograph courtesy of the Buckminster Fuller Institute, Los Angeles.

Figures 7.6, 7.9, 7.13, 7.56, Photographs by Wendy Klemyk.

Figures 7.7 and 7.53. From *Domebook 2* (Bolinas, Calif.: Shelter Publications, 1971). Reprinted by permission of Steve Baer.

Figure 7.8. Photograph by Steve Long. University of Massachusetts Photocenter.

Figures 7.10 and 7.11. Courtesy of Zvi Hecker, architect.

Figures 7.12, 7.24, 7.33 (top), 7.34, 7.39, 7.45, 7.50, 7.52, and 7.54. Photographs by Stan Sherer. Figure 7.24 is used by permission of the Trustees of the Smith College.

Figure 7.14. From Wolf Strache, *Forms and Patterns in Nature*, New York: Pantheon Books, a Division of Random House, Inc., 1956.

Figure 7.15. From Vincenzo de Michele, *Minerali* (Milan: Istituto Geografico de Agostini-Novara, 1971).

Figure 7.16. From Earl H. Pemberton, *Minerals of California*, New York: Van Nostrand Reinhold Company Inc., 1983.

Figures 7.17 and 7.43. From Viktor Goldschmidt, *Atlas der Krystallformen*, Heidelberg: Carl Winters, 1913–1923.

Figure 7.18. Photograph by Carl Roessler.

Figure 7.19. E. Haeckel, *The Voyage of H.M.S. Challenger* (Berlin: Georg Reimer, 1887), plates 12, 20, and 63.

Figure 7.20. Photograph by Lawrence Conner, Ph.D., entomologist.

Figure 7.21. Karl Blossfeldt, *Wundergarten der Natur*, Berlin: Verlag für Kunstwissenschaft, 1932.

Figure 7.22. Museo e Gallerie Nazionali di Capodimonte, Naples. Illustration by permission of Soprintendenza ai B.A.S. di Napoli.

Figure 7.23. Alinari/Art Resource, New York.

Figure 7.25. National Gallery of Art, Washington. Chester Dale Collection.

Figure 7.26. The Salvador Dali Foundation, Inc., St. Petersburg, Fla.

Figure 7.27. Photograph supplied by artist, Mary Bauermeister.

Figure 7.28. Reproduced by permission of Mrs. Anni Albers and the Josef Albers Foundation, Inc.

Figures 7.29, 7.30, and 7.51. ©M. C. Escher Heirs c/o Cordon Art-Baarn-Holland. Figure 7.51 is reproduced from Bruno Ernst, *The Magic Mirror of M. C. Escher*, New York: Ballantine Books, 1976.

Figure 7.31. Photograph by permission of the Marine Midland Bank, Corporate Communications Group.

Figure 7.32. Photograph by Jeremiah O. Bragstad.

Figure 7.33 (bottom). Photograph reproduced from Arthur L. Loeb, *Space Structures: Their Harmony and Counterpoint*, Reading, Mass.: Addison-Wesley, Advanced Book Program, 1976.

Figure 7.35. Sculpture by Hugo F. Verheyen.

Figure 7.36. From Gyorgy Kepes, *The New Landscape in Art and Science*, Chicago: Paul Theobald and Company, 1956.

Figure 7.40. Photograph by Bob Thayer. ©1984 by the Providence Journal Company.

Figure 7.41. Photograph supplied by the sculptor, Robinson Fredenthal.

Figure 7.42. From William Blackwell, *Geometry in Architecture*, p. 155. ©Copyright 1984 by John Wiley & Sons, Inc. Reprinted by permission of John Wiley & Sons, Inc.

Figure 7.46. From Peter S. Stevens, *Patterns in Nature*, Boston: Little, Brown, and Company, 1974.

Figure 7.47. Photographs by R. W. G. Wyckoff.

Figure 7.49. Cardboard models by Lucio Saffaro.

Figure 7.55. Lampshade by Bahman Negahban, architect, and Ezat O. Negahban, calligrapher.

Figure 7.57. Reprinted by permission of Dick A. Termes.

Figure 7.58. Reprinted from *Better Homes and Gardens Christmas Ideas*. Copyright Meredith Corportation, 1957. All rights reserved.

Chapter 8

Figure 8.2. Photograph by Stan Sherer.

Chapter 9

Figures 9.1 and 9.3 are Leonardo da Vinci's drawings for Luca Pacioli's *Divina Proportione*, Milan, 1509.

Chapter 10

Figures 10.8 and 10.11. ©M.C. Escher Heirs c/o Cordon Art–Baarn–Holland.

Chapter 11

Figure 11.4. From Ralph W. Rudolph, "Boranes and Heteroboranes: A Paradigm for the Electron Requirements of Clusters?" *Accounts of Chemical Research, 9* (1976):446–52. Copyright 1976 American Chemical Society.

Figure 11.21. Drawing by the artist Ferenc Lantos.

Chapter 12

Figure 12.1. After A. W. Hofmann, "On the Combining Power of Atoms," *Proceedings of the Royal Institution of Great Britian 4* (1865): 401–30.

Figure 12.2. Reprinted from *Philosophical Magazine,* series 4, *21*, (1861): Plate V, Figure 2.

Figures 12.3–12.5. From J. H. van 't Hoff, *The Arrangement of Atoms in Space*, 2nd ed. (London: Longmans, Green, 1898).

Figure 12.6. From Linus Pauling and Roger Hayward, *The Architecture of Molecules*. W. H. Freeman and Company. Copyright © 1964.

Figure 12.7. Reproduced by permission of Pergamon Press.

Figure 12.8. Reprinted from Linus Pauling, *The Nature of the Chemical Bond and the Structure of Molecules and Crystals: An Introduction to Modern Structural Chemistry*, 3rd edition. Copyright 1960 by Cornell University. Used by permission of Cornell University Press.

Figure 12.9. Used by permission of John C. Spurlino and Florentine A. Quiocho, Department of Biochemistry, Rice University.

Figures 12.10 and 12.11. Courtesy of D. L. D. Caspar.

Figure 12.12. From E. B. Matzke, "The Three-Dimensional Shape of Bubbles in Foam — An Analysis of the Rôle of Surface Forces in Three-Dimensional Cell Shape Determination," *American Journal of Botany 33*

Figures 12.15–12.16. Models made by Charles Ingersoll, Sr. Photographs by Fred Clow.

Figure 12.23. Model made by Charles Ingersoll, Sr. Photogarph by William Saunders.

Chapter 21

Figure 21.1, Hirshhorn Museum and Sculpture Garden, Washington, D.C.

All illustrations NOT listed above were created for Shaping Space by (or for) the authors of the chapters in which they appear.

Acknowledgments

I am grateful to Ann Kostant, long-time Mathematics Editor of Springer, New York, for encouraging me to prepare this volume and for making it possible. Smith College students Amy Wesolowski and Marissa Neal typed the LaTeX code and made helpful comments and suggestions.

Stan Sherer, many of whose photographs appear in this volume, scanned and enhanced the illustrations and helped in countless ways.

My Smith College colleague George Fleck's sound advice and eagle eye for detail have been, once again, invaluable.

Marjorie Senechal

Chapter 2

Jean Pedersen thanks Les Lange, Editor of *California Mathematics*, for giving permission to use in this article some of the ideas that were originally part of "Pop-up Polyhedra," *California Mathematics (April 1983):37–41.*

Chapter 4

Joe Malkevitch thanks D. M. Bailey of the British Museum and Dr. Maxwell Anderson of the Metropolitan Museum of Art for their cooperation in obtaining access to the information about early man-made polyhedra. J. Wills (Siegen) provided him with a copy of Lindemann's paper and Robert Machalow (York College Library) helped obtain copies of many articles, obscure and otherwise.

Chapter 10

Chung Chieh thanks Nancy McLean for redoing some of the drawings.

Chapter 11

This chapter was written when the authors were at the University of Connecticut (István Hargittai as Visiting Professor of Physics (1983–84) and Visiting Professor of Chemistry (1984–85) and Magdolna Hargittai as Visiting Scientist at the Institute of Materials Science), both on leave from the Hungarian Academy of Sciences, Budapest. They express their appreciation to the University of Connecticut and their colleagues there for hospitality, and to Professor Arthur Greenberg of the New Jersey Institute of Technology for many useful references on polycyclic hydrocarbons and for his comments on the manuscript.

Chapter 15

Research for this chapter was supported by the National Science Foundation Grant MCS8301971.

Chapter 16

Egon Schulte thanks Professors Ludwig Danzer and Branko Grünbaum for many helpful suggestions.

Chapter 18

Research for this chapter was supported in part by a grant to Henry Crapo from National Science and Engineering Research Council of Canada and in part by grants to Walter Whiteley from Fonds pour la Formation de Chercheurs et l'Aide à la Recherche Québec and NSERC, Canada. This chapter also grows out of previous joint work by Crapo and Whiteley, Ash and Bolker, and Bolker and Whiteley. The unification which is its theme developed while they were writing it.

Because time was short, the pleasure of shaping the chapter and working out the details fell to the final author, Henry Crapo. They all share the responsibility for any errors in the text.

Chapter 20

Tom Banchoff acknowledges with gratitude the help he received from correspondence and conversations with Professor Coxeter during the course of preparation of this paper. Computer images in this chapter were generated in collaboration with David Laidlaw and David Margolis, and with the cooperation of the entire graphics group at Brown University. Several of the illustrations are taken from the film *The Hypersphere: Foliation and Projection* by Huseyin Koçak, David Laidlaw, David Margolis, and the author.

Chapter 22

Moritz Schmitt thanks the DFG Research Center MATHEON, Institut für Mathematik, Freie Universität Berlin, Arnimallee 2, 14195 Berlin, Germany, for support. Günter M. Ziegler's work was partially supported by DFG, Research Training Group "Methods for Discrete Structures," also at the Institut für Mathematik.

Index